Lecture Notes in Computer Science 2948

Edited by G. Goos, J. Hartmanis, and J. van Leeuwen

Springer
Berlin
Heidelberg
New York
Hong Kong
London
Milan
Paris
Tokyo

Gary L. Mullen
Alain Poli
Henning Stichtenoth (Eds.)

Finite Fields and Applications

7th International Conference, Fq7
Toulouse, France, May 5-9, 2003
Revised Papers

 Springer

Series Editors

Gerhard Goos, Karlsruhe University, Germany
Juris Hartmanis, Cornell University, NY, USA
Jan van Leeuwen, Utrecht University, The Netherlands

Volume Editors

Gary L. Mullen
The Pennsylvania State University
Department of Mathematics, University Park, PA 16802, USA
E-mail: mullen@math.psu.edu

Alain Poli
University Paul Sabatier, AAECC/IRIT
118 route de Narbonne, 31300 Toulouse, France
E-mail: poli@cict.fr

Henning Stichtenoth
University Duisburg-Essen
Department of Mathematics, Campus Essen, 45117 Essen, Germany
E-mail: henning@sabanciuniv.edu

Cataloging-in-Publication Data applied for

A catalog record for this book is available from the Library of Congress.

Bibliographic information published by Die Deutsche Bibliothek
Die Deutsche Bibliothek lists this publication in the Deutsche Nationalbibliografie;
detailed bibliographic data is available in the Internet at <http://dnb.ddb.de>.

CR Subject Classification (1998): G.1, G.2, G.3, E.4, I.1, F.2.1

ISSN 0302-9743
ISBN 3-540-21324-4 Springer-Verlag Berlin Heidelberg New York

Springer-Verlag is a part of Springer Science+Business Media

springeronline.com

© Springer-Verlag Berlin Heidelberg 2004
Printed in Germany

Typesetting: Camera-ready by author, data conversion by PTP-Berlin, Protago-TeX-Production GmbH
Printed on acid-free paper   SPIN: 10990230   06/3142   5 4 3 2 1 0

Preface

This volume represents the refereed proceedings of the **7th International Conference on Finite Fields and Applications ($\mathbf{F_{q^7}}$)** held during May 5–9, 2003, in Toulouse, France. The conference was hosted by the Pierre Baudis Congress Center, downtown, and held at the excellent conference facility. This event continued a series of biennial international conferences on Finite Fields and Applications, following earlier meetings at the University of Nevada at Las Vegas (USA) in August 1991 and August 1993, the University of Glasgow (UK) in July 1995, the University of Waterloo (Canada) in August 1997, the University of Augsburg (Germany) in August 1999, and the Universidad Autónoma Metropolitana-Iztapalapa, in Oaxaca (Mexico) in 2001.

The Organizing Committee of F_{q^7} consisted of Claude Carlet (INRIA, Paris, France), Dieter Jungnickel (University of Augsburg, Germany), Gary Mullen (Pennsylvania State University, USA), Harald Niederreiter (National University of Singapore, Singapore), Alain Poli, Chair (Paul Sabatier University, Toulouse, France), Henning Stichtenoth (Essen University, Germany), and Horacio Tapia-Recillas (Universidad Autónoma Metropolitan-Iztapalapa, Mexico).

The program of the conference consisted of four full days and one half day of sessions, with eight invited plenary talks, and close to 60 contributed talks.

Finite fields have an inherently fascinating structure and they are important tools in discrete mathematics. Their applications range from combinatorial design theory, finite geometries, and algebraic geometry to coding theory, cryptology, and scientific computing. A particularly fruitful aspect is the interplay between theory and applications which has led to many new perspectives in research on finite fields. This interplay has been a dominant theme in earlier F_q conferences and was very much in evidence at F_{q^7}. Applied or applications-oriented topics accounted for a significant part of the program.

These proceedings reflect the wide variety of topics represented at the conference. Most invited talks and a good proportion of the contributed talks are on permanent record here. All contributed talks were screened before the conference and all full papers were carefully refereed. We would like to take this opportunity to thank the members of the Organizing Committee and all referees who helped in these tasks. These colleagues contributed enormously to the quality of the conference presentations and to guaranteeing high standards for these proceedings.

We greatly appreciate the generous financial support received for the conference. A fair portion of the funds was provided from the AAECC Lab. (Applied Algebra and Error Correcting Codes) of Alain Poli.

Regarding the present proceedings, we thank very much Prof. Yves Soulet (Irsamc, Paul Sabatier University, Toulouse, France) for his remarkable work in arranging the proceedings volume. We also thank Alfred Hofmann of Springer-Verlag who gave us the opportunity to publish this volume with a top publisher and in an attractive form. Working with him and Anna Kramer at Springer-Verlag has been a pleasure.

Finally, we are pleased to confirm that the F_q series will continue with F_{q^8} in Puerto Rico in August 2005. We expect another lively and stimulating meeting there, which should, like the previous conferences, serve as an important meeting place for theoretical as well as applied aspects of finite fields. We hope to see you there!

December 2003

Gary L. Mullen
Alain Poli
Henning Stichtenoth

Table of Contents

On the Autocorrelation of Cyclotomic Generators

Wilfried Meidl[1] and Arne Winterhof[1,2]

[1] Temasek Laboratories
National University of Singapore
10 Kent Ridge Crescent
Singapore 119260
tslmw@nus.edu.sg

[2] Johann Radon Institute for Computational and Applied Mathematics,
Austrian Academy of Sciences, c/o Johannes Kepler University Linz,
Altenbergerstraße 69, 4040 Linz, Austria.
arne.winterhof@oeaw.ac.at

Abstract. We extend a result of Ding and Helleseth on the autocorrelation of a cyclotomic generator in several ways. We define and analyze cyclotomic generators of arbitrary orders and over arbitrary finite fields, and we consider two, in general, different definitions of autocorrelation. Cyclotomic generators are closely related to the discrete logarithm. Hence, the results of this paper do not only describe interesting cryptographic properties of cyclotomic generators and their generalizations but also desirable features of the discrete logarithm.

1 Introduction

Let (s_n) be a q-periodic sequence over the residue class ring $\mathbb{Z}_d$. The *autocorrelation* of (s_n) is the complex-valued function defined by

$$A_d(q,t) := \frac{1}{q} \sum_{n=0}^{q-1} \varepsilon_d^{s_{n+t}-s_n}, \quad 1 \le t \le q-1, \tag{1}$$

where $\varepsilon_d = e^{2\pi\sqrt{-1}/d}$. The autocorrelation measures the amount of similarity between the sequence (s_n) and a shift of (s_n) by t positions. If (s_n) is a random sequence over $\mathbb{Z}_d$ of period q then $|A_d(q,t)|$ can be expected to be quite small for all values $1 \le t \le q-1$. The security of many cryptographic systems depends upon the generation of pseudorandom, i.e., unpredictable quantities and a low autocorrelation is a desirable feature for pseudorandom sequences.

Let q be a prime power, γ a primitive element of the finite field $\mathbb{F}_q$ of q elements, and $d > 1$ a divisor of $q - 1$. The *cyclotomic classes of order d* give a partition of $\mathbb{F}_q^* := \mathbb{F}_q \setminus \{0\}$ defined by

$$D_0 := \{\gamma^{id} : 0 \le i \le (q-1)/d - 1\} \quad \text{and} \quad D_j := \gamma^j D_0, \quad 1 \le j \le d-1.$$

G. Mullen, A. Poli and H. Stichtenoth (Eds.): Fq7 2003, LNCS 2948, pp. 1–11, 2003.

We order the elements of $\mathbb{F}_q = \{\xi_0, \xi_1, \ldots, \xi_{q-1}\}$ in the following way,

$$\xi_n := n_1\beta_1 + n_2\beta_2 + \ldots + n_r\beta_r$$

if

$$n = n_1 + n_2 p + \ldots + n_r p^{r-1}, \quad 0 \le n_1, n_2, \ldots, n_r \le p - 1,$$

where $q = p^r$, p is the characteristic of $\mathbb{F}_q$, and $\{\beta_1, \beta_2, \ldots, \beta_r\}$ is a basis of $\mathbb{F}_q$ over $\mathbb{F}_p$. We consider the q-periodic sequence (s_n) over $\mathbb{Z}_d$ defined by

$$s_n := \begin{cases} j, \text{ if } \xi_n \in D_j, \, 0 \le j \le d-1, \, 1 \le n \le q-1, \\ 0, \text{ if } n = 0, \end{cases} \tag{2}$$

and $s_{n+q} = s_n$, $n \ge 0$. For $r = 1$ this sequence is called *cyclotomic generator of order* d. In this case the autocorrelation is

$$A_d(p, t) = (-1 + \varepsilon_d^j + \varepsilon_d^{-i-j})/p, \tag{3}$$

if $t \in D_j$ and $-1 \in D_i$. See [8, Theorem 5] (see also [4, Chapter 10.3]) for a proof of the case $d = 3$. For $r > 1$ we prove an upper bound on $|A_d(q, t)|$ of the order of magnitude $q^{-1/2} \log(p)^{r-1}$ in Section 3, where the implied constant depends on r only.

Although this result is weaker it is also of high interest because of the close relation of the sequence (s_n) to the discrete logarithm in $\mathbb{F}_q$. The *discrete logarithm* (or *index*) $\mathrm{ind}_\gamma(\xi)$ of an element $\xi \in \mathbb{F}_q^*$ is the unique integer l with $\xi = \gamma^l$ and $0 \le l \le q-2$. With the convention $\mathrm{ind}_\gamma(0) := q-1$ we have

$$s_n = \mathrm{ind}_{\gamma,d}(\xi_n), \quad 0 \le n \le q-1,$$

where $\mathrm{ind}_{\gamma,d}(\xi)$ denotes the residue class of $\mathrm{ind}_\gamma(\xi)$ modulo d. Many cryptographic systems as the Diffie-Hellman key exchange depend on the intractability and unpredictability of the discrete logarithm (see e. g. [20]).

Moreover, we introduce a slight modification of the autocorrelation which might be better suited for the sequences (s_n) when $r > 1$ and coincides with (1) when $r = 1$,

$$A_d^\oplus(q, t) := \frac{1}{q} \sum_{n=0}^{q-1} \varepsilon_d^{s_{n \oplus t} - s_n}, \quad 0 \le t \le q-1, \tag{4}$$

where $n \oplus t := k$ if and only if $\xi_n + \xi_t = \xi_k$. Now the analogue of (3) for $A_d^\oplus(q, t)$ holds true. We prove this result in Section 2. We also consider the sequences (s_n) defined by

$$s_n := \mathrm{ind}_{\gamma,m}(\xi_n), \quad 0 \le n \le q-1, \tag{5}$$

and $s_{n+q} = s_n$, $n \ge 0$, where m is not a divisor of $q-1$. We reduce this general case to some extent to the case of divisors of $q-1$. For the most interesting partial case $m = q$ we get

$$|A_q^\oplus(q, t)| \le \frac{2\pi}{q} + |A_{q-1}^\oplus(q, t)| \le \frac{2\pi + 3}{q}$$

and

$$|A_q(q,t)| \le \frac{2\pi}{q} + |A_{q-1}(q,t)| = O(q^{-1/2}\log(p)^{r-1}).$$

Finally in Section 4, we prove similar results for parts of the period, i.e., upper bounds on the *aperiodic autocorrelation*.

2   Exact Autocorrelation Values

In this section we prove (3) in the following slightly more general form. Although the result is known [9, Theorem 7] we are not aware of a reference for the following short proof and present it here for completeness.

Theorem 1. *Let q be a prime power and $d > 1$ be a divisor of $q - 1$. For a sequence of the form (2) and $A_d^\oplus(q,t)$ defined by (4) we have*

$$A_d^\oplus(q,t) = (-1 + \varepsilon_d^j + \varepsilon_d^{-i-j})/q,$$

if $\xi_t \in D_j$ and $-1 \in D_i$.

Proof. By definition we have

$$q A_d^\oplus(q,t) = \sum_{n=0}^{q-1} \varepsilon_d^{s_{n\oplus t} - s_n} = \varepsilon_d^j + \varepsilon_d^{-i-j} + \sum_{\substack{n=1 \\ n\oplus t \ne 0}}^{q-1} \varepsilon_d^{s_{n\oplus t} - s_n}.$$

Verify that

$$\chi_d(\xi_n) := \varepsilon_d^{s_n}, \quad 1 \le n \le q - 1,$$

is a nontrivial multiplicative character of $\mathbb{F}_q$. We use the convention $\chi_d(0) := 0$. Thus with [11, Lemma 7.3.7] we have

$$\sum_{\substack{n=1 \\ n\oplus t \ne 0}}^{q-1} \varepsilon_d^{s_{n\oplus t} - s_n} = \sum_{n=0}^{q-1} \chi_d(\xi_n + \xi_t)\chi_d^{-1}(\xi_n) = -1,$$

where we used $\xi_{n\oplus t} = \xi_n + \xi_t$.   □

3   Bounds on the Autocorrelation for Arbitrary Finite Fields

In this section we establish upper bounds on the autocorrelation $A_d(q,t)$ for a sequence of the form (2) over arbitrary finite fields. In the case of a prime field Theorem 1 yields the exact value.

Theorem 2. *Let q be a prime power and $d > 1$ be a divisor of $q - 1$. For a sequence of the form (2) we have the following upper bound on the autocorrelation $A_d(q,t)$,*

$$|A_d(q,t)| = O(q^{-1/2}(\log p)^{r-1}), \quad 1 \le t \le q - 1,$$

where the implied constant depends on r only.

Proof. For $0 \leq t, n \leq q - 1$ let

$$t = t_1 + t_2 p + \ldots + t_r p^{r-1}, \quad 0 \leq t_1, t_2, \ldots, t_r \leq p - 1,$$

and

$$n = n_1 + n_2 p + \ldots + n_r p^{r-1}, \quad 0 \leq n_1, n_2, \ldots, n_r \leq p - 1,$$

be the p-adic expansions of t and n, respectively. Put $w_1 := 0$ and define for $1 \leq i \leq r$ recursively

$$w_{i+1} := \begin{cases} 1, & \text{if } t_i + n_i + w_i \geq p, \\ 0, & \text{otherwise.} \end{cases}$$

Then we have

$$n + t = z_1 + z_2 p + \ldots + z_r p^{r-1}, \quad 0 \leq z_1, z_2, \ldots, z_r \leq p - 1,$$

with

$$z_i = t_i + n_i + w_i - w_{i+1} p, \quad 1 \leq i \leq r,$$

and

$$\xi_{n+t} = \xi_n + \xi_t + \omega,$$

where

$$\omega = \sum_{i=2}^{r} w_i \beta_i.$$

Note that for fixed t we have at most 2^{r-1} possible choices for ω and the sets

$$P_\omega := \{\xi_n \in \mathbb{F}_q \ : \ \xi_{n+t} = \xi_n + \xi_t + \omega\}$$

define a partition of $\mathbb{F}_q$. For fixed $w_2, \ldots, w_r \in \{0, 1\}$ the set P_ω can be written in the form

$$P_\omega = \left\{ \alpha + \sum_{j=1}^{r} u_j \beta_j \ : \ 0 \leq u_j \leq k_j - 1, j = 1, \ldots, r \right\},$$

where

$$\alpha = \sum_{\substack{j=1 \\ w_{j+1}=1}}^{r-1} (p - (t_j + w_j)) \beta_j$$

and

$$k_j = \begin{cases} p - (t_j + w_j), & w_{j+1} = 0, \ 1 \leq j < r, \\ t_j + w_j, & w_{j+1} = 1, \ 1 \leq j < r, \\ p, & j = r. \end{cases}$$

Now the autocorrelation of (s_n) can be estimated,

$$q\,|AC_d(q,t)| \tag{6}$$

$$\leq 2 + \left| \sum_{\xi \in \mathbb{F}_q} \chi_d(\xi + \xi_t)\chi_d^{-1}(\xi) + \sum_{\omega \neq 0} \sum_{\xi \in P_\omega} (\chi_d(\xi + \xi_t + \omega) - \chi_d(\xi + \xi_t))\chi_d^{-1}(\xi) \right|$$

$$\leq 3 + \sum_{\omega \neq 0} \left(\left| \sum_{\xi \in P_\omega} \chi_d(\xi + \xi_t + \omega)\chi_d^{-1}(\xi) \right| + \left| \sum_{\xi \in P_\omega} \chi_d(\xi + \xi_t)\chi_d^{-1}(\xi) \right| \right).$$

With the method of Polya and Vinogradov and Weil's bound (see [6, 27–30]) we can prove that the absolute values of the inner sums are smaller than

$$2q^{1/2}(1 + \log(p))^{r-1}. \tag{7}$$

Hence, we have

$$|A_d(q,t)| < (3 + 4(2^{r-1} - 1)q^{1/2}(1 + \log(p))^{r-1})/q \tag{8}$$

and the result follows. For the convenience of the reader we add the most important steps for the deduction of the upper bound (7). For more details we refer to [6, 27–30]. From

$$\chi_d(\xi + \xi_t + \omega)\chi_d^{-1}(\xi) = \frac{1}{q} \sum_{\zeta \in \mathbb{F}_q} \chi_d(\zeta + \xi_t + \omega)\chi_d^{-1}(\zeta) \sum_{\eta \in \mathbb{F}_q} \psi(\eta(\xi - \zeta))$$

we get

$$\left| \sum_{\xi \in P_\omega} \chi_d(\xi + \xi_t + \omega)\chi_d^{-1}(\xi) \right|$$

$$\leq \frac{1}{q} \sum_{\eta \in \mathbb{F}_q} \left| \sum_{\zeta \in \mathbb{F}_q} \chi_d(\zeta + \xi_t + \omega)\chi_d^{-1}(\zeta)\psi(-\eta\zeta) \right| \left| \sum_{\xi \in P_\omega} \psi(\eta\xi) \right|$$

$$< \frac{2}{q^{1/2}} \sum_{\eta \in \mathbb{F}_q} \left| \sum_{\xi \in P_\omega} \psi(\eta\xi) \right|$$

by Weil's Theorem, where ψ denotes the additive canonical character of $\mathbb{F}_q$. Now we have

$$\left| \sum_{\xi \in P_\omega} \psi(\eta\xi) \right| = \prod_{j=1}^{r} \left| \sum_{u_j=0}^{k_j-1} \psi(\eta\beta_j)^{u_j} \right| \leq p \prod_{j=1}^{r-1} \min\left(k_j, \frac{1}{\sin(\mathrm{Tr}(\eta\beta_j)\pi/p)} \right),$$

if $\mathrm{Tr}(\eta\beta_r) = 0$, where Tr denotes the absolute trace of $\mathbb{F}_q$. Otherwise this sum vanishes since $k_r = p$. Now the mapping $\eta \mapsto (\mathrm{Tr}(\eta\beta_1), \ldots, \mathrm{Tr}(\eta\beta_r))$ is a bijection. Hence, we get

$$\sum_{\eta \in \mathbb{F}_q} \left| \sum_{\xi \in P_\omega} \psi(\eta\xi) \right| \leq p \prod_{j=1}^{r-1} \left(k_j + \sum_{u=1}^{p-1} \frac{1}{\sin(u\pi/p)} \right) \leq p \prod_{j=1}^{r-1} (k_j + p \log p),$$

which yields the desired upper bound.   □

Remark: Equation (6) can be easily modified to give an exact relation between $A_d(q,t)$ and $A_d^\oplus(q,t)$,

$$A_d(q,t) = A_d^\oplus(q,t)$$

$$1 - \frac{1}{q}\left(\varepsilon_d^{-k} - \varepsilon_d^{-i-j} + \sum_{w\neq 0}\sum_{\xi\in P_w}(\chi_d(\xi+\xi_t+w) - \chi_d(\xi+\xi_t))\chi_d^{-1}(\xi)\right),$$

if $\xi_t \in D_j$, $-1 \in D_i$, and $\xi_{q-t} \in D_k$. For $r = 1$ both values coincide and (8) yields the better bound $O(q^{-1})$.

Defining the sequence (2) we presupposed that the modulus d is a divisor of $q - 1$. Now we drop this condition and consider the sequences (s_n) defined by (5) over $\mathbb{Z}_m$ for an arbitrary integer m. For several moduli m that are close to a sufficiently large divisor d of $q - 1$, we can establish nontrivial upper bounds on the autocorrelation $A_m(q,t)$ of the corresponding sequence.

Corollary 1. *Let $m > 1$ be an integer. Then we have*

$$|A_m^\oplus(q,t)| \leq \min_{\substack{d|(q-1)\\d>1}} \frac{2\pi|m - d|(q - 1)}{md} + \frac{3}{q}$$

and

$$|A_m(q,t)| \leq \min_{\substack{d|(q-1)\\d>1}} \frac{2\pi|m - d|(q - 1)}{md} + O(q^{-1/2}\log(p)^{r-1}),$$

where the implied constant depends on r only.

Proof. We write $e(u) = e^{2\pi\sqrt{-1}u} \in \mathbb{C}$ for real u and $e_k(u) = e(u/k)$ for an integer k. We have for x with $|x| \leq q - 1$,

$$0 \leq \left|\frac{x}{d} - \frac{x}{m}\right| = \frac{|m - d||x|}{dm} \leq \frac{|m - d|(q - 1)}{dm}.$$

Since

$$|e(u) - 1| = 2|\sin(\pi u)| \leq 2\pi|u| \quad \text{for real } u,$$

we have

$$|e_m(x) - e_d(x)| \leq \frac{2\pi|m - d|(q - 1)}{dm}.$$

Hence,

$$q|A_m(q,t)| = \left|\sum_{n=0}^{q-1} e_m(s_{n+t} - s_n)\right|$$

$$\leq \left|\sum_{n=0}^{q-1}(e_m(\text{ind}_\gamma(\xi_{n+t}) - \text{ind}_\gamma(\xi_n)) - e_d(\text{ind}_\gamma(\xi_{n+t}) - \text{ind}_\gamma(\xi_n)))\right|$$

$$+ \left| \sum_{n=0}^{q-1} e_d(\text{ind}_\gamma(\xi_{n+t}) - \text{ind}_\gamma(\xi_n)) \right|$$
$$\leq q \left(\frac{2\pi|m-d|(q-1)}{md} + |A_d(q,t)| \right)$$

and

$$|A_m^\oplus(q,t)| = \frac{1}{q} \left| \sum_{n=0}^{q-1} e_m(s_{n\oplus t} - s_n) \right|$$
$$\leq \frac{2\pi|m-d|(q-1)}{md} + |A_d^\oplus(q,t)|,$$

respectively. Now the result follows from the previous theorems. □

4 Aperiodic Autocorrelation

In the previous sections we restricted ourselves to autocorrelations over the full period. Similar results for parts of the period are also cryptographically important and reflect local randomness. Actually for the generation of stream ciphers only a small part of the periodic sequence is used. The *aperiodic autocorrelations* $AA_d^\oplus(q,t,u,v)$ and $AA_d(q,t,u,v)$, $0 \leq u < v \leq q-1$, can be defined by

$$AA_d^\oplus(q,t,u,v) := \frac{1}{q} \sum_{n=u}^{v} \varepsilon_d^{s_{n\oplus t} - s_n}, \quad 1 \leq t \leq q-1,$$

and

$$AA_d(q,t,u,v) := \frac{1}{q} \sum_{n=u}^{v} \varepsilon_d^{s_{n+t} - s_n}, \quad 1 \leq t \leq q-1,$$

respectively. The character sum bounds of [22, 29, 31] yield nontrivial results on the aperiodic autocorrelations. For finite prime fields and fixed t the estimates of [3] can also be applied but the implied constant depends on t. See also [21] for an improvement. Without any restrictions on q and t we get the following result.

Theorem 3. *Let q be a prime power and $d > 1$ be a divisor of $q - 1$. For a sequence of the form (2) and for $0 \leq u < v \leq q-1$ we have*

$$AA_d^\oplus(q,t,u,v) = O(q^{-1/2} \log q), \quad 1 \leq t \leq q-1,$$

where the implied constant is absolute.

Proof. We have

$$q|AA_d^\oplus(q,t,u,v)| \leq 2 + \left| \sum_{n=u}^{v} \chi_d(\xi_n + \xi_t)\chi_d^{-1}(\xi_n) \right| < 2 + 2q^{1/2}(1 + \log q)$$

by [29, Theorem 3].   □

We can also prove an analog for $AA_d(q, t, u, v)$.

Theorem 4. *Let q be a prime power and $d > 1$ be a divisor of $q - 1$. For a sequence of the form (2) and for $0 \le u < v \le q - 1$ we have*

$$AA_d(q, t, u, v) = O(q^{-1/2}(\log p)^r), \quad 1 \le t \le q - 1,$$

where the implied constant depends on r only.

Proof. We may assume $r \ge 2$. As in the proof of Theorem 2 we can express $AA_d(q, t, u, v)$ in terms of character sums of the form

$$\sum_{\xi \in P'_\omega} \chi_d(\xi + \xi_t + \omega)\chi_d^{-1}(\xi),$$

where

$$P'_\omega = P_\omega \cap \{\xi_u, \xi_{u+1}, \ldots, \xi_v\}.$$

We split P'_ω in $2r - 1$ boxes,

$$V_{i,\omega} = \{\xi_n \in P'_\omega | n_i = v_i, n_{i+1} = v_{i+1}, \ldots, n_r = v_r, n_{i-1} \le v_{i-1} - 1\},$$

$$U_{i,\omega} = \{\xi_n \in P'_\omega | n_i = u_i, n_{i+1} = u_{i+1}, \ldots, n_r = u_r, n_{i-1} \ge u_{i-1} + 1\},$$

for $i = 3, 4, \ldots, r,$,

$$V_{2,\omega} = \{\xi_n \in P'_\omega | n_2 = v_2, n_3 = v_3, \ldots, n_r = v_r\},$$

$$U_{2,\omega} = \{\xi_n \in P'_\omega | n_2 = u_2, n_3 = u_3, \ldots, n_r = u_r\},$$

and

$$R_\omega = \{\xi_n \in P'_\omega | u_r + 1 \le n_r \le v_r - 1\}.$$

Now each character sum over $V_{i,\omega}$, $U_{i,\omega}$, and R_ω, respectively, can be estimated by

$$2q^{1/2}(\log p)^r$$

and the result follows.   □

5   Final Remarks

The *cyclotomic numbers* $(i, j)_d$ of order d are defined by

$$(i, j)_d = |(D_i + 1) \cap D_j|, \quad 0 \le i, j \le d - 1.$$

(For a monograph on cyclotomic numbers see [1].) The proof of [8, Theorem 5] for $A_3(q, t)$ depends on the knowledge of the cyclotomic numbers of order 3. It can be extended to $A_d(q, t)$ whenever the cyclotomic numbers of order d are known. In particular for $d = 2$ and $q = p$ a prime Theorem 1 follows already from the result of [23] (see also [28]). Here, we used the observation that the result does not depend on the knowledge of the cyclotomic numbers.

In contrast to Theorem 2 the result of Theorem 1 does not depend on the special choice of the ordering ξ_n, $n = 0, 1, \ldots, q - 1$, of the elements of $\mathbb{F}_q$.

Besides the autocorrelation, the linear complexity and the linear complexity profile are important measures for the randomness of sequences. For $r = 1$ and d a prime divisor of $q - 1$ in [7, 8] exact values for the linear complexity of the sequence (2) have been provided. In [18] lower bounds on the linear complexity of the sequence (2) for the general case were deduced. For lower bounds on the linear complexity profile see [12, 24, 25, 32]. Related sequences over $\mathbb{F}_d$, where d is a prime power divisor of $q - 1$, were investigated in [5].

The autocorrelation of similar sequences was determined in [14, 17, 26]. The *generalized Sidelńikov sequences* (σ_n) are the $q - 1$ periodic sequences defined by

$$\sigma_n = \chi_d(\gamma^n - 1), \quad n = 0, 1, \ldots, q - 2.$$

Equivalently, we can define $q - 1$ periodic sequences (s_n) over $\mathbb{Z}_d$ by

$$s_n = \mathrm{ind}_{\gamma,d}(\gamma^n - 1), \quad n = 0, 1, \ldots, q - 2.$$

First results on the linear complexity of these sequences in the binary case can be found in [10, 13]. Although Sidelńikov sequences and the sequences investigated in this paper look very similar at first glance, the latter sequences have some advantages. For large p and $r > 1$ we have two nice pseudorandom properties, Theorems 1 and 2, and the ordering ξ_n, $n = 0, 1, \ldots, q-1$, can be faster generated than the ordering $\gamma^n - 1$, $n = 0, 1, \ldots, q - 2$.

Theorem 2 yields nontrivial bounds only if the characteristic p is sufficiently large. Theorem 1 indicates that we have also good randomness properties in the case of small characteristic. However for some special choices of t our method also yields nontrivial bounds on $A_d(q, t)$ in the case of small characteristic p, where if $p = 2$ the sets P_ω are cosets of subgroups and a slightly better estimation can be performed involving some results of [29, 30]. It remains a challenging open problem to find a nontrivial upper bound on $A_d(q, t)$ for small characteristic and all t, as well.

The lower bounds on the linear complexity profile and the upper bounds on the autocorrelation confirm that the cyclotomic generator has good randomness properties. Because of the close relation of the sequence (2) to the discrete logarithm in $\mathbb{F}_q$ the good randomness properties also support the assumption of the intractability and unpredictability of the discrete logarithm, and thus of the hardness of the discrete logarithm problem.

Although for several cryptographic applications just the cases $q = p$ and $q = 2^r$ are important also the case that $q = p^r$ with small $r > 1$ and large p has gained increasing interest. In particular the *elliptic curve discrete logarithm problem* (see e.g. [2, 19]) and the *XTR discrete logarithm problem* (see [15, 16]) can be reduced to the discrete logarithm problem in a finite extension field.

Acknowledgment

The research of the authors is supported by DSTA research grant R-394-000-011-422. The second author is also supported by the Austrian Academy of Sciences and FWF research grant S8313.
The authors wish to thank Cunsheng Ding for useful discussion, in particular, for pointing to Sidelńikov sequences.

References

1. B. C. Berndt, R.J. Evans, and K. S. Williams, Gauss and Jacobi sums. Canadian Mathematical Society Series of Monographs and Advanced Texts. A Wiley-Interscience Publication. John Wiley & Sons, Inc., New York, 1998.

2. I. F. Blake, G. Seroussi, and N. P. Smart, Elliptic curves in cryptography. Reprint of the 1999 original. London Mathematical Society Lecture Note Series, 265. Cambridge University Press, Cambridge, 2000.

3. D. A. Burgess, On Dirichlet characters of polynomials. Proc. London Math. Soc. (3) 13 (1963), 537–548.

4. T. W. Cusick, C. Ding, and A. Renvall, Stream ciphers and number theory. North-Holland Mathematical Library, 55. North-Holland Publishing Co., Amsterdam, 1998.

5. Z. Dai, J. Yang, G. Gong, and P. Wang, On the linear complexity of generalised Legendre sequence. Sequences and their applications (Bergen, 2001), 145–153, Discrete Math. Theor. Comput. Sci. (Lond.), Springer, London, 2002.

6. H. Davenport and D. J. Lewis, Character sums and primitive roots in finite fields. Rend. Circ. Mat. Palermo (2) 12 (1963), 129–136.

7. C. Ding, T. Helleseth, and W. Shan, On the linear complexity of Legendre sequences, IEEE Transactions on Information Theory 44 (1998), 1276–1278.

8. C. Ding and T. Helleseth, On cyclotomic generator of order r. Inform. Process. Lett. 66 (1998), no. 1, 21–25.

9. T. Helleseth, On the crosscorrelation of m-sequences and related sequences with ideal autocorrelation. Sequences and their applications (Bergen, 2001), 34–45, Discrete Math. Theor. Sci. (Lond.), Springer, London, 2002.

10. T. Helleseth and K. Yang, On binary sequences of period $n = p^m - 1$ with optimal autocorrelation. Sequences and their applications (Bergen, 2001), 209–217, Discrete Math. Theor. Comput. Sci. (Lond.), Springer, London, 2002.

11. D. Jungnickel, Finite fields. Structure and arithmetics. Bibliographisches Institut, Mannheim, 1993.

12. S. Konyagin, T. Lange, and I. Shparlinski, Linear complexity of the discrete logarithm, Designs, Codes, and Cryptography 28 (2003), no. 2, 135–146.

13. G. M. Kyureghyan and A. Pott, On the linear complexity of the Sidelnikov-Lempel-Cohn-Eastman sequences, Designs, Codes, and Cryptography 29 (2003), 149–164.

14. A. Lempel, M. Cohn, and W. L. Eastman, A class of balanced binary sequences with optimal autocorrelation properties. IEEE Trans. Information Theory IT-23 (1977), no. 1, 38–42.

15. A. K. Lenstra and E. R. Verheul, The XTR public key system. Advances in cryptology—CRYPTO 2000 (Santa Barbara, CA), 1–19, Lecture Notes in Comput. Sci., 1880, Springer, Berlin, 2000.

16. A. K. Lenstra and E. R. Verheul, An overview of the XTR public key system. Public-key cryptography and computational number theory (Warsaw, 2000), 151–180, de Gruyter, Berlin, 2001.

17. H. D. Lüke, H. D. Schotten, and H. Hadinejad-Mahram, Generalized Sidelnikov sequences with optimal autocorrelation properties. Electronic Letters 36 (2000), no. 6, 525–527.

18. W. Meidl and A. Winterhof, Lower bounds on the linear complexity of the discrete logarithm in finite fields, IEEE Transactions on Information Theory 47 (2001), 2807–2811.

19. A. Menezes, Elliptic curve public key cryptosystems. The Kluwer International Series in Engineering and Computer Science, 234. Communications and Information Theory. Kluwer Academic Publishers, Boston, MA, 1993.

20. A. J. Menezes, P.C. van Oorschot, and S. A. Vanstone, Handbook of applied cryptography. With a foreword by Ronald L. Rivest. CRC Press Series on Discrete Mathematics and its Applications. CRC Press, Boca Raton, FL, 1997.

21. A. M. Naranjani, On Dirichlet characters of polynomials, Acta Arith. 43 (1984), 245–251.

22. G. I. Perelḿuter and I. E. Shparlinskiĭ, Distribution of primitive roots in finite fields. (Russian) Uspekhi Mat. Nauk 45 (1990), no. 1(271), 185–186; translation in Russian Math. Surveys 45 (1990), no. 1, 223–224.

23. O. Perron, Bemerkungen über die Verteilung der quadratischen Reste. Math. Z. 56 (1952), 122–130.

24. I. E. Shparlinski, Number Theoretic Methods in Cryptography. Basel, Birkhäuser, 1999.

25. I. E. Shparlinski, Cryptographic Applications of Analytic Number Theory. Basel, Birkhäuser, 2003.

26. V. M. Sidelńikov, Some k-valued pseudo-random sequences and nearly equidistant codes. Problems of Information Transmission 5 (1969), no. 1, 12–16.; translated from Problemy Peredači Informacii 5 (1969), no. 1, 16–22 (Russian).

27. A. Tietäväinen, Vinogradov's method and some applications. Number theory and its applications (Ankara, 1996), 261–282, Lecture Notes in Pure and Appl. Math., 204, Dekker, New York, 1999.

28. A. Winterhof, On the distribution of powers in finite fields. Finite Fields Appl. 4 (1998), no. 1, 43–54.

29. A. Winterhof, Some estimates for character sums and applications. Des. Codes Cryptogr. 22 (2001), no. 2, 123–131.

30. A. Winterhof, Incomplete additive character sums and applications. Finite fields and applications (Augsburg 1999), 462–476, Springer, Berlin, 2001.

31. A. Winterhof, Character sums, primitive elements, and powers in finite fields. J. Number Theory 91 (2001), no. 1, 153–163.

32. A. Winterhof, A note on the linear complexity profile of the discrete logarithm in finite fields, preprint.

The Weierstrass Semigroup of an m-tuple of Collinear Points on a Hermitian Curve

Gretchen L. Matthews

Department of Mathematical Sciences, Clemson University
Clemson SC 29634-0975, USA
gmatthe@clemson.edu,
WWW home page: www.math.clemson.edu/~gmatthe

Abstract. We examine the structure of the Weierstrass semigroup of an m-tuple of points on a smooth, projective, absolutely irreducible curve X over a finite field $\mathbb{F}$. A criteria is given for determining a minimal subset of semigroup elements which generate such a semigroup where $2 \leq m \leq |\mathbb{F}|$. For all $2 \leq m \leq q+1$, we determine the Weierstrass semigroup of any m-tuple of collinear $\mathbb{F}_{q^2}$-rational points on a Hermitian curve $y^q + y = x^{q+1}$.

1 Introduction

Let X be a smooth, projective, absolutely irreducible curve of genus $g > 1$ over a finite field $\mathbb{F}$. Let $\mathbb{F}(X)$ denote the field of rational functions on X defined over $\mathbb{F}$. The divisor of a rational function $f \in \mathbb{F}(X)$ will be denoted by (f) and the divisor of poles of f will be denoted by $(f)_\infty$.

Given m distinct $\mathbb{F}$-rational points $P_1, \ldots, P_m$ on X, the Weierstrass semigroup $H(P_1, \ldots, P_m)$ of the m-tuple $(P_1, \ldots, P_m)$ is defined by

$$H(P_1, \ldots, P_m) = \left\{ (\alpha_1, \ldots, \alpha_m) \in \mathbb{N}_0^m : \exists f \in \mathbb{F}(X) \text{ with } (f)_\infty = \sum_{i=1}^m \alpha_i P_i \right\},$$

and the Weierstrass gap set $G(P_1, \ldots, P_m)$ of the m-tuple $(P_1, \ldots, P_m)$ is defined by

$$G(P_1, \ldots, P_m) = \mathbb{N}_0^m \setminus H(P_1, \ldots P_m),$$

where $\mathbb{N}_0 := \mathbb{N} \cup \{0\}$ denotes the set of nonnegative integers. If $m = 1$, the Weierstrass gap set is the classically studied gap sequence. In [1], the authors generalized the notion of the semigroup of a point to the semigroup of a pair of points on a curve. This study was carried on by S. J. Kim [7] and M. Homma [5]. The Weierstrass gap set of an m-tuple of points where $m \geq 2$ has been examined by E. Ballico and Kim [2], and more recently, by C. Carvalho and F. Torres [3]. Weierstrass gap sets play an interesting role in the construction and analysis of algebraic geometry codes (see [4], [9], [6], [3]). While $| G(P_1) |= g$ for any $\mathbb{F}$-rational point P_1 on X, the cardinality of the set $G(P_1, \ldots, P_m)$ where $m \geq 2$ depends on the choice of points $P_1, \ldots, P_m$ [1]. However, any pair of $\mathbb{F}_{q^2}$-rational

G. Mullen, A. Poli and H. Stichtenoth (Eds.): Fq7 2003, LNCS 2948, pp. 12–24, 2003.

points on a Hermitian curve $y^q + y = x^{q+1}$ has the same Weierstrass semigroup [9]. The analogous result does not hold for triples of $\mathbb{F}_{q^2}$-rational points on a Hermitian curve [10].

In this paper, we consider the notion of a minimal generating subset of a Weierstrass semigroup of an m-tuple of points on an arbitrary (smooth, projective, absolutely irreducible) curve over a finite field $\mathbb{F}$. In Section 2, we discuss properties of minimal elements of the Weierstrass semigroup. This section concludes with a useful characterization of the elements of the minimal generating set of the Weierstrass semigroup of an m-tuple of points for $2 \leq m \leq |\mathbb{F}|$. An interesting application of this is found in Section 3 where we see that any m-tuple of collinear $\mathbb{F}_{q^2}$-rational points on a Hermitian curve $y^q + y = x^{q+1}$ has the same Weierstrass semigroup. In addition, we determine this Weierstrass semigroup and its minimal generating set.

2   Results for Arbitrary Curves

Let X be a smooth, projective, absolutely irreducible curve of genus $g > 1$ over a finite field $\mathbb{F}$. Fix m distinct $\mathbb{F}$-rational points $P_1, \ldots, P_m$ on X, where $2 \leq m \leq |\mathbb{F}|$. For $1 \leq l \leq m$, set $H_l := H(P_1, \ldots, P_l)$. Define a partial order $\preceq$ on $\mathbb{N}_0^m$ by $(n_1, \ldots, n_m) \preceq (p_1, \ldots, p_m)$ if and only if $n_i \leq p_i$ for all i, $1 \leq i \leq m$. It is convenient to collect here two results from [3] that will be used in this section.

Lemma 1. [3] If $(n_1, \ldots, n_m), (p_1, \ldots, p_m) \in H_m$ and $n_j = p_j$ for some j, $1 \leq j \leq m$, then there exists $\mathbf{q} = (q_1, \ldots, q_m) \in H_m$ whose coordinates satisfy the following properties:

1. $q_i = max(n_i, p_i)$ for $i \neq j$ and $n_i \neq p_i$.
2. $q_i \leq n_i$ for $i \neq j$ and $n_i = p_i$.
3. $q_j = n_j = 0$ or $q_j < n_j$.

Lemma 2. [3] Suppose that there exists i, $1 \leq i \leq m$, such that $(n_1, \ldots, n_m)$ is a minimal element of the set $\{\mathbf{p} \in H_m : p_i = n_i\}$ with respect to $\preceq$. If $n_i > 0$ and $n_j > 0$ for some j, $1 \leq j \leq m$, $j \neq i$, then $n_i \in G(P_i)$.

Proposition 3. Let $\mathbf{n} \in \mathbb{N}^m$. Then $\mathbf{n}$ is minimal in $\{\mathbf{p} \in H_m : p_i = n_i\}$ with respect to $\preceq$ for some i, $1 \leq i \leq m$, if and only if $\mathbf{n}$ is minimal in the set $\{\mathbf{p} \in H_m : p_i = n_i\}$ with respect to $\preceq$ for all i, $1 \leq i \leq m$.

Proof. Suppose $\mathbf{n} \in \mathbb{N}^m$ is minimal in $\{\mathbf{p} \in H_m : p_i = n_i\}$ with respect to $\preceq$ for some i, $1 \leq i \leq m$. Without loss of generality, we may assume that $i = 1$. Suppose there exists j, $2 \leq j \leq m$, such that $\mathbf{n}$ is not minimal in $\{\mathbf{p} \in H_m : p_j = n_j\}$. Then there exists $\mathbf{v} \in H_m$ such that $\mathbf{v} \preceq \mathbf{n}$, $\mathbf{v} \neq \mathbf{n}$, and $v_j = n_j$. Note that $v_1 < n_1$ as otherwise $\mathbf{v} \in \{\mathbf{p} \in H_m : p_1 = n_1\}$ contradicting the minimality of $\mathbf{n}$. Applying Lemma 1, we see that there exists $\mathbf{q} \in H_m$ with $q_1 = n_1$, $q_j < n_j$, and $q_i \leq n_i$ for all $1 \leq i \leq m$. Thus, $\mathbf{q} \preceq \mathbf{n}$, $\mathbf{q} \neq \mathbf{n}$, and $\mathbf{q} \in \{\mathbf{p} \in H_m : p_1 = n_1\}$. This contradicts the minimality of $\mathbf{n} \in \{\mathbf{p} \in H_m : p_1 = n_1\}$. Thus, $\mathbf{n}$ is minimal in $\{\mathbf{p} \in H_m : p_j = n_j\}$ for all j, $1 \leq j \leq m$.

Using these ideas, we set out to describe a subset of H_m that generates the entire semigroup H_m. To begin, set $\Gamma_1^+ = H(P_1)$, the Weierstrass semigroup of the point P_1. For $2 \leq l \leq m$, define

$$\Gamma_l^+ := \{\mathbf{n} \in \mathbb{N}^l : \mathbf{n} \text{ is minimal in } \{\mathbf{p} \in H_l : p_i = n_i\} \text{ for some } i, 1 \leq i \leq l\}.$$

The notion of Γ_2^+ is due to Kim [7]. As an immediate consequence of Proposition 3 and Lemma 2, we obtain the following result.

Lemma 4. *For* $2 \leq l \leq m$, $\Gamma_l^+ \subseteq G(P_1) \times \cdots \times G(P_l)$.

Using Γ_l^+, we will now describe a subset Γ_l of H_l for $1 \leq l \leq m$. First, set $\Gamma_1 = \Gamma_1^+ = H(P_1)$. For $2 \leq l \leq m$, define

$$\Gamma_l := \Gamma_l^+ \cup \left\{ \mathbf{n} \in \mathbb{N}_0^l : \begin{array}{l} (n_{i_1}, \ldots, n_{i_k}) \in \Gamma_k^+ \text{ for some } \{i_1, \ldots, i_m\} = \{1, \ldots, m\} \\ \text{such that } i_1 < \cdots < i_k \text{ and } n_{i_{k+1}} = \cdots = n_{i_m} = 0 \end{array} \right\}.$$

Clearly, Γ_m is completely determined by $\{\Gamma_l^+ : 1 \leq l \leq m\}$.

Example 5. Consider the curve defined by $y^8 + y = x^9$ over $\mathbb{F}_{64}$. Let $P_1 = P_\infty$ denote the point at infinity and $P_2 = P_{00}$ denote the common zero of x and y. It is well known that the Weierstrass gap set of the point P_1 (and P_2) is

$$
\begin{array}{ccccccc}
1 & 2 & 3 & 4 & 5 & 6 & 7 \\
10 & 11 & 12 & 13 & 14 & 15 \\
19 & 20 & 21 & 22 & 23 \\
28 & 29 & 30 & 31 \\
37 & 38 & 39 \\
46 & 47 \\
55
\end{array} \qquad .
$$

Equivalently, the Weierstrass semigroup of the point P_1 is the additive subsemigroup of $\mathbb{N}_0$ generated by 8 and 9; that is, $H(P_1) = \langle 8, 9 \rangle := \{8a + 9b : a, b \in \mathbb{N}_0\}$. Hence, $\Gamma_1 = \langle 8, 9 \rangle$. According to [9],

$$
\Gamma_2^+ = \left\{
\begin{array}{l}
(1, 55), (2, 47), (3, 39), (4, 31), (5, 23), (6, 15), (7, 7), (10, 46), \\
(11, 38), (12, 30), (13, 22), (14, 14), (15, 6), (19, 37), (20, 29), \\
(21, 21), (22, 13), (23, 5), (28, 28), (29, 20), (30, 12), (31, 4), \\
(37, 19), (38, 11), (39, 3), (46, 10), (47, 2), (55, 1)
\end{array}
\right\} .
$$

Then

$$\Gamma_2 = \Gamma_2^+ \cup \{(n, 0), (0, n) : n \in \langle 8, 9 \rangle\}.$$

We will show that Γ_m generates H_m by taking least upper bounds. Given $\mathbf{u}_1, \ldots, \mathbf{u}_l \in \mathbb{N}_0^m$, define the least upper bound of $\mathbf{u}_1, \ldots, \mathbf{u}_l$ by

$$\text{lub}\{\mathbf{u}_1, \ldots, \mathbf{u}_l\} = (\max\{u_{1_1}, \ldots, u_{l_1}\}, \ldots, \max\{u_{1_m}, \ldots, u_{l_m}\}) \in \mathbb{N}_0^m$$

In [7], Kim proved that $H_2 = \{\text{lub}\{\mathbf{u}_1, \mathbf{u}_2\} \in \mathbb{N}_0^2 : \mathbf{u}_1, \mathbf{u}_2 \in \Gamma_2\}$. To obtain a similar result for H_m where $m \geq 3$, we use the next fact which follows immediately from [3].

Proposition 6. *Suppose that* $1 \leq l \leq m \leq |\mathbb{F}|$ *and* $\mathbf{u_1}, \ldots, \mathbf{u_l} \in H_m$. *Then* $\mathrm{lub}\{\mathbf{u_1}, \ldots, \mathbf{u_l}\} \in H_m$.

Proof. Let $\mathbf{q_2} := \mathrm{lub}\{\mathbf{u_1}, \mathbf{u_2}\}$. For $3 \leq i \leq l$, define $\mathbf{q_i} := \mathrm{lub}\{\mathbf{q_{i-1}}, \mathbf{u_i}\}$. According to [3], $\mathbf{q_2} \in H_m$. Repeated application gives $\mathbf{q_i} \in H_m$ for all $i \in \{2, \ldots, l\}$. This completes the proof as $\mathrm{lub}\{\mathbf{u_1}, \ldots, \mathbf{u_l}\} = \mathbf{q_l} \in H_m$.

Theorem 7. *If* $1 \leq m \leq |\mathbb{F}|$, *then*

$$H_m = \{\mathrm{lub}\{\mathbf{u_1}, \ldots, \mathbf{u_m}\} \in \mathbb{N}_0^m : \mathbf{u_1}, \ldots, \mathbf{u_m} \in \Gamma_m\}.$$

Proof. The fact that $\{\mathrm{lub}\{\mathbf{u_1}, \ldots, \mathbf{u_m}\} \in \mathbb{N}_0^m : \mathbf{u_1}, \ldots, \mathbf{u_m} \in \Gamma_m\} \subseteq H_m$ follows from Proposition 6.

Suppose $\mathbf{n} \in H_m \setminus \Gamma_m$. Without loss of generality, we may assume that $\mathbf{n} \in \mathbb{N}^m$. (Otherwise, $(n_{i_1}, \ldots n_{i_l}) \in \mathbb{N}^l$ for some $\{i_1, \ldots, i_m\} = \{1, \ldots, m\}$ such that $i_1 < \cdots < i_l$ and $n_{i_{l+1}} = \cdots = n_{i_m} = 0$, and the same argument applies to $(n_{i_1}, \ldots n_{i_l})$). Then, according to Proposition 3, $\mathbf{n}$ is not minimal in $\{\mathbf{p} \in H_m : p_i = n_i\}$ for any i, $1 \leq i \leq m$. Hence, there exists $\mathbf{u_i} \in \Gamma_m$ with $u_{i_i} = n_i$, $\mathbf{u_i} \preceq \mathbf{n}$, and $\mathbf{u_i} \neq \mathbf{n}$ for each i, $1 \leq i \leq m$. Then $\mathbf{n} = \mathrm{lub}\{\mathbf{u_1}, \ldots, \mathbf{u_m}\}$, completing the proof.

According to Theorem 7 and the definition of Γ_m, the Weierstrass semigroup H_m is completely determined by $\{\Gamma_l^+ : 1 \leq l \leq m\}$. We conclude this section with a useful characterization of elements of the sets Γ_l^+, $1 \leq l \leq m$. To do this, it is helpful to consider dimensions of certain divisors. For a divisor D on X defined over $\mathbb{F}$, let $L(D)$ denote the set of rational functions $f \in \mathbb{F}(X)$ with divisor $(f) \geq -D$ together with the zero function. Then $L(D)$ is a finite dimensional vector space over $\mathbb{F}$. Let $l(D)$ denote the dimension of the vector space $L(D)$ over $\mathbb{F}$. The Riemann-Roch Theorem states that $l(D) = \deg D + 1 - g + l(K - D)$, where K is any canonical divisor on X. This gives a characterization of elements of the Weierstrass semigroup of an m-tuple $(P_1, \ldots, P_m)$ according to dimensions of divisors supported by the points $P_1, \ldots, P_m$. This is an easy generalization of a lemma due to Kim [7].

Lemma 8. *For* $(\alpha_1, \ldots, \alpha_m) \in \mathbb{N}^m$, *the following are equivalent:*
(i) $(\alpha_1, \ldots, \alpha_m) \in H(P_1, \ldots, P_m)$.
(ii) $l(\sum_{i=1}^m \alpha_i P_i) = l((\alpha_j - 1)P_j + \sum_{i=1, i \neq j}^m \alpha_i P_i) + 1$ *for all* j, $1 \leq j \leq m$.

Proposition 9. *Let* $1 \leq l \leq m \leq |\mathbb{F}|$ *and* $\mathbf{n} \in \mathbb{N}^l$. *Then* $\mathbf{n} \in \Gamma_l^+$ *if and only if* $\mathbf{n} \in H_l$ *and* $l(\sum_{j=1}^l (n_j - 1)P_j) = l((n_k - 1)P_k + \sum_{j=1, j \neq k}^l n_j P_j)$ *for all* k, $1 \leq k \leq l$.

Proof. Suppose $\mathbf{n} \in \Gamma_l^+$. If $l(\sum_{j=1}^l (n_j - 1)P_j) \neq l((n_k - 1)P_k + \sum_{j=1, j \neq k}^l n_j P_j)$ for some k, $1 \leq k \leq l$, then there exists $\mathbf{v} \in H_l$ with $\mathbf{v} \preceq \mathbf{n}$, $v_k \leq n_k - 1$, and $v_t = n_t$ for some t, $1 \leq t \leq l$. This contradicts the assumption that $\mathbf{n}$ is minimal in $\{\mathbf{p} \in H_l : p_t = n_t\}$. Thus, $l(\sum_{j=1}^l (n_j - 1)P_j) = l((n_k - 1)P_k + \sum_{j=1, j \neq k}^l n_j P_j)$ for all k, $1 \leq k \leq l$.

Suppose $\mathbf{n} \in H_l$ and $l(\sum_{j=1}^{l}(n_j - 1)P_j) = l((n_k - 1)P_k + \sum_{j=1, j\neq k}^{l} n_j P_j)$ for all k, $1 \leq k \leq l$. This implies

$$L\left((n_1 - 1)P_1 + \sum_{j=2}^{l} n_j P_j\right) = L\left(\sum_{j=1}^{l}(n_j - 1)P_j\right) = L\left((n_k - 1)P_k + \sum_{\substack{j=1 \\ j \neq k}}^{l} n_j P_j\right)$$

for all k, $1 \leq k \leq l$, as $L(\sum_{j=1}^{l}(n_j - 1)P_j) \subseteq L((n_k - 1)P_k + \sum_{j=1, j\neq k}^{l} n_j P_j)$. If $\mathbf{n} \notin \Gamma_l^+$, then there exists $\mathbf{u} \in H_l$ with $u_1 = n_1$, $\mathbf{u} \preceq \mathbf{n}$, and $\mathbf{u} \neq \mathbf{n}$. In particular, $u_k < n_k$ for some k, $2 \leq k \leq l$. Thus, there exists a rational function $f \in L((n_k - 1)P_k + \sum_{j=1, j\neq k}^{l} n_j P_j)$ such that $f \notin L((n_1 - 1)P_1 + \sum_{j=2}^{l} n_j P_j)$, which is a contradiction.

3   Computation of $H(P_1, \ldots, P_m)$ for Collinear Points $P_1, \ldots, P_m$ on a Hermitian Curve

In this section, we restrict our attention to the curve X defined by $y^q + y = x^{q+1}$ over $\mathbb{F}_{q^2}$. Given $a, b \in \mathbb{F}_{q^2}$ with $b^q + b = a^{q+1}$, let P_{ab} denote the common zero of $x - a$ and $y - b$. Fix $a \in \mathbb{F}_{q^2}$. Then there are exactly q elements $b_2, \ldots, b_{q+1} \in \mathbb{F}_{q^2}$ such that $b_i^q + b_i = a^{q+1}$. Set $P_1 = P_\infty, P_2 = P_{ab_2}, P_3 = P_{ab_3}, \ldots, P_{q+1} = P_{ab_{q+1}}$. For $1 \leq m \leq q+1$, let $H_m := H(P_1, \ldots, P_m)$. We set out to determine Γ_m for all $1 \leq m \leq q+1$.

Notice that the divisors of $x - a$ and y are given by

$$(x - a) = \sum_{i=2}^{q+1} P_{ab_i} - q P_\infty \qquad \text{and} \qquad (y) = (q+1)(P_{00} - P_\infty).$$

It will also be useful to consider functions $h_{ab_i} := y - b_i - a^q(x - a)$ where $2 \leq i \leq q+1$. Note that the divisor of h_{ab_i} is given by

$$(h_{ab_i}) = (q+1)(P_{ab_i} - P_\infty)$$

(see [8]). Using the functions x and y and the fact that X is a curve of genus $\frac{q(q-1)}{2}$, one can check $H(P_1) = \langle q, q+1 \rangle$ and that the Weierstrass gap set $G(P_1)$ is

$$
\begin{array}{ccccc}
1 & 2 & \cdots & q-2 & q-1 \\
(q+1)+1 & (q+1)+2 & \cdots & (q+1)+(q-2) & \\
\vdots & \vdots & \cdot^{\cdot^{\cdot}} & & \cdot^{\cdot^{\cdot}} \\
(q-3)(q+1)+1 & (q-3)(q+1)+2 & & & \\
(q-2)(q+1)+1 & & & &
\end{array}
$$

In fact, the above set is the Weierstrass gap set of any $\mathbb{F}_{q^2}$-rational point on X. Given $\alpha \in G(P)$ where P is any $\mathbb{F}_{q^2}$-rational point, α can be written uniquely as $\alpha = (t - j)(q + 1) + j$ with $1 \leq j \leq t \leq q - 1$. Here, j denotes the column containing α and t denotes the diagonal containing α in the above diagram.

From above, $\Gamma_1^+ = H(P_1) = \langle q, q + 1 \rangle$. According to [9, Theorem 3.7],

$$\Gamma_2^+ = \left\{ ((t_1 - j)(q + 1) + j, (t_2 - j)(q + 1) + j) : \begin{array}{l} 1 \leq j \leq t_1, t_2 \leq q - 1, \\ t_1 + t_2 = q + j - 1 \end{array} \right\}.$$

To describe Γ_m^+ for $3 \leq m \leq q + 1$, we must set up some notation. Given $1 \leq m \leq q + 1$, $\mathbf{t} = (t_1, \ldots, t_m) \in \mathbb{N}^m$, and $j \in \mathbb{N}$, define

$$\boldsymbol{\gamma}_{\mathbf{t},j} := ((t_1 - j)(q + 1) + j, (t_2 - j)(q + 1) + j, \ldots, (t_m - j)(q + 1) + j) \in \mathbb{N}_0^m.$$

Notice that if $1 \leq j \leq t_i \leq q - 1$ for all $1 \leq i \leq m$, then

$$\boldsymbol{\gamma}_{\mathbf{t},j} \in G(P_1) \times G(P_2) \times \cdots \times G(P_m).$$

We next show that certain $\boldsymbol{\gamma}_{\mathbf{t},j}$ form a generating set for the Weierstrass semigroup H_m.

Theorem 10. *Let $a \in \mathbb{F}_{q^2}$ and $P_1 = P_\infty, P_2 = P_{ab_2}, P_3 = P_{ab_3}, \ldots, P_{q+1} = P_{ab_{q+1}}$ be $q + 1$ distinct $\mathbb{F}_{q^2}$-rational points on the Hermitian curve X defined by $y^q + y = x^{q+1}$. For $2 \leq m \leq q + 1$,*

$$\Gamma_m^+ = \left\{ \boldsymbol{\gamma}_{\mathbf{t},j} : \begin{array}{l} \sum_{i=1}^m t_i = q + (m - 1)(j - 1), \\ 1 \leq j \leq t_i \leq q - 1 \text{ for all } 1 \leq i \leq m \end{array} \right\}.$$

In particular, the Weierstrass semigroup $H(P_1, \ldots, P_m)$ is generated by

$$\left\{ \begin{array}{c} \mathbf{n} \in \mathbb{N}_0^m : (n_{i_1}, \ldots n_{i_l}) = \boldsymbol{\gamma}_{\mathbf{t},j} \in \Gamma_l^+ \text{ and } n_{i_{l+1}} = \cdots = n_{i_m} = 0 \\ \text{for some } l \in \mathbb{N} \text{ and } \{i_1, \ldots, i_m\} = \{1, \ldots, m\} \end{array} \right\}.$$

Proof. We begin by setting up some notation. For $2 \leq m \leq q + 1$, set

$$S_m := \left\{ \boldsymbol{\gamma}_{\mathbf{t},j} : \begin{array}{l} \sum_{i=1}^m t_i = q + (m - 1)(j - 1), \\ 1 \leq j \leq t_i \leq q - 1 \text{ for all } 1 \leq i \leq m \end{array} \right\}.$$

For each $2 \leq i \leq q + 1$, let $h_i := h_{ab_i} \in \mathbb{F}_{q^2}(X)$ be as above so that

$$(h_i) = (q + 1)P_i - (q + 1)P_1.$$

Given $\mathbf{v} := (v_1, \ldots, v_m) \in \mathbb{Z}^m$, let $\mathbf{v}^+ := (v_{i_1}, \ldots, v_{i_l}) \in \mathbb{N}^l$ where $i_1 < \cdots < i_l$ and $v_i > 0$ if and only if $i = i_r$ for some $1 \leq r \leq l$; that is, $\mathbf{v}^+$ is the vector formed from $\mathbf{v}$ by deleting each coordinate of $\mathbf{v}$ containing a negative or zero entry.

We will prove that $\Gamma_m^+ = S_m$ by induction on m. By [9, Theorem 3.7],

$$\Gamma_2^+ = \{ \boldsymbol{\gamma}_{(t_1, t_2), j} : t_1 + t_2 = q + j - 1, 1 \leq j \leq t_1, t_2 \leq q - 1 \} = S_2,$$

which settles the case where $m = 2$. We now proceed by induction on $m \geq 3$. Assume that $\Gamma_l^+ = S_l$ holds for all $2 \leq l \leq m - 1$.

First, we claim that $S_m \subseteq \Gamma_m^+$. Let $\gamma_{\mathbf{t},j} \in S_m$. Then

$$\left(\frac{(x-a)^{q-j+1}}{h_2^{t_2-j+1} h_3^{t_3-j+1} \cdots h_m^{t_m-j+1}} \right)_\infty = \sum_{i=1}^{m} ((t_i - j)(q+1) + j) P_i.$$

Hence, $\gamma_{\mathbf{t},j} \in H_m$.

In order to show that $\gamma_{\mathbf{t},j} \in \Gamma_m^+$, it suffices to prove that $\gamma_{\mathbf{t},j}$ is minimal in $\{\mathbf{p} \in H_m : p_1 = (t_1 - j)(q+1) + j\}$. Suppose $\gamma_{\mathbf{t},j}$ is not minimal in

$$\{\mathbf{p} \in H_m : p_1 = (t_1 - j)(q+1) + j\}.$$

Then there exists $\mathbf{u} \in H_m$ with $u_1 = (t_1 - j)(q+1) + j$, $\mathbf{u} \preceq \gamma_{\mathbf{t},j}$, and $\mathbf{u} \neq \gamma_{\mathbf{t},j}$. Let $f \in \mathbb{F}_{q^2}(X)$ be such that $(f)_\infty = u_1 P_1 + \cdots + u_m P_m$. Without loss of generality, we may assume that $u_m < (t_m - j)(q+1) + j$ as $\mathbf{u} \neq \gamma_{\mathbf{t},j}$ gives $u_i < (t_i - j)(q+1) + j$ for some $2 \leq i \leq m$ and a similar argument holds if $2 \leq i \leq m-1$. Hence,

$$u_m = (t_m - j)(q+1) + j - k$$

for some $k \geq 1$. There are two cases to consider:

$$(1)\ j > k.$$
$$(2)\ j \leq k.$$

Case (1): Suppose $j > k$. Then

$$\left(f h_m^{t_m-j}(x-a)^{j-k} \right)_\infty = ((t_1 + t_m - j - k)(q+1) + k) P_1 + \sum_{i=2}^{m-1} \max\{u_i - (j-k), 0\} P_i.$$

Therefore,

$$\mathbf{v} := ((t_1 + t_m - j - k)(q+1) + k, v_2, \ldots, v_{m-1}) \in H_{m-1},$$

where $v_i = \max\{u_i - (j-k), 0\}$ for $2 \leq i \leq m-1$. Set

$$\mathbf{w} := \gamma_{(t_1+t_m-j, t_2-j+1+k, t_3-j+k, \ldots, t_{m-1}-j+k), k}.$$

Clearly,

$$\mathbf{v} \preceq \mathbf{w}.$$

Note that

$$\mathbf{w} \in S_{m-1}$$

since $t_1 + t_m - j + t_2 - j + 1 + k + \sum_{i=3}^{m-1}(t_i - j + k) = q + (m-2)(k-1)$, $k \leq t_2 - j + 1 + k \leq t_2 \leq q-1$ as $j - k > 0$, $k \leq t_i - j + k \leq t_i \leq q - 1$ for $3 \leq i \leq m-1$, and $k \leq j \leq t_1 + t_m - j \leq q - 1$ (otherwise, $\sum_{i=2}^{m-1} t_i \leq (m-2)(j-1) < (m-2)j$). By the induction hypothesis, $S_{m-1} = \Gamma_{m-1}^+$, and so

$$\mathbf{w} \in \Gamma_{m-1}^+.$$

By Proposition 3, $\mathbf{w}$ is minimal in $\{\mathbf{p} \in H_{m-1} : p_1 = (t_1 + t_m - j - k)(q+1) + k\}$. This leads to a contradiction as

$$\mathbf{v} \in \{\mathbf{p} \in H_{m-1} : p_1 = (t_1 + t_m - j - k)(q+1) + k\},$$
$$\mathbf{v} \preceq \mathbf{w}, \text{ and}$$
$$\mathbf{v} \neq \mathbf{w}.$$

Case (2): Suppose $j \leq k$. Then

$$\left(fh_m^{t_m - j}\right)_\infty = ((t_1 + t_m - 2j)(q+1) + j)P_1 + \sum_{i=2}^{m-1} u_i P_i$$

which implies that

$$\mathbf{v} := ((t_1 + t_m - j - j)(q+1) + j, u_2, \ldots, u_{m-1}) \in H_{m-1}.$$

Note that there exists i, $2 \leq i \leq m-1$, such that $t_i < q - 1$ since otherwise $2j \leq t_1 + t_m = q + (m-1)(j-1) - (m-2)(q-1)$ implies that $0 \leq 2 - m$ contradicting the assumption that $m \geq 3$. We may assume that $i = 2$ as a similar argument holds in the case $2 < i \leq m-1$. Set

$$\mathbf{w} := \gamma_{(t_1 + t_m - j, t_2 + 1, t_3 \ldots, t_{m-1}), j}.$$

Clearly,

$$\mathbf{v} \preceq \mathbf{w}.$$

Also note that

$$\mathbf{w} \in S_{m-1}$$

since $t_1 + t_m - j + t_2 + 1 + \sum_{i=3}^{m-1} t_i = q + (m-2)(j-1)$, $j \leq t_2 + 1 \leq q - 1$ as $t_2 < q - 1$, $j \leq t_i \leq q - 1$ for $3 \leq i \leq m - 1$, and $j \leq t_1 + t_m - j \leq q - 1$. By the induction hypothesis, $S_{m-1} = \Gamma_{m-1}^+$, and so

$$\mathbf{w} \in \Gamma_{m-1}^+.$$

By Proposition 3, $\mathbf{w}$ is minimal in $\{\mathbf{p} \in H_{m-1} : p_1 = (t_1 + t_m - j - j)(q+1) + j\}$. This leads to a contradiction as

$$\mathbf{v} \in \{\mathbf{p} \in H_{m-1} : p_1 = (t_1 + t_m - j - j)(q+1) + j\},$$
$$\mathbf{v} \preceq \mathbf{w}, \text{ and}$$
$$\mathbf{v} \neq \mathbf{w}.$$

Since both cases (1) and (2) yield a contradiction, it must be the case that $\gamma_{\mathbf{t}, j}$ is minimal in $\{\mathbf{p} \in H_m : p_1 = (t_1 - j)(q+1) + j\}$.

Therefore, by the definition of Γ_m^+, we have that $\gamma_{\mathbf{t}, j} \in \Gamma_m^+$. This completes the proof of the claim that

$$S_m \subseteq \Gamma_m^+.$$

Next, we will show that $\Gamma_m^+ \subseteq S_m$.

Suppose not; that is, suppose that there exists $\mathbf{n} \in \Gamma_m^+ \setminus S_m$. Then there exists $f \in \mathbb{F}_{q^2}(X)$ with pole divisor $(f)_\infty = n_1 P_1 + \cdots + n_m P_m$. By Lemma 4,

$$\mathbf{n} \in \Gamma_m^+ \subseteq G(P_1) \times G(P_2) \times \cdots \times G(P_m).$$

Thus,

$$\mathbf{n} = ((t_1 - j_1)(q + 1) + j_1, (t_2 - j_2)(q + 1) + j_2, \ldots, (t_m - j_m)(q + 1) + j_m)$$

where $1 \leq j_i \leq t_i \leq q - 1$ for all $1 \leq i \leq m$.

Without loss of generality, we may assume that $j_m = \max\{j_i : 2 \leq i \leq m\}$ as a similar argument holds if $j_r = \max\{j_i : 2 \leq i \leq m\}$ for some $2 \leq r \leq m-1$. Then

$$(fh_m^{t_m - j_m + 1})_\infty = (n_1 + (t_m - j_m + 1)(q + 1))P_1 + \sum_{i=2}^{m-1} n_i P_i,$$

which implies that $(n_1 + (t_m - j_m + 1)(q+1), n_2, \ldots, n_{m-1}) \in H_{m-1}$. Then there exists $\mathbf{u} \in \Gamma_{m-1}$ such that

$$\mathbf{u} \preceq (n_1 + (t_m - j_m + 1)(q + 1), n_2, \ldots, n_{m-1})$$

and $u_2 = n_2 = (t_2 - j_2)(q+1) + j_2$. If $u_1 \leq n_1$, then $(u_1, \ldots, u_{m-1}, 0) \preceq \mathbf{n}$ which contradicts the minimality of $\mathbf{n}$ in $\{\mathbf{p} \in H_m : p_2 = n_2\}$. Thus, $u_1 > n_1 > 0$. By the induction hypothesis,

$$\mathbf{u}^+ = \boldsymbol{\gamma}_{(T_{i_1}, \ldots, T_{i_l}), j'} \in S_l = \Gamma_l^+$$

for some l, $2 \leq l \leq m - 1$, and some $(T_{i_1}, \ldots, T_{i_l})$ and j' satisfying $1 \leq j' \leq T_{i_r} \leq q - 1$ for $1 \leq r \leq l$ and $\sum_{r=1}^l T_{i_r} = q + (l-1)(j' - 1)$.

Hence, there exists an index set $\{i_1, \ldots, i_{m-1}\} = \{1, \ldots, m-1\}$ such that $i_1 < i_2 < \cdots < i_l$ and

$$u_{i_r} = \begin{cases} (T_{i_r} - j')(q + 1) + j' & \text{if } 1 \leq r \leq l \\ 0 & \text{if } l + 1 \leq r \leq m - 1 \end{cases}.$$

Since $u_1 > n_1 > 0$, $i_1 = 1$. Similarly, $i_2 = 2$ because $u_2 = n_2 \neq 0$. Since

$$(T_2 - j')(q + 1) + j' = u_{i_2} = u_2 = (t_2 - j_2)(q + 1) + j_2$$

implies that $(q + 1) \mid (j' - j_2)$, we must have that $j' = j_2$ as $-(q-1) \leq j' - j_2 \leq q - 1$. In addition, $T_2 = t_2$. As a result,

$$\mathbf{u}^+ = \boldsymbol{\gamma}_{(T_1, T_2, T_{i_3}, \ldots, T_{i_l}), j_2},$$

$$u_{i_r} = \begin{cases} (T_{i_r} - j_2)(q + 1) + j_2 & \text{if } 1 \leq r \leq l \\ 0 & \text{if } l + 1 \leq r \leq m - 1 \end{cases},$$

$T_1 + T_2 + T_{i_3} + \cdots + T_{i_l} = q + (l-1)(j_2 - 1)$, and $j_2 \leq T_{i_r} \leq q-1$ for all $1 \leq r \leq l$. At this point, we separate the remainder of the proof into two cases:

$$(1) \ u_1 - (t_m - j_m + 1)(q + 1) \geq 0$$
$$(2) \ u_1 - (t_m - j_m + 1)(q + 1) < 0$$

Case (1): Suppose $u_1 - (t_m - j_m + 1)(q + 1) \geq 0$.
Since $q + 1 \nmid j_2$, it follows that $u_1 - (t_m - j_m + 1)(q + 1) > 0$. Set

$$\mathbf{v} := (u_1 - (t_m - j_m + 1)(q+1), u_2, u_3, \ldots, u_{m-1}, (t_m - j_m + j_2 - j_2)(q+1) + j_2).$$

Notice that $\mathbf{v} \preceq \mathbf{n}$ since $u_1 \leq n_1 + (t_m - j_m + 1)(q+1)$, $u_i \leq n_i$ for $2 \leq i \leq m-1$, and $j_2 \leq j_m = \max\{j_i : 2 \leq i \leq m\}$. We claim that $\mathbf{v}^+ \in S_{l+1}$. To see this, it is helpful to express $\mathbf{v}^+$ as

$$\mathbf{v}^+ = \boldsymbol{\gamma}_{(T_1 - t_m + j_m - 1, T_2, T_{i_3}, \ldots, T_{i_l}, t_m - j_m + j_2), j_2}.$$

It is easy to see that $T_1 - t_m + j_m - 1 + T_2 + T_{i_3} + \cdots + T_{i_l} + t_m - j_m + j_2 = q + l(j_2 - 1)$, $T_1 - (t_m - j_m) - 1 \leq T_1 \leq q - 1$, $j_2 \leq T_{i_r} \leq q - 1$ for $2 \leq r \leq l$, and $j_2 \leq t_m - j_m + j_2 \leq t_m \leq q - 1$ as $j_2 \leq j_m$. If $T_1 - t_m + j_m - 1 < j_2$, then $u_1 - (t_m - j_m + 1)(q+1) = (T_1 - j_2 - (t_m - j_m + 1))(q+1) + j_2 < 0$ which is not the case.

Thus, $j_2 \leq T_1 - t_m + j_m - 1$, establishing the claim that $\mathbf{v}^+ \in S_{l+1}$. Since $S_{l+1} \subseteq \Gamma_{l+1}^+ \subseteq H_{l+1}$, it follows that $\mathbf{v} \in \Gamma_m \subseteq H_m$. Now, $\mathbf{v} \preceq \mathbf{n}$ and $\mathbf{n} \in \Gamma_m^+$ force $\mathbf{n} = \mathbf{v}$ as otherwise $\mathbf{n}$ is not minimal in $\{\mathbf{p} \in H_m : p_2 = n_2\}$. Hence, $l + 1 = m$ and $\mathbf{n} = \mathbf{v} = \mathbf{v}^+ \in S_m$, which is a contradiction.

Case (2): Suppose that $u_1 - (t_m - j_m + 1)(q + 1) < 0$.
There are two subcases to consider:

$$(a) \ j_1 < t_1.$$
$$(b) \ j_1 = t_1.$$

Subcase (a): Suppose $j_1 < t_1$. Set

$$\mathbf{v} := ((t_1 - j_1 + j_2 - 1 - j_2)(q+1) + j_2, u_2, \ldots, u_{m-1}, (T_1 - t_1 + j_1 - j_2)(q+1) + j_2).$$

Notice that $\mathbf{v} \preceq \mathbf{n}$ and $\mathbf{v} \neq \mathbf{n}$ since $(t_1 - j_1 - 1)(q+1) + j_2 \leq (t_1 - j_1)(q+1) \leq (t_1 - j_1)(q+1) + j_1$, $u_i \leq n_i$ for $2 \leq i \leq m - 1$, and $u_1 < (t_m - j_m + 1)(q+1)$ implies that $T_1 - j_2 \leq t_m - j_m$ which leads to $(T_1 - t_1 + j_1 - j_2)(q+1) + j_2 \leq (t_m - j_m)(q+1) + j_m$ as $j_2 \leq j_m$. The fact that $j_1 < t_1$ gives $\mathbf{v}^+ \in \mathbb{N}^{l+1}$. We claim that $\mathbf{v}^+ \in S_{l+1}$. To see this, it is helpful to express $\mathbf{v}^+$ as

$$\mathbf{v}^+ = \boldsymbol{\gamma}_{(t_1 - j_1 + j_2 - 1, T_2, T_{i_3}, \ldots, T_{i_l}, T_1 - t_1 + j_1), j_2}.$$

It is easy to see that $t_1 - j_1 + j_2 - 1 + T_2 + T_{i_3} + \cdots + T_{i_l} + T_1 - t_1 + j_1 = q + l(j_2 - 1)$, $j_2 \leq T_{i_r} \leq q - 1$ for $2 \leq r \leq l$, $j_2 \leq t_1 - j_1 + j_2 - 1$ as $j_1 < t_1$, and $T_1 - (t_1 - j_1) \leq q - 1$. In order to conclude that $\mathbf{v}^+ \in S_{l+1}$, it only remains to show that $t_1 - j_1 + j_2 - 1 \leq q - 1$ and $j_2 \leq T_1 - t_1 + j_1$. It suffices to show that $j_2 \leq$

$T_1 - t_1 + j_1$ since this implies that $j_2 \le q - (t_1 - j_1)$ and so $t_1 - j_1 + j_2 - 1 \le q - 1$. If $j_2 > T_1 - t_1 + j_1$, then $(T_1 - j_2)(q+1) < (t_1 - j_1)(q+1) + j_1 - j_2$, contradicting the fact that $u_1 > n_1$. Hence, $j_2 \le T_1 - t_1 + j_1$ and $\mathbf{v}^+ \in S_{l+1} \subseteq \Gamma_{l+1}^+ \subseteq H_{l+1}$. It follows that $\mathbf{v} \in H_m$ and so $\mathbf{v} \in \{\mathbf{p} \in H_m : p_2 = n_2\}$. This yields a contradiction as $\mathbf{n}$ is minimal in $\{\mathbf{p} \in H_m : p_2 = n_2\}$, concluding the proof in this subcase.

Subcase (b): Suppose that $j_1 = t_1$. Set

$$\mathbf{v} := (0, u_2, \dots, u_{m-1}, (T_1 - j_2)(q+1) + j_2).$$

Then $\mathbf{v} \preceq \mathbf{n}$ and $\mathbf{v} \ne \mathbf{n}$ since $0 < n_1$, $u_i \le n_i$ for $2 \le i \le m - 1$, and $u_1 \prec (t_m - j_m + 1)(q+1)$ implies $T_1 - j_2 \le t_m - j_m$ which means $(T_1 - j_2)(q+1) + j_2 \le (t_m - j_m)(q+1) + j_m$ as $j_2 \le j_m$.

It is easy to see that $\mathbf{v}^+ \in S_l$ as $\sum_{r=1}^{l} T_{i_r} = q + (l-1)(j_2 - 1)$ and $j_2 \le T_{i_r} \le q - 1$ for all $1 \le r \le l$. As before, it follows that $\mathbf{v} \in H_m$ and $\mathbf{v} \in \{\mathbf{p} \in H_m : p_2 = n_2\}$. Since $\mathbf{v} \ne \mathbf{n}$, this contradicts the minimality of $\mathbf{n}$ in the set $\{\mathbf{p} \in H_m : p_2 = n_2\}$, concluding the proof in this subcase.

Since both cases (1) and (2) yield a contradiction, it must be the case that no such $\mathbf{n}$ exists. Hence, $\Gamma_m^+ \setminus S_m = \emptyset$. This establishes that $\Gamma_m^+ \subseteq S_m$, concluding the proof that $\Gamma_m^+ = S_m$.

To illustrate Theorem 10, we provide an example.

Example 11. As in Example 5, consider the curve X defined by $y^8 + y = x^9$ over $\mathbb{F}_{64} = \mathbb{F}_2(\omega)$ where $\omega^6 + \omega^4 + \omega^3 + \omega + 1 = 0$.

Let $P_1 = P_\infty$, $P_2 = P_{00}$, $P_3 = P_{01}$, $P_4 = P_{0\omega^9}$. Since $\Gamma_1 = \langle 8, 9 \rangle$ and Γ_2^+ is described in Example 5, to determine $H(P_1, P_2, P_3)$ it only remains to find Γ_3^+. By Theorem 10, $\Gamma_3^+ =$

$$\left\{ \begin{array}{l}
(1,1,46), (1,10,37), (1,19,28), (1,28,19), (1,37,10), (1,46,1), \\
(2,2,38), (2,11,29), (2,20,20), (2,29,11), (2,38,2), \\
(3,3,30), (3,12,21), (3,21,12), (3,30,3), \\
(4,4,22), (4,13,13), (4,22,4), \\
(5,5,14), (5,14,5), (6,6,6), \\
(10,1,37), (10,10,28), (10,19,19), (10,28,10), (10,37,1), \\
(11,2,29), (11,11,20), (11,20,11), (11,29,2), \\
(12,3,21), (12,12,12), (12,21,3), \\
(13,4,13), (13,13,4), \\
(14,5,5), \\
(19,1,28), (19,10,19), (19,19,10), (19,28,1), \\
(20,2,20), (20,11,11), (20,20,2), \\
(21,3,12), (21,12,3), \\
(22,4,4), \\
(28,1,19), (28,10,10), (28,19,1), \\
(29,2,11), (29,11,2), \\
(30,3,3), \\
(37,1,10), (37,10,1), \\
(38,2,2), \\
(46,1,1)
\end{array} \right\}.$$

To find $H(P_1, P_2, P_3, P_4)$, we only need to apply Theorem 10 to see that $\Gamma_4^+ =$

$$\left\{ \begin{array}{l}
(1,1,1,37), (1,1,10,28), (1,1,19,19), (1,1,28,10), (1,1,37,1), (1,10,1,28), \\
(1,10,10,19), (1,10,19,10), (1,10,28,1), (1,19,1,19), (1,19,10,10), (1,19,19,1), \\
(1,28,1,10), (1,28,10,1), (1,37,1,1), \\
(2,2,2,29), (2,2,11,20), (2,2,20,11), (2,2,29,2), (2,11,2,20), (2,11,11,11), \\
(2,11,20,2), (2,20,2,11), (2,20,11,2), (2,29,2,2), \\
(3,3,3,21), (3,3,12,12), (3,3,21,3), (3,12,3,12), (3,12,12,3), (3,21,3,3), \\
(4,4,4,13), (4,4,13,4), (4,13,4,4), \\
(5,5,5,5), \\
(10,1,1,28), (10,1,10,19), (10,1,19,10), (10,1,28,1), (10,10,1,19), (10,10,10,10), \\
(10,10,19,1), (10,19,1,10), (10,19,10,1), (10,28,1,1), \\
(11,2,2,20), (11,2,11,11), (11,2,20,2), (11,11,2,11), (11,11,11,2), (11,20,2,2), \\
(12,3,3,12), (12,3,12,3), (12,12,3,3), \\
(13,4,4,4), \\
(19,1,1,19), (19,1,10,10), (19,1,19,1), (19,10,1,10), (19,10,10,1), (19,19,1,1), \\
(20,2,2,11), (20,2,11,2), (20,11,2,2), \\
(21,3,3,3), \\
(28,1,1,10), (28,1,10,1), (28,10,1,1), \\
(29,2,2,2), \\
(37,1,1,1)
\end{array} \right\}.$$

Similarly, one can use Theorem 10 to find Γ_5^+, Γ_6^+, Γ_7^+, Γ_8^+, and Γ_9^+.

4   Acknowledgements

The author wishes to thank the anonymous referee whose careful reading and detailed comments led to numerous improvements in the proof of Theorem 10. This project was supported by NSF grant DMS-0201286 and an ORAU Ralph E. Powe Junior Faculty Enhancement Award.

References

1. E. Arbarello, M. Cornalba, P. Griffiths, and J. Harris, *Geometry of Algebraic Curves*, Springer-Verlag, 1985.
2. E. Ballico and S. J. Kim, *Weierstrass multiple loci of n-pointed algebraic curves*, J. Algebra **199** (1998), 455–471.
3. C. Carvalho and F. Torres, *On Goppa codes and Weierstrass gaps at several points*, preprint.
4. A. Garcia, S. J. Kim, and R. F. Lax, *Consecutive Weierstrass gaps and minimum distance of Goppa codes*, J. Pure Appl. Algebra **84** (1993), 199–207.
5. M. Homma, *The Weierstrass semigroup of a pair of points on a curve*, Arch. Math. **67** (1996), 337–348.
6. M. Homma and S. J. Kim, *Goppa codes with Weierstrass pairs*, J. Pure Appl. Algebra **162** (2001), 273–290.
7. S. J. Kim, *On the index of the Weierstrass semigroup of a pair of points on a curve*, Arch. Math. **62** (1994), 73–82.

8. H. Maharaj, G. L. Matthews, and G. Pirsic, *Riemann-Roch spaces for the Hermitian function field with applications to low-discrepency sequences and algebraic geometry codes*, preprint.

9. G. L. Matthews, *Weierstrass pairs and minimum distance of Goppa codes*, Des. Codes and Cryptog. **22** (2001), 107–121.

10. G. L. Matthews, *Weierstrass pairs and minimum distance of Goppa codes*, Ph.D. dissertation, Louisiana State University, Baton Rouge, Louisiana, USA, May 1999.

On Cyclic top-associative
Generalized Galois Rings

Santos González[1], Viktor T. Markov[2], Consuelo Martínez[1],
Aleksandr A. Nechaev[2], and Ignacio F. Rúa[1]

[1] Departamento de Matemáticas, Universidad de Oviedo,
33007 Oviedo, Spain
santos@pinon.ccu.uniovi.es
[2] Center of New Information Technologies, Moscow State University,
119899 Moscow, Russia
nechaev@cnit.msu.ru

Abstract. A Generalized Galois Ring (GGR) S is a finite nonassociative ring with identity of characteristic p^n, for a prime number p, such that its top-factor $\overline{S} = S/pS$ is a finite semifield. It is well known that if S is an associative Galois Ring (GR) then the set $S^* = S \setminus pS$ is a finite multiplicative abelian group. This group is cyclic if and only if S is either a finite field, or a residual integer ring of odd characteristic or the ring $\mathbb{Z}_4$. A GGR is called top-associative if $\overline{S}$ is a finite field. In this paper we study the conditions for a top-associative not associative GGR S to be cyclic.

1 Introduction

1.1 Generalized Galois Rings and Finite Semifields

Classical (associative) *Galois Rings* (GR) have been extensively studied in the literature [5, 7, 15–18, 20], and several applications of these rings to coding theory and cryptography have been considered [8–13, 18]. A GR is an associative and commutative ring with identity, and it is uniquely determined by its characteristic p^n and its cardinality p^{rn}. Moreover, for any natural numbers r and n and for any prime number p there exists a unique, up to isomorphism, GR S of characteristic p^n and cardinality p^{rn}. It is denoted by $GR(p^{rn}, p^n)$ [15, 20]. In [2] the notion of *Generalized Galois Ring (GGR)* was introduced and defined as a *finite not necessarily associative ring S with identity such that the set of all its zero divisors has the form pS for a natural number p*. Many properties of associative Galois Rings, related to characteristic, cardinality and ideal lattice structure are preserved by the new definition.

A finite nonassociative ring D is called a *finite semifield* if the set $D^* = D \setminus \{0\}$ is closed under the multiplication and it is a loop, i.e. D is a finite ring with identity such that for any pair of elements $a, b \in D$, $a \neq 0$, there exists a unique solution of the equation $ax = b$ (resp. $xa = b$). The characteristic of a finite semifield D is a prime number p, its associative-commutative centre $P = Z(D)$

G. Mullen, A. Poli and H. Stichtenoth (Eds.): Fq7 2003, LNCS 2948, pp. 25–39, 2003.
© Springer-Verlag Berlin Heidelberg 2003

is a finite field of $q = p^c$ elements ($c \in \mathbb{N}$), and D is a P-algebra of dimension d, for a natural number d. If D is associative then it is a finite field ($d = 1$), whereas if it is not associative then $d \geq 3$ [1, 6].

Finite semifields appear in a natural way in the study of GGR: a finite nonassociative ring S with identity is a GGR if and only if there exist a natural number n and a prime number p such that the characteristic of S is p^n and the top-factor $\overline{S} = S/pS$ is a finite semifield [2]. If $\overline{S}$ has p^d elements then the set of nonzero divisors $S^* = S \setminus pS$ of a GGR S is a loop of cardinality $(p^d - 1)(p^d)^{n-1}$. In [2] we proved that for any finite semifield D of characteristic p and for any natural number n there exists a (not necessarily unique) GGR S of characteristic p^n and top-factor $\overline{S} = S/pS$ isomorphic to D. We will denote the class of GGR of characteristic p^n and with a given finite semifield D as quotient ring by $GGR(D, p^n)$. The case $D = GF(p^d)$ a finite field is especially interesting, and rings in the class $GGR(GF(p^d), p^n)$ are called *top-associative*.

1.2   Multiplicative Structure of Galois Rings

One of the main tools in the study of a GR S is the *Teichmüller coordinate set (TCS)*. It is a subset of S, closed under the multiplication and isomorphic to the multiplicative semigroup of the finite field $\overline{S} = S/pS$ under the natural epimorphism $S \to \overline{S}$. Any associative GR $S = GR(q^n, p^n)$ ($q = p^r$) contains a unique subset $\Gamma(S)$ with this property: $\Gamma(S) = \{\alpha \in S \mid \alpha^q = \alpha\}$; it induces a p-adic decomposition $S = \Gamma(S) + p\Gamma(S) + \dots + p^{n-1}\Gamma(S)$ [15, 16, 18, 20]. Many properties of the GR can be obtained from the existence of the TCS and the mentioned p-adic decomposition, for instance, the full description of the lattice of Galois subrings or the structure of the group of automorphisms of S.

Moreover, given a Galois Ring $S = GR(q^n, p^n)$ ($q = p^r$), the multiplicative structure of the group of units $S^* = S \setminus pS$ can be obtained using the existence of the TCS. The group S^* is abelian of cardinality $(q - 1)q^{n-1}$, and it can be decomposed as a direct sum of the multiplicative group $\Gamma(S)^* = \Gamma(S) \setminus \{0\}$ and the p-group $e + pS = \{e + ps \mid s \in S\}$, where e denotes the identity of S. The first group is cyclic of cardinality $q - 1$, whereas the second one decomposes as a direct sum of cyclic groups [20]:

$$e + pS \cong \begin{cases} C_{p^{n-1}} \times \overset{(r}{\dots} \times C_{p^{n-1}} & \text{if } p > 2, \\ C_2 \times C_{2^{n-2}} \times C_{2^{n-1}} \times \overset{(r-1}{\dots} \times C_{2^{n-1}} & \text{if } p = 2. \end{cases}$$

Hence S^* is cyclic if and only if S is a finite field ($n = 1$), or a residual integer ring of odd characteristic ($p > 2, r = 1$) or the ring $\mathbb{Z}_4$ ($p = n = 2, r = 1$).

1.3   Multiplicative Structure of Finite Semifields and Generalized Galois Rings

The structure of the multiplicative loop of nonzero elements $D^* = D \setminus \{0\}$ of an arbitrary finite semifield D has not been described up to now. In [23], G.P.

Wene introduced the notion of *right primitive semifield* as a finite semifield D having an element w (a *right primitive element*) that generates D^*, that is, D^* consists of all right principal powers of w: $D^* = \{w^{i)} \mid i = 0, \ldots, \#D - 2\}$, where if e is the identity of the semifield, the (right) principal powers of an arbitrary element a are defined by: $a^{0)} = e$, $\forall i \in \mathbb{N}: a^{i+1)} = a^{i)} * a$. If D is a right primitive semifield of cardinality p^d then the set $\{e, w, w^{2)}, \ldots, w^{d-1)}\}$ is a basis of D over $\mathbb{Z}_p$. The notion of *left primitive semifield* can be defined in the obvious way. Wene proved that some important classes of finite semifields are right primitive, and conjectured that *any finite semifield is right primitive, i.e. it possesses a right primitive element* [23, 24]. Recently, I.F. Rúa [21] has proved that the answer to this conjecture is negative. There exists a semifield of 32 elements that is neither right nor left primitive. Nevertheless, there are many semifields for which assertion of the conjecture remains true [21].

In [3] the existence of Teichmüller Coordinate Sets in GGR S whose top-factor $\overline{S}$ is a right (or left) primitive semifield was considered. It was proved that the existence of a TCS in a GGR guarantees its associativity. This fact suggests that the study of the multiplicative structure of an arbitrary nonassociative GGR will not follow the lines of the associative case.

In this paper we study the cyclic condition of a GGR S, i.e. we look for conditions for the loop $S^* = S \setminus pS$ to be generated by a single element $\alpha \in S^*$. First we will consider *right cyclic* GGR, when there exists an element α such that the set of all its right principal powers is the set S^* of nonzero elements of S. Then we will consider *right-left cyclic* top-associative GGR, when there exists an element α such that the set of all elements generated from α by multiplications on the right or on the left by α is the set S^* of nonzero elements of S. In our study, properties of matrices over GR and finite fields will be extensively used, together with maximal order matrices over finite fields. So, most of the problems will be eventually reduced to the study of properties of matrices over finite fields. In particular, a result on conjugates of a matrix by powers of a maximal order matrix, that has its own interest, is included.

2   Right Cyclic GGR

In what follows *ring* will mean nonassociative ring [22], and D will denote a finite semifield of characteristic p and cardinality p^d. Furthermore, $S \in GGR(D, p^n)$ will denote a GGR of characteristic p^n and top-factor $\overline{S}$ isomorphic to D. The *centre* of a ring T is the subring:

$$Z(T) = \{a \in T \mid \forall b, c \in T : (ab)c = a(bc) = b(ac), \ ab = ba\}.$$

The subring R generated by the identity e of S is contained in the centre of S and is a Galois Ring $GR(p^n, p^n) = \mathbb{Z}_{p^n}$ (notice that $\overline{R} = R/pR = \mathbb{Z}_p$ and that $\overline{S}$ is an $\overline{R}$-vector space of dimension d). It was proved in [2] that S is a free module of rank d over R, and a subset $\{x_1, \ldots, x_d\} \subseteq S$ is an R-basis of S if and only if $\{\overline{x_1}, \ldots, \overline{x_d}\} \subseteq \overline{S}$ is an $\overline{R}$-basis of $\overline{S}$. Let us fix an R-basis of S, $\mathcal{B} = \{x_1, \ldots, x_d\} \subseteq S$.

In this section we shall introduce and study the concepts of right and left order of an element.

Definition 1. *Let α be an element of the loop $S^* = S \backslash pS$. Then the right order of α is the smallest natural number k such that $\alpha^{k)} = e$, and it will be denoted by $ord_r(\alpha)$. The left order of α, $ord_l(\alpha)$, is defined in an analogous way. The GGR S is called* right (resp. left) cyclic *if there exists an element $\alpha \in S^*$ such that $ord_r(\alpha) = |S^*|$ (resp. $ord_l(\alpha) = |S^*|$).*

Notice that the right and left order of an element are well defined because S^* is a finite loop. If S is an associative GGR, i.e. a GR, the right order of an element α coincides with the left order, and it is the order of the multiplicative subgroup generated by α. Therefore S is right (left) cyclic if and only if the group S^* is cyclic. In the nonassociative case the situation is rather different. The right order of an element $\alpha \in S^*$ can be smaller than the cardinality of the subloop generated by α, that contains elements obtained by finitely many multiplications of the element α with any possible arrangement of brackets.

To study the right order of an element $\alpha \in S^*$ we will consider the linear transformation $R_\alpha : S \to S$, given by $R_\alpha(x) = x\alpha$. This map is a bijection and induces a permutation of the set S^*. The order of this permutation will be denoted by $T(R_\alpha)$. Clearly $ord_r(\alpha)$ divides $T(R_\alpha)$, since $ord_r(\alpha)$ is the length of the cycle containing α in the permutation induced by R_α. The linear transformation R_α has a matrix $A \in GL(d, R)$ with respect to the basis $\mathcal{B}$ fixed above. Since $T(R_\alpha) = T(A)$ (the order of the matrix A), the study of the right order of α will use properties of matrices over the ring $\mathbb{Z}_{p^n}$. The following properties of matrices over associative Galois Rings were proved by A.A. Nechaev in [19].

Theorem 1. *If $R = \mathbb{Z}_{p^n}$, d is a natural number and $A \in GL(d, R)$, then $T(A) \leq (p^d - 1)p^{n-1}$.*

A matrix $A \in GL(d, R)$ whose order satisfies $T(A) = (p^d - 1)p^{n-1}$ is called maximal order matrix.

For any $A \in M_d(R)$, $d \times d$-matrix over the ring R, we shall denote by a the matrix over $\overline{R} = R/pR = \mathbb{Z}_p$ given by $a_{ij} = \overline{A_{ij}}$ for any $i, j \in \{1, \ldots, d\}$. Given $\alpha^\downarrow$ in the set $R^{(d)}$ of all column vectors of length d over R, $\alpha_\downarrow$ will denote its image in the ring $\mathbb{Z}_p^{(d)}$. Any matrix $A \in GL(d, R)$ induces a permutation φ_A in $R^{(d)}$ ($\varphi_A(x^\downarrow) = Ax^\downarrow$). The next result characterizes maximal order matrices in $GL(d, R)$.

Theorem 2. *Let $R = \mathbb{Z}_{p^n}$ and let d be a natural number ($d \geq 2$ in the case $p = 2, n > 2$). The matrix $A \in GL(d, R)$ is a maximal order matrix if and only if it has the form*

$$A = A_*(E + pH(A_*)),$$

where $A_* \in GL(d, R)$ is a matrix of order $p^d - 1$, E is the identity matrix and $H(x) \in R[x]$ is a polynomial of degree less than d satisfying

$$\overline{0} \neq \overline{H}(x) \in \mathbb{Z}_p[x],$$

and, if $p = 2 < n$, satisfying

$$\overline{e} \neq \overline{H}(x),$$

where e is the identity of R and $\overline{H}(x)$ denotes the image of $H[x]$ in the ring $\mathbb{Z}_p[x]$. In such a case, the matrix $a = a_* \in GL(d, \mathbb{Z}_p)$ is a maximal order matrix $(T(a) = p^d - 1)$ and for any vector $\alpha^{\downarrow} \in R^{(d)}$ such that $\alpha_{\downarrow} \in \mathbb{Z}_p^{(d)}$ is not the vector $0_{\downarrow}$, the length of the cycle containing $\alpha^{\downarrow}$ in the permutation φ_A is $(p^d - 1)p^{n-1}$.

Remark 1. If $a \in GL(d, \mathbb{Z}_p)$ is a maximal order matrix then its characteristic polynomial is an irreducible polynomial of degree d over $\mathbb{Z}_p$ whose roots are primitive elements of the field $GF(p^d)$.

Proposition 1. *Let $A \in GL(d, R)$ $(R = \mathbb{Z}_{p^n})$ be a matrix such that $T(a) = p^d - 1$. The matrix $B \in M_d(R)$ commutes with A if and only if B is a polynomial in A.*

Now we can easily get the following corollaries concerning the right order of elements in a GGR.

Corollary 1. *If $S \in GGR(D, p^n)$, $|D| = p^d$ and $\alpha \in S^*$ is an arbitrary element, then $ord_r(\alpha) \leq (p^d - 1)p^{n-1}$. Furthermore, $ord_r(\alpha) = (p^d - 1)p^{n-1}$ if and only if $T(R_\alpha) = (p^d - 1)p^{n-1}$. In this case D is a right primitive semifield.*
An element α satisfying the equality $ord_r(\alpha) = (p^d - 1)p^{n-1}$ is called maximal right order *element.*

Corollary 2. *Let $S \in GGR(D, p^n)$ be a GGR. Then S is right cyclic if and only if S is a cyclic associative GGR or a right primitive semifield.*

A non right cyclic ring $S \in GGR(D, p^n)$ may contain elements of maximal right order. According to Corollary 1 this is only possible if D is a right primitive semifield. The following result guarantees the existence of maximal right order elements in S when D is a right primitive semifield of odd characteristic.

Proposition 2. *Let $S \in GGR(D, p^n)$ be a GGR with D a right primitive semifield of odd characteristic. Then S contains an element of maximal right order.*

Proof. Let $w \in D^*$ be a right primitive element of the semifield D, i.e. $ord_r(w) = p^d - 1$ where $|D| = p^d$, and take $\alpha \in S^*$ such that $\overline{\alpha} = w$. If $R = \mathbb{Z}_{p^n}$ is the subring of S generated by the identity then $\overline{R} = R/pR = \mathbb{Z}_p$. Since $\mathcal{C} = \{e, w, w^{2)}, \ldots, w^{d-1)}\}$ is a $\mathbb{Z}_p$-basis of D we have that $\mathcal{B} = \{e, \alpha, \alpha^{2)}, \ldots, \alpha^{d-1)}\}$ is an R-basis of S. Let us denote by $A \in GL(d, R)$ the matrix of the linear transformation R_α with respect to $\mathcal{B}$. Then the matrix $a \in GL(d, \mathbb{Z}_p)$ is the coordinate matrix of the linear transformation R_w with respect to $\mathcal{C}$ and, since $ord_r(w) = p^d - 1$, we have that $T(a) = p^d - 1$ and so $p^d - 1 \mid T(A) \mid (p^d - 1)p^{n-1}$. If $T(A) = (p^d - 1)p^{n-1}$ then, by Corollary 1, the element α is an element of maximal right order $(p^d - 1)p^{n-1}$.

Otherwise, $T(A) = (p^d - 1)p^t$, where $t < n - 1$. Since $a^{p^{dt}} = a$ we have that $A = A^{p^{dt}}(E + pC)$ for some matrix C, since A is invertible. So the matrix

pC commutes with A. By Proposition 1, $pC = pQ(A)$ for some $Q(x) \in R[x]$. Furthermore, $T(A^{p^{dt}}) = p^d - 1$, and we will denote $A^{p^{dt}}$ by A_*. Using the equality $A = A_*(E + pQ(A))$ it is not difficult to prove that $pQ(A) = pQ'(A_*)$ for some polynomial $Q'(x) \in R[x]$, and so $A = A_*(E + pQ'(A_*))$. Since the characteristic polynomial of A_* has degree d we can assume that the degree of Q' is smaller than d.

Since $T(A) \neq (p^d - 1)p^{n-1}$, according to Theorem 2, we have that $\overline{Q'}(x) = \overline{0}$. Now consider an element $\lambda \in R$ such that $\overline{\lambda}$ is not equal to $\overline{0}$. The matrix of the linear transformation R_λ with respect to $\mathcal{B}$ is equal to λE and commutes with A. The matrix of the linear transformation $R_{\alpha(e+p\lambda)} = R_\alpha + pR_{\alpha\lambda} = R_\alpha(E + pR_\lambda)$ with respect to $\mathcal{B}$ is

$$A(E + p\lambda E) = A_*(E + pQ'(A_*))(E + p\lambda E)$$

$$= A_*(E + p(Q'(A_*) + \lambda E) + p^2\lambda Q'(A_*)).$$

Then the polynomial $Q''(x) = Q'(x) + \lambda + p\lambda Q'(x) \in R[x]$ satisfies $\overline{Q''}(x) \neq \overline{0}$. By Theorem 2, $A(E + p\lambda E)$ is a maximal order matrix and, by Corollary 1, $\alpha(e + p\lambda e)$ is an element of maximal right order $(p^d - 1)p^{n-1}$.

In view of this result we can conclude that many GGR S have a maximal right order element α, but the set of all right principal powers of α may not exhaust the loop S^*. In the following section we will consider the case when S^* coincides with the set of elements generated from α by multiplications on the right and on the left by α. We will restrict ourselves to the top-associative case, i.e. we will assume that the top-factor $\overline{S}$ is a finite field (right-left cyclic top-associative GGR).

3   Right-Left Cyclic Top-Associative GGR

The study of the problem of the right-left cyclic condition of top-associative GGR can be divided in two different parts. Firstly we will prove that it suffices to solve the problem for GGR of small characteristic. Then we will find a solution of the problem in the case of small characteristic, translating the problem to another one of matrices over finite fields. In the process we will obtain an interesting result on the conjugates of a matrix by the powers of a maximal order matrix (always matrices over finite fields).

3.1   Lifting Property of the Right-Left Cyclic Condition

Definition 2. *Let $S \in GGR(GF(p^d), p^n)$ be a top-associative GGR with identity e and let $\alpha \in S^*$. Consider the group RL_α generated by the maps $R_\alpha, L_\alpha : S \to S$, given by $R_\alpha(x) = x\alpha$ and $L_\alpha(x) = \alpha x$. Then the set $RL_\alpha(e) = \{C(e) \mid C \in RL_\alpha\}$ is a subset of S^* containing α and the identity. We will say that S is right-left cyclic if there exists an element $\alpha \in S^*$ such that $\overline{\alpha}$ is a primitive element of $GF(p^d)$ and $RL_\alpha(e) = S^*$.*

Definition 3. *Let $S \in GGR(D, p^n)$ $(n > 1)$ be a (not necessarily top-associative) GGR. We define the* critical factor *of S by the following way:*

$$S_{crit} = \begin{cases} S/p^2 S & \text{if } p > 2 \text{ or } n = 2, \\ S/2^3 S & \text{if } p = 2 < n \end{cases}$$

Notice that the critical factor of a GGR is also a GGR. The characteristic of S_{crit} is p^2 or 2^3, depending on its definition. Our aim is to prove that a top-associative GGR of odd characteristic is right-left cyclic if and only if its critical factor is right-left cyclic, that is, the right-left cyclic condition can be lifted from the critical factor.

In what follows we will use the following notation: if α and β are elements in a GGR S such that $\alpha = \beta + p^t \gamma$ for a $\gamma \in S$ and $t \in \mathbb{N}$, then we will write $\alpha \underset{p^t}{\equiv} \beta$.

Lemma 1. *Let $S \in GGR(GF(p^d), p^n)$ be a top-associative GGR with identity e and let $\alpha \in S^*$ such that $\bar{\alpha}$ is a primitive element of $GF(p^d)$. If $H \in RL_\alpha$ satisfies the condition $H(e) \underset{p}{\equiv} e$ then $H(\beta) \underset{p}{\equiv} \beta$ for any element $\beta \in S$.*

Proof. Since S is a top-associative GGR the maps $\overline{R_\alpha}, \overline{L_\alpha} : GF(p^d) \to GF(p^d)$, given by $\overline{R_\alpha}(x) = \overline{\alpha x}$ and $\overline{L_\alpha}(x) = \overline{x \alpha}$, are the same. Hence, the map $\overline{H} : GF(p^d) \to GF(p^d)$, given by $\overline{H}(x) = \overline{H(x)}$, is equal to $(\overline{R_\alpha})^h$ for a natural number h. Then, since $(\overline{R_\alpha})^h = R_{\overline{\alpha}^h}$, the condition $H(e) \underset{p}{\equiv} e$ implies $\bar{\alpha}^h = \bar{e}$, that is, $p^d - 1$ divides h, and so $\overline{H} = \overline{Id}$. Therefore $H(\beta) \underset{p}{\equiv} \beta$ for any element $\beta \in S$.

Definition 4. *Let $S \in GGR(D, p^n)$ be a GGR (not necessarily top-associative). A subset $\Gamma \subseteq S$ is called a* coordinate set *if $|\Gamma| = |D|$ and $\overline{\Gamma} = \{\bar{a} \mid a \in \Gamma\} = D$.*

Given a coordinate set $\Gamma \subseteq S$ it is not difficult to prove that for any element $\beta \in S$ there exist unique elements $\beta_0, \beta_1, \ldots, \beta_{n-1} \in \Gamma$ such that $\beta = \beta_0 + p\beta_1 + \cdots + p^{n-1}\beta_{n-1}$. Furthermore, $\beta \in S^*$ if and only if $\overline{\beta_0} \neq \overline{0}$ [3].

Lemma 2. *Let $S \in GGR(GF(p^d), p^n)$ be a top-associative GGR with identity e and let $\alpha \in S^*$ be an element such that $\bar{\alpha}$ is a primitive element of $GF(p^d)$. Then $S^* = RL_\alpha(e)$ if and only if for any $r \in \{0, \ldots, p^d - 2\}$ and for any $t < n$ there exists an element $C_{t+1}^r \in RL_\alpha$ such that $C_{t+1}^r(e) \underset{p^{t+1}}{\equiv} e + p^t \alpha^{r)}$.*

Proof. The condition is necessary, since the set of elements $\{e + p^t a^{r)} \mid r \in \{0, \ldots, p^d - 2\}, t < n\}$ is contained in S^*. So, let us assume that the condition of the theorem holds. Since $\bar{\alpha}$ is a primitive element of $GF(p^d)$ the set $\Gamma = \{0, e, \alpha, \alpha^{2)}, \ldots, \alpha^{p^d - 2)}\}$ is a coordinate set of S. We shall prove, by induction on t, that for any elements $\beta_0, \ldots, \beta_{n-1} \in \Gamma$ such that $\beta_0 \neq 0$, there exists $B_t \in RL_\alpha$ such that $B_t(e) \underset{p^t}{\equiv} \beta = \beta_0 + p\beta_1 + \cdots + p^{n-1}\beta_{n-1}$. In the case $t = 1$, since $\overline{\beta_0} = \overline{\alpha^{r)}}$ for some $r \in \{0, \ldots, p^d - 2\}$, the map $B_1 = (R_\alpha)^r \in RL_\alpha$ verifies $B_1(e) \underset{p}{\equiv} \beta$. Let us now suppose that there exists $B_t \in RL_\alpha$ such that $B_t(e) \underset{p^t}{\equiv} \beta$.

Then there exists $\gamma_t \in \Gamma$ such that $B_t(e) \underset{p^{t+1}}{\equiv} \beta_0 + p\beta_1 + \cdots + p^{t-1}\beta_{t-1} + p^t\gamma_t$.
If $\gamma_t = \beta_t$ take $B_t = B_{t+1}$. Otherwise there exists $s \in \{0,\ldots,p^d-2\}$ such that $\overline{\alpha^s)} = \overline{\beta_t - \gamma_t}$. We consider the map B_t^{-1}, then $B_t^{-1}(\alpha^{s)}) \underset{p}{\equiv} \alpha^{r)}$ for some $r \in \{0,\ldots,p^d-2\}$ and, by hypothesis, there exists $C_{t+1}^r \in RL_\alpha$ such that $C_{t+1}^r(e) \underset{p^{t+1}}{\equiv} e + p^t\alpha^{r)}$. Now the map $B_{t+1} = B_t C_{t+1}^r \in RL_\alpha$ verifies $B_{t+1}(e) \underset{p^{t+1}}{\equiv} \beta$.
Finally, if we consider $B_n \in RL_\alpha$ then $B_n(e) = \beta$ and so $S^* = RL_\alpha(e)$.

Given $S \in GGR(D,p^n)$ $(n > 1)$ we shall denote by $\widetilde{\ } : S \to S_{crit}$ the canonical epimorphism from S onto its critical factor S_{crit}.

Proposition 3. *Let $S \in GGR(GF(p^d),p^n)$ $(n > 1)$ be a top-associative GGR of odd characteristic with identity e, and let $\alpha \in S^*$ be an element such that $\overline{\alpha}$ is a primitive element of $GF(p^d)$. Then $S^* = RL_\alpha(e)$ if and only if $S_{crit}^* = \widetilde{RL_\alpha(e)}$.*

Proof. It is clear that if $S^* = RL_\alpha(e)$ then $S_{crit}^* = \widetilde{RL_\alpha(e)}$. Now let us assume that $S_{crit}^* = \widetilde{RL_\alpha(e)}$. In view of the previous lemma it is enough to prove that for any $r \in \{0,\ldots,p^d-2\}$ and for any $t < n$ there exists an element $C_{t+1}^r \in RL_\alpha$ such that $C_{t+1}^r(e) \underset{p^{t+1}}{\equiv} e + p^t\alpha^{r)}$. Let $r \in \{0,\ldots,p^d-2\}$, since $S_{crit}^* = \widetilde{RL_\alpha(e)}$ there exists $C_2^r \in RL_\alpha$ such that $C_2^r(e) \underset{p^2}{\equiv} e + p\alpha^{r)}$. So there exists $\gamma \in \Gamma = \{0,e,\alpha,\alpha^{2)},\ldots,\alpha^{p^d-2)}\}$ such that $C_2^r(e) \underset{p^3}{\equiv} e + p\alpha^{r)} + p^2\gamma$. According to Lemma 1 the congruence $C_2^r(\beta) \underset{p^2}{\equiv} \beta + p\gamma_\beta$ holds for any element $\beta \in \Gamma$ (for suitable $\gamma_\beta \in \Gamma$). Let us prove by induction that for any natural number $k \geq 2$ the following equations are true:

$$(C_2^r)^k(e) \underset{p^3}{\equiv} e + kp\alpha^{r)} + kp^2\gamma + \binom{k}{2}p^2\gamma_{\alpha^{r)}}$$

and

$$(C_2^r)^k(\alpha^{r)}) \underset{p^2}{\equiv} \alpha^{r)} + kp\gamma_{\alpha^{r)}}.$$

Indeed:

$$(C_2^r)^2(e) \underset{p^3}{\equiv} C_2^r(e + p\alpha^{r)} + p^2\gamma) \underset{p^3}{\equiv} e + 2p\alpha^{r)} + p^2(2\gamma + \gamma_{\alpha^{r)}})$$

and

$$(C_2^r)^2(\alpha^{r)}) \underset{p^2}{\equiv} C_2^r(\alpha^{r)} + p\gamma_{\alpha^{r)}}) \underset{p^2}{\equiv} \alpha^{r)} + 2p\gamma_{\alpha^{r)}}.$$

So, if the congruences are true for k then:

$$(C_2^r)^{k+1}(e) \underset{p^3}{\equiv} C_2^r\left(e + kp\alpha^{r)} + kp^2\gamma + \binom{k}{2}p^2\gamma_{\alpha^{r)}}\right)$$

$$\underset{p^3}{\equiv} e + (k+1)p\alpha^{r)} + (k+1)p^2\gamma + \binom{k+1}{2}p\gamma_{\alpha^{r)}}$$

and

$$(C_2^r)^{k+1}(\alpha^r)) \underset{p^2}{\equiv} C_2^r(\alpha^r) + kp\gamma_{\alpha^r})) \underset{p^2}{\equiv} \alpha^r) + (k+1)p\gamma_{\alpha^r}).$$

When $k = p$ we have: $(C_2^r)^p(e) \underset{p^3}{\equiv} e + p^2\alpha^r)$ and $(C_2^r)^p(\alpha^r)) \underset{p^2}{\equiv} \alpha^r)$. Finally, we can

prove by induction that for any $t \geq 2$ the equality $(C_2^r)^{p^{t-1}}(e) \underset{p^{t+1}}{\equiv} e + p^t\alpha^r)$

holds. The case $t = 2$ is true, as we have just proved. If the congruence holds for t then there exists $\beta \in \Gamma$ such that $(C_2^r)^{p^{t-1}}(e) \underset{p^{t+2}}{\equiv} e + p^t\alpha^r) + p^{t+1}\beta$. Notice

that $(C_2^r)^{p^{t-1}}(\alpha^r)) \underset{p^2}{\equiv} \alpha^r)$, so:

$$(C_2^r)^{p^t}(e) = ((C_2^r)^{p^{t-1}})^p(e) \underset{p^{t+2}}{\equiv} ((C_2^r)^{p^{t-1}})^{p-1}(e + p^t\alpha^r) + p^{t+1}\beta)$$

$$\underset{p^{t+2}}{\equiv} ((C_2^r)^{p^{t-1}})^{p-2}(e + 2p^t\alpha^r) + 2p^{t+1}\beta) \underset{p^{t+2}}{\equiv} \cdots \underset{p^{t+2}}{\equiv} e + p^{t+1}\alpha^r) + p^{t+2}\beta$$

$$\underset{p^{t+2}}{\equiv} e + p^{t+1}\alpha^r).$$

The result follows from Lemma 2 taking $C_{t+1}^r = (C_2^r)^{p^{t-1}}$.

This result shows that if the critical factor of a top-associative GGR S is right-left cyclic, so is the ring S. Whether a similar property holds when the characteristic of S is even is an open problem.

Next we shall study the problem of the right-left cyclic condition of top-associative GGR of characteristic p^2.

3.2   Right-Left Cyclic Condition of Top-Associative GGR of Characteristic p^2

In this section $S \in GGR(GF(p^d), p^2)$ will be a top-associative GGR of characteristic p^2 and $\alpha \in S^*$ will be an element such that $\overline{\alpha}$ is a primitive element of $GF(p^d)$, so the right order of the element α is equal to $p^d - 1$ or $(p^d - 1)p$. If $R = \mathbb{Z}_{p^2}$ is the subring of S generated by the identity e, then the set $\mathcal{B} = \{e, \alpha, \ldots, \alpha^{d-1}\}$ is an R-basis of S. We shall denote by $A \in GL(d, R)$ the matrix of the linear transformation $R_\alpha : S \to S$ with respect to $\mathcal{B}$. Since $T(A) = (p^d - 1)$ or $T(A) = (p^d - 1)p$, according to Theorem 2, there exists a polynomial $H(x) \in R[x]$ of degree less than d such that $A = A_*(E + pH(A_*))$ where $A_* \in GL(d, R)$ is a matrix of order $p^d - 1$ (if $T(A) = p^d - 1$ take $H(x) = 0$). Notice that $A_* = A^{p^d}$ and $pH(A) = pH(A_*)$, since the characteristic of S is p^2. So $A^{p^d - 1} = E - pH(A_*) = E - pH(A)$. On the other hand, since $\overline{R_\alpha} = \overline{L_\alpha}$, the matrix of the linear transformation L_α with respect to $\mathcal{B}$ is equal to $A(E + pB)$ for some $B \in M_d(R)$, since A is invertible. For any element $\beta \in S$ we shall denote by $\beta^\downarrow \in R^{(d)}$ the coordinate vector of β in the basis $\mathcal{B}$. We shall denote by $\beta_\downarrow$ the image of $\beta^\downarrow$ in $\mathbb{Z}_p^{(d)}$ under the natural epimorphism $\pi : R \to R/pR = \mathbb{Z}_p$. Given a matrix $C \in M_d(R)$ we will denote by $c \in M_d(\mathbb{Z}_p)$ the image of C under π. In particular, the image of the matrix $H(A)$ will be denoted by $h(a)$. The next result presents a matrix characterization of right-left cyclic top-associative GGR.

Proposition 4. *The set $RL_\alpha(e)$ is equal to S^* if and only if the set of vectors $W = \{h(a)e_\downarrow, \{a^t ba^{-t}e_\downarrow\}_{t\in\mathbb{N}}\} \subseteq \mathbb{Z}_p^{(d)}$ is a $\mathbb{Z}_p$-generator system of $\mathbb{Z}_p^{(d)}$.*

Proof. In terms of matrices with respect to the basis $\mathcal{B}$, the elements in RL_α are given by

$$\{A^{t_0}(E+pB)\ldots A^{t_{k-1}}(E+pB)A^{t_k} \mid k, t_i \in \mathbb{N}_0\}$$

$$= \{A^l + sp\sum_{i=0}^{k-1} A^{r_i} BA^{l-r_i} \mid s \in \{0,1\}, k, l, r_i \in \mathbb{N}_0\}$$

since the characteristic of S is p^2. Let us assume that $RL_\alpha(e) = S^*$ then, in view of Lemma 2, for any $r \in \{0,\ldots,p^d-2\}$ there exists $C_2^r \in RL_\alpha$ such that $C_2^r(e) = e + p\alpha^{r)}$. If the matrix associated to C_2^r is $A^l + sp\sum_{i=0}^{k-1} A^{r_i} BA^{l-r_i}$ $(k, l, r_i \in \mathbb{N}_0)$, then the following equality holds:

$$(e + p\alpha^{r)})^\downarrow = \left(A^l + sp\sum_{i=0}^{k-1} A^{r_i} BA^{l-r_i}\right) e^\downarrow$$

So

$$e_\downarrow = a^l e_\downarrow$$

and, since $T(a) = p^d - 1$, we have that $l = (p^d - 1)t$, for some natural number t. Then $A^l = A^{(p^d-1)t} = (E - pH(A))^t = E - ptH(A)$ and $pA^l = pE$, so:

$$(e + p\alpha^{r)})^\downarrow = \left(E - ptH(A) + ps\left(\sum_{i=0}^{k-1} A^{r_i} BA^{-r_i}\right)\right) e^\downarrow.$$

Therefore

$$\alpha_\downarrow^{r)} = \left(s\sum_{i=0}^{k-1} a^{r_i} ba^{-r_i} - th(a)\right) e_\downarrow.$$

Since

$$\{\alpha_\downarrow^{r)} \mid r \in \{0,\ldots,p^d-2\}\} = \mathbb{Z}_p^{(d)} \setminus \{0_\downarrow\},$$

the set of vectors W is a $\mathbb{Z}_p$-generator system of $\mathbb{Z}_p^{(d)}$.

Reciprocally, let $r \in \{0,\ldots,p^d-2\}$, then there exists natural numbers $s, s_0,\ldots,s_{p^d-2} \in \{0,\ldots,p-1\}$ such that

$$\alpha_\downarrow^{r)} = -sh(a)e_\downarrow + \sum_{t=0}^{p^d-2} s_t a^t ba^{-t} e_\downarrow$$

since W is a $\mathbb{Z}_p$-generator system of $\mathbb{Z}_p^{(d)}$. Consider the following element of RL_α:

$$T = (R_\alpha^{-1}L_\alpha)^{s_0} R_\alpha (R_\alpha^{-1}L_\alpha)^{s_1} R_\alpha \ldots (R_\alpha^{-1}L_\alpha)^{s_{p^d-2}} R_\alpha R_\alpha^{(p^d-1)(s-1)}.$$

Then the matrix of T with respect to $\mathcal{B}$ has the form:

$$C = E + p\left(-sH(A) + \sum_{t=0}^{p^d-2} s_t A^t B A^{-t}\right),$$

so $Ce^{\downarrow} = (e + p\alpha^r)^{\downarrow}$ and $T(e) = e + p\alpha^r)$.

According to this proposition it is necessary to study the $\mathbb{Z}_p$-vector subspace generated by the elements of the form $\{a^{-t}ba^t e_{\downarrow}\}_{t\in\mathbb{N}}$, that is, to study conjugates of a given matrix over a finite field by powers of a maximal order matrix. So we have reduced our original problem of right-left cyclic top-associative GGR to a problem of matrices over finite fields. In order to provide a solution we need to prove the following Lemma.

Lemma 3. *Let $L = GF(p^d)$ be the finite field of p^d elements, let $\alpha \in L$ be a primitive element and let $f_i(x) \in \mathbb{Z}_p[x]$ be the minimal polynomial over $\mathbb{Z}_p$ of the element $\alpha^{p^i-1} \in L$ (where $i \in \{1,\ldots,d-1\}$). Then $\mathbb{Z}_p(\alpha^{p^i-1}) = L$, and $f_i(x) \neq f_j(x)$ if $i \neq j$.*

Proof. We need to prove that the degree of $f_i(x)$ is d, so let us assume that the degree of $f_i(x)$ is $t < d$. Then, since $t \mid d$, we have that $t \leq \frac{d}{2}$ and the multiplicative order of the element α^{p^i-1} divides $p^t - 1$. But $ord(\alpha^{p^i-1}) = \frac{p^d-1}{(p^d-1,p^i-1)} = \frac{p^d-1}{p^{(d,i)}-1}$ and, since $(d,i) \mid d$ and $i < d$, we have that $(d,i) \leq \frac{d}{2}$. So $ord(\alpha^{p^i-1}) \mid p^t - 1$ implies that $p^d - 1 \mid (p^{(d,i)} - 1)(p^t - 1) \leq (p^{\frac{d}{2}} - 1)(p^{\frac{d}{2}} - 1) = p^d - 2p^{\frac{d}{2}} + 1$, a contradiction. Therefore the degree of $f_i(x)$ is d.

Let us now assume that for some $1 \leq i < j \leq d-1$ the equality $f_i(x) = f_j(x)$ holds. Then there exists $1 \leq s \leq d-1$ such that $(\alpha^{p^i-1})^{p^s} = \alpha^{p^j-1}$, i.e. $(p^i - 1)p^s \equiv p^j - 1 \pmod{p^d - 1}$. Since $i \leq (d-2)$ and $s \leq d-1$ we have that $i + s \leq 2d - 3$. We have two different possibilities: $i + s < d$ or $i + s = d + k$ with $0 \leq k < s \leq d-1$.

If $i + s < d$ then $0 \leq (p^i - 1)p^s = p^{i+s} - p^s < p^d - 1$. Since $0 \leq p^j - 1 < p^d - 1$ and $(p^i - 1)p^s \equiv p^j - 1 \pmod{p^d - 1}$, we have that $(p^i - 1)p^s = p^j - 1$, i.e. p divides 1, a contradiction.

If $i + s = d + k$ with $0 \leq k < s \leq d-1$ then $(p^i - 1)p^s = p^{i+s} - p^s = p^{d+k} - p^s = (p^d - 1)p^k + (p^k - p^s)$, i.e. $(p^i - 1)p^s \equiv p^k - p^s \pmod{p^d - 1}$. Hence $p^j - 1 \equiv (p^i - 1)p^s \equiv p^k - p^s \equiv p^d - p^s + p^k - 1 \pmod{p^d - 1}$. Since $1 \leq p^k < p^s < p^d$ and $0 \leq p^j - 1 < p^d - 1$, we conclude that $p^j - 1 = p^d - p^s + p^k - 1$. Therefore $p^j = p^d - p^s + p^k = p^k(p^{d-k} - p^{s-k} + 1)$ and p divides 1, a contradiction.

Now we can prove the following theorem.

Theorem 3. *Let $K = \mathbb{Z}_p$ and let d be a natural number greater than 2. If $A \in GL(d, K)$ and $B \in M_d(K)$ are matrices such that $T(A) = p^d - 1$, and $e^{\downarrow}$ denotes the vector $(1, 0, \ldots, 0)^{\downarrow}$, then the K-vector subspace of $K^{(d)}$ generated by the vectors $\{A^{-t}BA^t e^{\downarrow}\}_{t\in\mathbb{N}}$ has dimension less than or equal to 1 if B commutes with A, and d otherwise.*

Proof. If B commutes with A it is clear that the dimension is, at most, equal to 1. Let us suppose that B does not commute with A. Notice that it is enough to prove the result when A is a companion matrix of the form

$$S_f = \begin{pmatrix} 0 & 0 & \ldots & 0 & f_0 \\ 1 & 0 & \ldots & 0 & f_1 \\ 0 & 1 & \ldots & 0 & f_2 \\ \vdots & \vdots & \ddots & \vdots & \vdots \\ 0 & 0 & \ldots & 1 & f_{d-1} \end{pmatrix}$$

where the roots of $f(x) = x^d - f_{d-1}x^{d-1} - \cdots - f_1 x - f_0 \in K[x]$ are primitive elements of $L = GF(p^d)$. Indeed, if A is a maximal order matrix then $\{e^{\downarrow}, Ae^{\downarrow}, \ldots, A^{d-1}e^{\downarrow}\}$ is a K-basis of $K^{(d)}$ and so there exists $C \in GL(d, K)$ such that $C^{-1}AC = S_f$, for some $f(x) \in K[x]$. It is not difficult to see that $\dim_K(\{A^{-t}BA^te^{\downarrow}\}_{t\in\mathbb{N}}) = d$ if and only if $d = \dim_K(\{S_f^{-t}(C^{-1}BC)S_f^te^{\downarrow}\}_{t\in\mathbb{N}})$, since $Ce^{\downarrow} = A^ke^{\downarrow}$ for some natural number k and so

$$\dim_K(\{A^{-t}BA^te^{\downarrow}\}_{t\in\mathbb{N}}) = \dim_K(\{A^{-(t+k)}BA^{t+k}e^{\downarrow}\}_{t\in\mathbb{N}})$$

$$= \dim_K(\{A^{-k}(CS_f^{-t}C^{-1})B(CS_f^tC^{-1})A^ke^{\downarrow}\}_{t\in\mathbb{N}})$$

$$= \dim_K(\{(A^{-k}C)S_f^{-t}(C^{-1}BC)S_f^te^{\downarrow}\}_{t\in\mathbb{N}}) = \dim_K(\{S_f^{-t}(C^{-1}BC)S_f^te^{\downarrow}\}_{t\in\mathbb{N}}).$$

Therefore, let us assume that $A = S_f$, take $\alpha \in L$ such that $f(\alpha) = 0$ and consider the following matrices in $GL(d, L)$:

$$P = \begin{pmatrix} 1 & \alpha & \alpha^2 & \ldots & \alpha^{d-1} \\ 1 & \alpha^p & \alpha^{2p} & \ldots & \alpha^{(d-1)p} \\ \vdots & \vdots & \vdots & \ddots & \vdots \\ 1 & \alpha^{p^{d-1}} & \alpha^{2p^{d-1}} & \ldots & \alpha^{(d-1)p^{d-1}} \end{pmatrix}, \quad D = \begin{pmatrix} \alpha & 0 & \ldots & 0 \\ 0 & \alpha^p & \ldots & 0 \\ \vdots & \vdots & \ddots & \vdots \\ 0 & 0 & \ldots & \alpha^{p^{d-1}} \end{pmatrix}.$$

It is not difficult to check that $PAP^{-1} = D$. If we set $PBP^{-1} = C$ and $x^{\downarrow} = Pe^{\downarrow} = (1, \ldots, 1)^{\downarrow}$ we have that $A^{-t}BA^te^{\downarrow} = P^{-1}(D^{-t}CD^t)x^{\downarrow}$ for any natural number t. Then it is clear that

$$\dim_K(\{A^{-t}BA^te^{\downarrow}\}_{t\in\mathbb{N}}) = \dim_K(\{D^{-t}CD^tx^{\downarrow}\}_{t\in\mathbb{N}}).$$

We shall prove that this dimension is equal to d.

Since B does not commute with A, then the matrix C does not commute with D and so it is not a diagonal matrix. Therefore there exists some $C_{kj} \neq 0$ ($k \neq j$). We shall associate, to a polynomial $\lambda(x) = \sum_{t=0}^{p^d-2} \lambda_t x^t \in K[x]$, the matrix $M_\lambda = \sum_{t=0}^{p^d-2} \lambda_t D^{-t}CD^t$. Since $\lambda(x) \in K[x]$ we have:

$$M_\lambda = \begin{pmatrix} C_{11}\lambda(1) & C_{12}\lambda(\alpha^{p-1}) & C_{13}\lambda(\alpha^{p^2-1}) & \ldots & C_{1d}\lambda(\alpha^{p^{d-1}-1}) \\ C_{21}\lambda(\alpha^{1-p}) & C_{22}\lambda(1) & C_{23}\lambda(\alpha^{p^2-p}) & \ldots & C_{2d}\lambda(\alpha^{p^{d-1}-p}) \\ \vdots & \vdots & \vdots & \ddots & \vdots \\ C_{d1}\lambda(\alpha^{1-p^{d-1}}) & C_{d2}\lambda(\alpha^{p-p^{d-1}}) & C_{d3}\lambda(\alpha^{p^2-p^{d-1}}) & \ldots & C_{dd}\lambda(1) \end{pmatrix}$$

$$= \begin{pmatrix} C_{11}\lambda(1) & C_{12}\lambda(\alpha^{p-1}) & \cdots & C_{1d}\lambda(\alpha^{p^{d-1}-1}) \\ C_{21}\lambda(\alpha^{p^{d-1}-1})^p & C_{22}\lambda(1)^p & \cdots & C_{2d}\lambda(\alpha^{p^{d-2}-1})^p \\ \vdots & \vdots & \ddots & \vdots \\ C_{d1}\lambda(\alpha^{p-1})^{p^{d-1}} & C_{d2}\lambda(\alpha^{p^2-1})^{p^{d-1}} & \cdots & C_{dd}\lambda(1)^{p^{d-1}} \end{pmatrix}$$

Now we shall use the notation of the previous lemma. For any $i \in \{1, \ldots, d-1\}$ consider the polynomial:

$$F_i(x) = (x-1) \prod_{\substack{j=1 \\ j \neq i}}^{d-1} f_j(x) \in K[x].$$

By Lemma 3, any element $g(x)$ in the ideal $(F_i(x))$ verifies that $g(\alpha^{p^j-1}) = 0$ if $j \in \{0, \ldots, d-1\}$ with $j \neq i$, and the set $\{g(\alpha^{p^i-1}) \mid g(x) \in (F_i(x))\}$ is equal to L. Now consider a polynomial $\mu(x) \in (F_i(x))$ with $i \equiv j - k \pmod{d}$, then:

$$M_\mu = \begin{pmatrix} 0 \cdots & 0 & C_{1(i+1)}\mu(\alpha^{p^i-1}) & 0 & \cdots & 0 \\ 0 \cdots & 0 & 0 & C_{2(i+2)}\mu(\alpha^{p^i-1})^p & \cdots & 0 \\ \vdots & \vdots & \vdots & \vdots & \ddots & \vdots \\ 0 \cdots & C_{di}\mu(\alpha^{p^i-1})^{p^{d-1}} & 0 & 0 & \cdots & 0 \end{pmatrix}$$

The k-th row of the matrix M_μ has the form:

$$(0, \ldots, 0, C_{kj}\mu(\alpha^{p^i-1})^{p^{k-1}}, 0, \ldots, 0)$$

and

$$(0, \ldots, 0, C_{kj}\mu(\alpha^{p^i-1})^{p^{k-1}}, 0, \ldots, 0)x^\downarrow = C_{kj}\mu(\alpha^{p^i-1})^{p^{k-1}}.$$

Therefore the set

$$\{(0, \ldots, 0, C_{kj}\mu(\alpha^{p^i-1})^{p^{k-1}}, 0, \ldots, 0)x^\downarrow \mid \mu(x) \in (F_i(x))\}$$

is equal to L. But this is the set of all possible elements in the k-th coordinate of the K-linear combinations of $\{D^{-t}CD^tx^\downarrow\}_{t \in \mathbb{N}}$, so $\dim_K(\{D^{-t}CD^tx^\downarrow\}_{t \in \mathbb{N}}) = d$.

As a corollary to this theorem we can give an answer to our problem of right-left cyclic top-associative GGR.

Corollary 3. Let $S \in GGR(GF(p^d), p^n)$ $(n > 1)$ be a top-associative not associative GGR, char $S = 2^2$ or odd, and let $\alpha \in S^*$ be an element such that $\overline{\alpha}$ is a primitive element of $GF(p^d)$. Then $S^* = RL_\alpha(e)$ if and only if $\widetilde{R_\alpha} \neq \widetilde{L_\alpha}$.

Proof. If $S^* = RL_\alpha(e)$ then it is clear that $\widetilde{R_\alpha} \neq \widetilde{L_\alpha}$, in view of Corollary 2. On the other hand, if $\widetilde{R_\alpha} \neq \widetilde{L_\alpha}$ then $L_\alpha = R_\alpha(E + p\varphi)$ such that $\overline{\varphi} \neq \overline{0}$. If $\overline{\varphi}$ commutes with $\overline{R_\alpha}$ then, by Proposition 1, it is a polynomial in $\overline{R_\alpha}$, so $\overline{\varphi(e)} = \overline{\alpha^s} \neq \overline{0}$ for some $s \in \mathbb{N}$ and, therefore, $\alpha = L_\alpha(e) = R_\alpha(E + p\varphi)(e) \underset{p^2}{\equiv} \alpha + p\alpha^{s+1})$, a contradiction. Therefore $\overline{\varphi}$ does not commute with $\overline{R_\alpha}$. According to Propositions 3 and 4 and Theorem 3, we conclude that $S^* = RL_\alpha(e)$.

Corollary 4. *Let* $S \in GGR(GF(p^d), p^n)$ *(n > 1) be a top-associative not associative GGR, char* $S = 2^2$ *or odd. Then* S *is right-left cyclic if and only if* S_{crit} *is not commutative.*

Proof. If S is right-left cyclic then, by Corollary 3, S_{crit} is not commutative. Reciprocally, let us assume that S_{crit} is not commutative. Using a refinement of the normal basis theorem (Theorem 2.40 [14]) there exists a $\mathbb{Z}_p$-basis of $GF(p^d)$, $\{x_1, \ldots, x_d\}$, consisting of primitive elements of $GF(p^d)$. Taking elements $\alpha_1, \ldots, \alpha_d \in S^*$, such that $\overline{\alpha_i} = x_i$ (for $i \in \{1, \ldots, d\}$), the set $\mathcal{B} = \{\alpha_1, \ldots, \alpha_d\}$ is a $\mathbb{Z}_{p^n}$-basis of S. Moreover, since S_{crit} is not commutative, there exists $i \in \{1, \ldots, d\}$ such that $\widetilde{R_{\alpha_i}} \neq \widetilde{L_{\alpha_i}}$, and we can apply the previous corollary to conclude that S is right-left cyclic.

We have shown that a top-associative GGR S is right-left cyclic if and only if the critical factor ring of S is not commutative. Following similar techniques and, in particular, using Theorem 3, it is possible to characterize cyclic GGR whose top-factor is a right primitive finite semifield by the fact that its critical factor ring is not associative [4].

References

1. Cordero, M., Wene, G.P.: A survey of semifields. Discrete Math. **208/209** (1999) 125-137.
2. González, S., Markov, V.T., Martínez, C., Nechaev, A.A., Rúa, I.F.: Nonassociative Galois Rings. Discrete Math. Appl. **12** (2002) 591-606.
3. González, S., Markov, V.T., Martínez, C., Nechaev, A.A., Rúa, I.F.: Coordinate Sets of Generalized Galois Rings. Journal of Algebra and its Applications (to appear).
4. González, S., Markov, V.T., Martínez, C., Nechaev, A.A., Rúa, I.F.: Cyclic Generalized Galois Rings (preprint).
5. Janusz, G.J.: Separable algebras over commutative rings. Trans. Amer. Math. Soc. **122** (1966) 461-479.
6. Knuth, D.E.: Finite semifields and projective planes. J. Algebra **2** (1965) 182-217.
7. Krull, W.: Algebraische Theorie der Ringe. II. Math. Ann. **91** (1924) 1-46.
8. Kurakin, V.L., Kuzmin, A.S., Markov, V.T., Mikhalev, A.V., Nechaev, A.A.: Linear codes and polylinear recurrences over finite rings and modules (a survey). Applied Algebra, Algebraic Algorithms and Error-Correcting Codes 13, Lect. Notes Comput. Sci. **1719** (Springer, Berlin, 1999) 363-391.
9. Kurakin, V.L., Kuzmin, A.S., Mikhalev, A.V., Nechaev, A.A.: Linear recurring sequences over rings and modules. J. of Math. Sciences **76** (1995) 2793-2915.
10. Nechaev, A.A., Kuzmin, A.S.: Trace-function on a Galois ring in coding theory. Lect. Notes Comput. Sci. **1255** (1997) 277-290.
11. Kuzmin, A.S., Nechaev, A.A.: Complete weight enumerators of generalized Kerdock code and linear recursive codes over Galois rings. Proceedings of the WCC'99 Workshop on Coding and Cryptography (January 11-14), Paris, France. (1999) 332-336.
12. Kuzmin, A.S., Nechaev, A.A.: Linear recurrent sequences over Galois rings. Algebra i Logika **34** (1995) 169-189.

13. Kuzmin, A.S., Kurakin, V.L., Nechaev, A.A.: Pseudorandom and polylinear sequences. Proceedings in discrete mathematics **1** (1997) 139-202.
14. Lidl, R., Niederreiter, H.: Finite fields. Encyclopedia of Mathematics and its Applications **20** (Cambridge University Press, Cambridge, 1997).
15. McDonald, B.R.: Finite rings with identity. Pure and Applied Mathematics **28** (Marcel Dekker, New York, 1974).
16. Nechaev, A.A.: Finite rings of principal ideals. Matemat. Sbornik **91** (1973) 350-366.
17. Nechaev, A.A.: A basis of generalized identities of a finite commutative local principal ideal ring. Algebra i Logika **18** (1979) 186-193.
18. Nechaev, A.A.: Kerdock's code in a cyclic form. Diskret. Mat. **1** (1989) 123-139.
19. Nechaev, A.A.: Cycle types of linear maps over finite commutative rings. Russian Acad. Sci. Sb. Math. **78** (1994) 283-311.
20. Raghavendran, R.: Finite associative rings. Compositio Math. **21** (1969) 195-229.
21. Rúa, I.F.: Primitive and non primitive finite semifields. Communications in Algebra (to appear).
22. Schafer, R.D.: An introduction to nonassociative algebras. Pure and Applied Mathematics **22** (Academic Press, New York, London, 1966).
23. Wene, G.P.: On the multiplicative structure of finite division rings. Aequationes Math. **41** (1991) 222-233.
24. Wene, G.P.: Semifields of dimension $2n, n \geq 3$, over $GF(p^m)$ that have left primitive elements. Geom. Dedicata **41** (1992) 1-3.

Linear Recurrences with Polynomial Coefficients and Computation of the Cartier-Manin Operator on Hyperelliptic Curves

Alin Bostan[1], Pierrick Gaudry[2], Éric Schost[3]

[1] Laboratoire STIX, École polytechnique, France and IMAR, Romania
Alin.Bostan@stix.polytechnique.fr
[2] Laboratoire LIX, École polytechnique, France
gaudry@lix.polytechnique.fr
[3] Laboratoire STIX, École polytechnique, France
Eric.Schost@polytechnique.fr

Abstract. We improve an algorithm originally due to Chudnovsky and Chudnovsky for computing one selected term in a linear recurrent sequence with polynomial coefficients. Using baby-steps / giant-steps techniques, the nth term in such a sequence can be computed in time proportional to $\sqrt{n}$, instead of n for a naive approach.

As an intermediate result, we give a fast algorithm for computing the values taken by an univariate polynomial P on an arithmetic progression, taking as input the values of P on a translate on this progression.

We apply these results to the computation of the Cartier-Manin operator of a hyperelliptic curve. If the base field has characteristic p, this enables us to reduce the complexity of this computation by a factor of order $\sqrt{p}$. We treat a practical example, where the base field is an extension of degree 3 of the prime field with $p = 2^{32} - 5$ elements.

1 Introduction

In this paper, we investigate some complexity questions related to linear recurrent sequences. Specifically, we concentrate on recurrences with polynomial coefficients; our main focus is on the complexity of computing one selected term in such a recurrence.

A well-known particular case is that of recurrences with constant coefficients, where the nth term can be computed with a complexity that is logarithmic in n, using binary powering techniques.

In the general case, there is a significant gap, as for the time being no algorithm with complexity polynomial in $\log(n)$ is known. Yet, in [10], Chudnovsky and Chudnovsky proposed an algorithm that allows to compute one selected term in such a sequence without computing all intermediate ones. This algorithm appears as a generalization of those of Pollard [23] and Strassen [30] for integer factorization; using baby-steps / giant-steps techniques, it requires a number of operations which is roughly linear in $\sqrt{n}$ to compute the nth term in the sequence.

G. Mullen, A. Poli and H. Stichtenoth (Eds.): Fq7 2003, LNCS 2948, pp. 40–58, 2003.
© Springer-Verlag Berlin Heidelberg 2003

Our main contribution is an improvement of the algorithm of [10]; for simplicity, we only give the details in the case when all coefficients are polynomials of degree 1, as the study in the general case would follow in the same manner. The complexity of our algorithm is still (roughly) linear in $\sqrt{n}$; Chudnovsky and Chudnovsky actually suggested that this bound might be essentially optimal. We improve the time and space complexities by factors that are logarithmic in n; in practice, this is far from negligible, since in the application detailed below, n has order 2^{32}. A precise comparison with Chudnovsky and Chudnovsky's algorithm is made in Section 3.

Along the way, we also consider a question of basic polynomial arithmetic: given the values taken by a univariate polynomial P on a set of points, how fast can we compute the values taken by P on a translate of this set of points? An obvious solution is to make use of fast interpolation and evaluation techniques, but we show that one can do better when the evaluation points form an arithmetic sequence.

Computing the Cartier-Manin operator. Our initial motivation is an application to point-counting procedures in hyperelliptic curve cryptography, related to the computation of the Cartier-Manin operator of curves over finite fields. We now present these matters in more detail.

The Cartier-Manin operator of a curve defined over a finite field, together with the Hasse-Witt matrix, are useful tools to study the arithmetic properties of the Jacobian of that curve. Indeed, the supersingularity, and more generally the p-rank, can be read from the invariants of the Hasse-Witt matrix. In the case of hyperelliptic curves, this matrix was used in [13, 21] as part of a point-counting procedure for cryptographic-oriented applications.

Indeed, thanks to a result of Manin, computing the Cartier-Manin operator gives the coefficients of the Zeta function modulo p; this partial information can then be completed by some other algorithms. However, in [13] and [21], the method used to compute the Hasse-Witt matrix has a complexity which is essentially linear in p.

It turns out that one can do better. The entries of the Hasse-Witt matrix of a hyperelliptic curve $y^2 = f(x)$ defined over a finite field of characteristic p are coefficients of the polynomial $h = f^{(p-1)/2}$, so they satisfy a linear recurrence with rational function coefficients. Using our results on linear recurrences, this remark yields an algorithm to compute the Hasse-Witt matrix whose complexity now grows like $\sqrt{p}$, up to logarithmic factors, instead of p.

We demonstrate the interest of these techniques by a point-counting example, for a curve of genus 2 defined over a finite field whose characteristic just fits in one 32-bit machine word; this kind of fields have an interest for efficiency reasons [3].

Note finally that other point-counting algorithms, such as the p-adic methods used in Kedlaya's algorithm [18], also provide efficient point-counting procedures in small characteristic, but their complexity remains at least linear in p [12]. On the other hand, Kedlaya's algorithm outputs the whole Zeta function and should be preferred if available. Therefore, the range of application of our algorithm is when the characteristic is too large for Kedlaya's algorithm to be run.

Organization of the paper. We start in Section 2 with our algorithm for shifting a polynomial given by its values on some evaluation points. This building block is used in Section 3 to describe our improvement on Chudnovsky and Chudnovsky's algorithm. In Section 4 we apply these results to the computation of the Cartier-Manin operator of a hyperelliptic curve. We conclude in Section 5 with a numerical example.

Notation. In what follows, we give complexity estimates in terms of number of base ring operations (additions, subtractions, multiplications and inversions of unit elements) and of storage requirements; this last quantity is measured in terms of number of elements in the ring. We pay particular attention to polynomial and matrix multiplications and use the following notation.

- Let R be a commutative ring; we suppose that R is unitary, its unit element being denoted by 1_R, or simply 1. Let φ be the map $\mathbb{N} \to R$ sending n to $n \cdot 1_R = 1_R + \cdots + 1_R$ (n times); the map φ extends to a map $\mathbb{Z} \to R$. When the context is clear, we simply denote the ring element $\varphi(n)$ by n.
- We denote by $\mathsf{M} : \mathbb{N} \to \mathbb{N}$ a function that represents the complexity of univariate polynomial multiplication, *i.e.* such that over any ring R, the product of two degree d polynomials can be computed within $\mathsf{M}(d)$ base ring operations. Using the algorithms of [25, 24, 8], $\mathsf{M}(d)$ can be taken in $O(d \log(d) \log(\log(d)))$.
  We suppose that the function M verifies the inequality $\mathsf{M}(d_1) + \mathsf{M}(d_2) \leq \mathsf{M}(d_1 + d_2)$ for all positive integers d_1 and d_2; in particular, the inequality $\mathsf{M}(d) \leq \frac{1}{2} \mathsf{M}(2d)$ holds for all $d \geq 1$. On the other hand, we make the (natural) hypothesis that $\mathsf{M}(cd) \in O(\mathsf{M}(d))$ for all $c \geq 1$.
  We also assume that the product of two degree d polynomials can be computed in space $O(d)$; this is the case for all classical algorithms, such as naive, Karatsuba and Schönhage-Strassen multiplications.
- We let ω be a real number such that for every commutative ring R, all $n \times n$ matrices over R can be multiplied within $O(n^\omega)$ operations in R. The classical multiplication algorithm gives $\omega = 3$. Using Strassen's algorithm [29], we can take $\omega = \log_2(7) \simeq 2.81$. We assume that the product of two $n \times n$ matrices can be computed in space $O(n^2)$, which is the case for classical as well as Strassen's multiplications.

In the sequel, we need the following classical result on polynomial arithmetic over R. The earliest references we are aware of are [22, 4], see [31] for a detailed account. We also refer to [6] for a solution that is in the same complexity class, but where the constant hidden in the $O(\ )$ notation is actually smaller than that in [31].

Multipoint evaluation. If P is a polynomial of degree d in $R[X]$ and $r_0, \ldots, r_d$ are points in R, then the values $P(r_0), \ldots, P(r_d)$ can be computed using $O(\mathsf{M}(d)\log(d))$ operations in R and $O(d\log(d))$ space.

2   Shifting Evaluation Values

In this section, we address a particular case of the question of *shifting evaluation values* of a polynomial. The question reads as follows: Let P be a polynomial of degree d in $R[X]$, where R is a commutative unitary ring. Let a and $r_0, \ldots, r_d$ be in R. Given $P(r_0), \ldots, P(r_d)$, how fast can we compute $P(r_0+a), \ldots, P(r_d+a)$?

A reasonable condition for this question to make sense is that all differences $r_i - r_j$, $i \neq j$, are units in R; otherwise, uniqueness of the answer might be lost. Under this assumption, using fast interpolation and fast multipoint evaluation, the problem can be answered within $O(\mathsf{M}(d)\log(d))$ operations in R. We now show that the cost reduces to $\mathsf{M}(2d)+O(d)$ operations in R, in the particular case when $r_0, \ldots, r_d$ are in arithmetic progression, so we gain a logarithmic factor.

Our solution reduces to the multiplication of two suitable polynomials of degree at most $2d$; $O(d)$ additional operations come from additional pre- and post-processing operations. As mentioned in Section 1, all operations made below on integer values actually take place in R.

The algorithm underlying Proposition 1 is given in Figure 1; we use the notation $\mathtt{coeff}(Q, k)$ to denote the coefficient of degree k of a polynomial Q. We stress the fact that the polynomial P is *not* part of the input of our algorithm.

Input $P(0), \ldots, P(d)$ and a in R
Output $P(a), \ldots, P(a+d)$

- Compute

$$\delta(0,d) = \prod_{j=1}^{d}(-j), \quad \delta(i,d) = \frac{i}{i-d-1}\delta(i-1,d) \quad i = 1,\ldots,d$$

$$\Delta(a,0,d) = \prod_{j=0}^{d}(a-j), \quad \Delta(a,k,d) = \frac{a+k}{a+k-d-1}\Delta(a,k-1,d) \quad k = 1,\ldots,d$$

- Let

$$\widetilde{P} = \sum_{i=0}^{d}\frac{P(i)}{\delta(i,d)}X^i, \quad S = \sum_{i=0}^{2d}\frac{1}{a+i-d}X^i, \quad Q = \widetilde{P}S.$$

- Return the sequence $\Delta(a,0,d)\cdot\mathtt{coeff}(Q,d), \ldots, \Delta(a,d,d)\cdot\mathtt{coeff}(Q,2d)$.

Fig. 1. Shifting evaluation values

Proposition 1 *Let R be a commutative ring with unity, and $d \in \mathbb{N}$ such that $1,\ldots,d$ are units in R. Let P be in $R[X]$ of degree d, such that the sequence*

$$P(0),\ldots,P(d)$$

is known. Let a be in R, such that $a - d,\ldots,a + d$ are units in R. Then the sequence

$$P(a),\ldots,P(a+d)$$

can be computed within $\mathsf{M}(2d) + O(d)$ base ring operations, using space $O(d)$.

PROOF. Our assumption on R enables to write the Lagrange interpolation formula:

$$P = \sum_{i=0}^{d} P(i) \frac{\prod_{j=0,j\neq i}^{d}(X - j)}{\prod_{j=0,j\neq i}^{d}(i - j)}.$$

From now on, we denote by $\delta(i,d)$ the denominator $\prod_{j=0,j\neq i}^{d}(i - j)$ and by $\widetilde{P}_i$ the ratio $P(i)/\delta(i,d)$.

First note that all $\delta(i,d), i = 0,\ldots,d$, can be computed in $O(d)$ operations in R. Indeed, computing the first value $\delta(0,d) = \prod_{j=1}^{d}(-j)$ takes d multiplications. Then for $i = 1,\ldots,d$, $\delta(i,d)$ can be deduced from $\delta(i - 1, d)$ for two ring operations using the formula

$$\delta(i,d) = \frac{i}{i - d - 1}\delta(i - 1, d),$$

so their inductive computation requires $O(d)$ multiplications as well. Thus the sequence $\widetilde{P}_i, i = 0,\ldots,d$, can be computed in admissible time and space $O(d)$ from the input sequence $P(i)$. Accordingly, we rewrite the above formula as

$$P = \sum_{i=0}^{d} \widetilde{P}_i \prod_{j=0,j\neq i}^{d} (X - j).$$

For k in $0,\ldots,d$, let us evaluate P at $a + k$:

$$P(a + k) = \sum_{i=0}^{d} \widetilde{P}_i \prod_{j=0,j\neq i}^{d} (a + k - j).$$

Using our assumption on a, we can complete each product by the missing factor $a + k - i$:

$$P(a + k) = \sum_{i=0}^{d} \widetilde{P}_i \frac{\prod_{j=0}^{d}(a + k - j)}{a + k - i} = \left(\prod_{j=0}^{d}(a + k - j) \right) \cdot \left(\sum_{i=0}^{d} \widetilde{P}_i \frac{1}{a + k - i} \right). \tag{1}$$

Just as we introduced the sequence $\delta(i,d)$ above, we now introduce the sequence $\Delta(a,k,d)$ defined by $\Delta(a,k,d) = \prod_{j=0}^{d}(a + k - j)$. In a parallel manner, we deduce that all $\Delta(a, k, d), k = 0, \ldots, d$ can be computed in time and space $O(d)$, using the formulas:

$$\Delta(a, 0, d) = \prod_{j=0}^{d}(a - j), \quad \Delta(a, k, d) = \frac{a + k}{a + k - d - 1}\Delta(a, k - 1, d).$$

Let us denote $Q_k = P(a + k)/\Delta(a, k, d)$. We now show that knowing $\widetilde{P}_i, i = 0, \ldots, d$, we can compute $Q_k, k = 0, \ldots, d$ in $\mathsf{M}(2d)$ base ring operations and space $O(d)$; this is enough to conclude, by the above reasoning.

Using the coefficients $\Delta(a, k, d)$, Equation (1) reads

$$Q_k = \sum_{i=0}^{d} \widetilde{P}_i \frac{1}{a + k - i}. \tag{2}$$

Let $\widetilde{P}$ and S be the polynomials:

$$\widetilde{P} = \sum_{i=0}^{d} \widetilde{P}_i X^i, \quad S = \sum_{i=0}^{2d} \frac{1}{a + i - d} X^i;$$

then by Equation (2), for $k = 0, \ldots, d$, Q_k is the coefficient of degree $k + d$ in the product $\widetilde{P}S$. This concludes the proof. $\qquad\square$

We will conclude this section by an immediate corollary of this proposition; we first give a few comments.

- An alternative $O(\mathsf{M}(d))$ algorithm which does *not* require any inversibility hypotheses can be designed in the special case when $a = d+1$. The key fact is that for any degree d polynomial P, the sequence $P(0), P(1), \ldots$ is linearly recurrent, of characteristic polynomial $Q(X) = (1 - X)^{d+1}$. Thus, if the first terms $P(0), \ldots, P(d)$ are known, the next $d+1$ terms $P(d+1), \ldots, P(2d+1)$ can be recovered in $O(\mathsf{M}(d))$ using the algorithm in [27, Theorem 3.1].
- The general case when the evaluation points form an arbitrary arithmetic progression reduces to the case treated in the above proposition. Indeed, suppose that $r_0, \ldots, r_d$ form an arithmetic progression of difference δ, that $P(r_0), \ldots, P(r_d)$ are known and that we want to compute the values $P(r_0 + a), \ldots, P(r_d + a)$, where $a \in R$ is divisible by δ. Introducing the polynomial $Q(X) = P(\delta X + r_0)$, we are under the hypotheses of the above proposition, and it suffices to determine the shifted evaluation values of Q by a/δ.
- The reader may note the similarity of our problem with the question of computing the Taylor expansion of a given polynomial P at a given point in R. The algorithm of [2] solves this question with a complexity of $\mathsf{M}(d) + O(d)$ operations in R and space $O(d)$. The complexity results are thus quite similar; it turns out that analogous generating series techniques are used in that algorithm.

- In [15], an operation called *middle product* is defined: Given a ring R, and A, B in $R[X]$ of respective degrees d and $2d$, write $AB = C_0 + C_1 X^{d+1} + C_2 X^{2d+2}$, with all C_i of degree at most d; then the middle product of A and B is the polynomial C_1. This is precisely what is needed in the above algorithm.

  Up to considering the reciprocal polynomial of A, the middle product by A can be seen as the transpose of the map of multiplication by A. General program transformation techniques [7, 15] then show that it can be computed in time $\mathsf{M}(d) + O(d)$, but with a possible loss in space complexity. In [6], it is shown how to keep the same space complexity, at the cost of a constant increase in time complexity. Managing both requirements remains an open question, already stated in [17, Problem 6].

Corollary 1 *Let R be a commutative ring with unity, and $d \in \mathbb{N}$ such that $1, \ldots, 2d + 1$ are units in R. Let P be a degree d polynomial in $R[X]$ such that the sequence*

$$P(0), \ldots, P(d)$$

is known. For any s in $\mathbb{N}$, the sequence

$$P(0), P(2^s), \ldots, P(2^s d)$$

can be computed in time $s\,\mathsf{M}(2d) + O(sd) \in O(s\,\mathsf{M}(d))$ and space $O(d)$.

PROOF. For any $s \in \mathbb{N}$, let us denote by $P_s(X)$ the polynomial $P(2^s X)$. We prove by induction that all values $P_s(0), \ldots, P_s(d)$ can be computed in time $s\,\mathsf{M}(2d) + O(sd)$ and space $O(d)$, which is enough to conclude. The case $s = 0$ is obvious, as there is nothing to compute. Suppose then that $P_s(0), \ldots, P_s(d)$ can be computed in time $s\,\mathsf{M}(2d) + O(sd)$ and using $O(d)$ temporary space allocation.

Under our assumption on R, Proposition 1 shows that the values $P_s(d + 1), \ldots, P_s(2d + 1)$ can be computed in time $\mathsf{M}(2d) + O(d)$, using again $O(d)$ temporary space allocation. The values $P_s(0), P_s(2), \ldots, P_s(2d)$ coincide with $P_{s+1}(0), P_{s+1}(1), \ldots, P_{s+1}(d)$, so the corollary is proved.   □

3   Computing one Selected Term of a Linear Sequence

In this section, we recall and improve the complexity of an algorithm due to Chudnovsky and Chudnovsky [10] for computing selected terms of linear recurrent sequences with polynomial coefficients. The results of the previous section are used as a basic subroutine for these questions.

As in the previous section, R is a commutative ring with unity. Let A be a $n \times n$ matrix of polynomials in $R[X]$. For simplicity, in what follows, we only treat the case of degree 1 polynomials, since this is what is needed in the sequel. Nevertheless, all results extend *mutatis mutandis* to arbitrary degree.

For r in R, we denote by $A(r)$ the matrix over R obtained by specializing all coefficients of A at r. In particular, for k in $\mathbb{N}$, $A(k \cdot 1_R)$ is simply denoted by

$A(k)$, following the convention used up to now. Given a vector of initial conditions $U_0 = [u_1, \ldots, u_n]^t \in R^n$ and given k in $\mathbb{N}$, we consider the question of computing the kth term of the linear sequence defined by the relation $U_i = A(i)U_{i-1}$ for $i > 0$, that is, the product

$$U_k = A(k)A(k-1) \cdots A(1)U_0.$$

For simplicity, we write

$$U_k = \left(\prod_{i=1}^{k} A(i) \right) U_0,$$

performing all successive matrix products, $i = 1, \ldots, k$, on the left side. We use this convention hereafter.

In the particular case when A is a matrix of constant polynomials, and taking only the dependence on k into account, the binary powering method gives a time complexity of order $O(\log(k))$ base ring operations.

In the general case, the naive solution consists in evaluating all matrices $A(i)$ and performing all products. With respect to k only, the complexity of this approach is of order $O(k)$ base ring operations. In [10], Chudnovsky and Chudnovsky propose an algorithm that reduces this cost to essentially $O(\sqrt{k})$. We first recall the main lines of this algorithm; we then present some improvements in both time and space complexities.

The algorithm of Chudnovsky and Chudnovsky. The original algorithm uses baby-step / giant-step techniques, so for simplicity we assume that k is a square in $\mathbb{N}$. Let C be the $n \times n$ matrix over $R[X]$ defined by

$$C = \prod_{i=1}^{\sqrt{k}} A(X + i),$$

where $A(X + i)$ denotes the matrix A with all polynomials evaluated at $X + i$. By assumption on A, the entries of C have degree at most $\sqrt{k}$. For r in R, we denote by $C(r)$ the matrix C with all entries evaluated at r. Then the requested output U_k can be obtained by the equation

$$U_k = \left(\prod_{j=0}^{\sqrt{k}-1} C(j\sqrt{k}) \right) U_0. \tag{3}$$

Here are the main steps of the algorithm underlying Equation (3), originally due to [10].

Baby steps. The "baby steps" part of the algorithm consists in computing the polynomial matrix C. In [10], this is done within $O(n^\omega \mathsf{M}(\sqrt{k}))$ base ring operations, as products of polynomial matrices with entries of degree $O(\sqrt{k})$ are required.

Giant steps. In the second part the matrix C is evaluated on the arithmetic progression $0, \sqrt{k}, 2\sqrt{k}, \ldots, (\sqrt{k} - 1)\sqrt{k}$ and the value of U_k is obtained using Equation (3). Using fast evaluation techniques, all evaluations are done within $O(n^2 \, \mathsf{M}(\sqrt{k}) \log(k))$ base ring operations, while performing the $\sqrt{k}$ successive matrix-vector products in Equation (3) adds a negligible cost of $O(n^2 \sqrt{k})$ operations in R.

Summing all the above costs gives an overall complexity bound of

$$O\big(n^\omega \mathsf{M}(\sqrt{k}) + n^2 \, \mathsf{M}(\sqrt{k}) \log(k)\big)$$

base ring operations for computing a selected term of a linear sequence. Due to the use of fast evaluation algorithms in degree $\sqrt{k}$, the space complexity is $O(n^2 \sqrt{k} + \sqrt{k} \log(k))$.

In the particular case when A is the 1×1 matrix $[X]$, the question reduces to the computation of $\prod_{j=1}^{k} j$ in the ring R. For this specific problem, note that the ideas presented above were already used in [23, 30], for the purpose of factoring integers.

Avoiding multiplications of polynomial matrices. In what follows, we show how to avoid the multiplication of polynomial matrices, and reduce the cost of the above algorithm to $O(n^\omega \sqrt{k} + n^2 \, \mathsf{M}(\sqrt{k}) \log(k))$ base ring operations, storing only $O(n^2 \sqrt{k})$ elements of R.

Our improvements are obtained through a modification of the baby steps phase; the underlying idea is to work with the *values* taken by the polynomial matrices instead of their representation on the monomial basis. This idea is encapsulated in the following proposition.

Proposition 2 *Let A be a $n \times n$ matrix with entries in $R[X]$, of degree at most 1. Let $N \geq 1$ be an integer and let C be the $n \times n$ matrix over $R[X]$ defined by*

$$C = \prod_{i=1}^{N} A(X + i).$$

Then one can compute all scalar matrices $C(0), C(1), \ldots, C(N)$ within $O(n^\omega N)$ operations in R and with a memory requirement of $O(n^2 N)$ elements in R.

PROOF. We first compute the scalar matrices $[A(1), A(2), \ldots, A(2N)]$. Since all entries of A are linear in X, the complexity of this preliminary step is $O(n^2 N)$, both in time and space.

Then, we construct the matrices $(C'_j)_{0 \leq j \leq N}$ and $(C''_j)_{0 \leq j \leq N}$, which are defined as follows: we let C'_0 and C''_0 equal the identity matrix I_n and we recursively define

$$\begin{aligned} C'_j &= A(N+j)C'_{j-1} & \text{for } 1 \leq j \leq N, \\ C''_j &= C''_{j-1}A(N-j+1) & \text{for } 1 \leq j \leq N. \end{aligned}$$

Explicitly, for $0 \leq j \leq N$, we have

$$C'_j = A(N + j) \cdots A(N + 1)$$

and

$$C''_j = A(N) \cdots A(N - j + 1),$$

thus

$$C''_{N-j} = A(N) \cdots A(j + 1).$$

Computing all the scalar matrices (C'_j) and (C''_j) requires $2N$ matrix multiplications with entries in R; their cost is bounded by $O(n^\omega N)$ in time and by $O(n^2 N)$ in space. Lastly, the formula

$$C(j) = A(N + j) \cdots A(N + 1)A(N) \cdots A(j + 1) = C'_j C''_{N-j}, \quad 0 \leq j \leq N$$

enables to recover $C(0), C(1), \ldots, C(N)$ in time $O(n^\omega N)$ and space $O(n^2 N)$. $\square$

From this proposition, we deduce the following corollary, which shows how to compute the scalar matrices used in the giant steps.

Corollary 2 *Let A and C be polynomial matrices as in Proposition 2. If the elements $1, \ldots, 2N + 1$ are units in R, then for any integer $s \geq 1$, the sequence*

$$C(0), C(2^s), \ldots, C(2^s(N - 1))$$

can be computed using $O(n^\omega N + n^2 s\, \mathsf{M}(N))$ operations in R and $O(n^2 N)$ memory space.

PROOF. This is an immediate consequence of Proposition 2 and Corollary 1. $\square$

The above corollary enables us to perform the "giant steps" phase of Chudnovsky and Chudnovsky's algorithm in the special case when $N = 2^s$; this yields the 4^sth term in the recurrent sequence. Using this intermediate result, the following theorem shows how to compute the kth term, for arbitrary k, using the 4-adic expansion of k.

Theorem 1 *Let A be a $n \times n$ matrix with linear entries in $R[X]$ and let U_0 be in R^n. Suppose that (U_i) is the sequence of elements in R^n defined by the linear recurrence*

$$U_{i+1} = A(i + 1)U_i, \quad \text{for all } i \geq 0.$$

Let $k > 0$ be an integer and suppose that $1, \ldots, 2\lceil\sqrt{k}\rceil + 1$ are units in R. Then the vector U_k can be computed within $O(n^\omega \sqrt{k} + n^2 \,\mathsf{M}(\sqrt{k}) \log(k))$ operations in R and using memory space $O(n^2 \sqrt{k})$.

The proof of Theorem 1 is divided in two steps. We begin by proving the proposition in the particular case when k is a power of 4, then we treat the general case.

The case k is a power of 4. Let us suppose that $N = 2^s$ and $k = N^2$, so that $k = 4^s$. With this choice of k, Corollary 2 shows that the values $C(0), C(N), \ldots, C((N-1)N)$ can be computed within the required time and space complexities. Then we go on to the giant step phase described at the beginning of the section, and summarized in Equation (3). It consists in performing $\sqrt{k}$ successive matrix-vector products, which has a cost in both time and space of $O(n^2\sqrt{k})$.

The general case. We now consider the general case. Let $k = \sum_{i=0}^{s} k_i 4^i$ be the 4-adic expansion of k, with $k_i \in \{0, 1, 2, 3\}$ for all i. Given any t, we will denote by $\lceil k \rceil^t$ the integer $\sum_{i=0}^{t-1} 4^i k_i$. Using this notation, we define a sequence $(V_t)_{0 \le t \le s}$ as follows: we let $V_0 = U_0$ and, for $0 \le t \le s$ we set

$$V_{t+1} = A(\lceil k \rceil^t + 4^t k_t) \cdots A(\lceil k \rceil^t + 1)V_t. \tag{4}$$

It is easy to verify that $V_{s+1} = U_k$. Therefore, it suffices to compute the sequence (V_t) within the desired complexities.

Supposing that the term V_t has been determined, we estimate the cost of computing the next term V_{t+1}. If k_t is zero, we have nothing to do. Otherwise, we let $V_{t+1}^{(0)} = V_t$, and, for $1 \le j \le k_t$, we let $A^{(j)}(X) = A\left(X + \lceil k \rceil^t + 4^t(j-1)\right)$. Then we define $V_{t+1}^{(j)}$ by

$$V_{t+1}^{(j)} = A^{(j)}(4^t) \cdots A^{(j)}(1)V_{t+1}^{(j-1)}, \quad j = 1, \ldots, k_t.$$

By Equation (4), we have $V_{t+1}^{k_t} = V_{t+1}$. Thus, passing from V_t to V_{t+1} amounts to computing k_t selected terms of a linear recurrence of the special form treated in the previous paragraph. Using the complexity result therein and the fact that all k_t are bounded by 3, the total cost of the general case is thus

$$O\left(\sum_{t=0}^{s} \left(n^\omega 2^t + n^2 t\, \mathsf{M}(2^t)\right)\right) = O\left(n^\omega 2^s + n^2 s \left(\sum_{t=0}^{s} \mathsf{M}(2^t)\right)\right).$$

Using the fact that $2^s \le \sqrt{k} \le 2^{s+1}$ and the assumptions on the function M, we easily deduce that the whole complexity fits into the bound $O\left(n^\omega \sqrt{k} + n^2\, \mathsf{M}(\sqrt{k}) \log(k)\right)$, as claimed. Similar considerations also yield the bound concerning the memory requirements. This concludes the proof of Theorem 1.

Comments. The question of a lower time bound for computing U_k is still open. The simpler question of reducing the cost to $O\left(n^\omega \sqrt{k} + n^2\, \mathsf{M}(\sqrt{k})\right)$ base ring operations, that is gaining a logarithmic factor, already raises challenging problems.

As the above paragraphs reveal, this improvement could be obtained by answering the following question: Let P be a polynomial of degree d in $R[X]$. Given r in R, how fast can we compute $P(0), P(r), \ldots, P(rd)$ from the data of $P(0), P(1), \ldots, P(d)$? A complexity of order $O(\mathsf{M}(d))$ would immediately give the improved bound mentioned above. We leave it as an open question.

4   The Cartier-Manin Operator on Hyperelliptic Curves

We finally show how to apply the above results to the computation of the Cartier-Manin operator, and start by reviewing some known facts on this operator.

Let $\mathcal{C}$ be a hyperelliptic curve of genus g defined over the finite field $\mathbb{F}_{p^d}$ with p^d elements, where p is the characteristic of $\mathbb{F}_{p^d}$. We suppose that $p > 2$ and that the equation of $\mathcal{C}$ is of the form $y^2 = f(x)$, where $f \in \mathbb{F}_{p^d}[X]$ is a monic squarefree polynomial of degree $2g+1$. The generalization to hyperelliptic curves of the Hasse invariant for elliptic curves is the so-called Hasse-Witt matrix, which is defined as follows:

Definition 1 *Let h_k be the coefficient of degree k in the polynomial $f^{(p-1)/2}$. The Hasse-Witt matrix is the $g \times g$ matrix with coefficients in $\mathbb{F}_{p^d}$ given by*

$$H = (h_{ip-j})_{1 \leq i,j \leq g}.$$

This matrix was introduced in [16]; in a suitable basis, it represents the operator on differential forms that was introduced by Cartier in [9]. Manin then showed in [20] that this matrix is strongly related to the action of the Frobenius endomorphism on the p-torsion part of the Jacobian of $\mathcal{C}$. The article [33] provides a complete survey about those facts; they can be summarized by the following theorem:

Theorem 2 (Manin) *Let $\mathcal{C}$ be a hyperelliptic curve of genus g defined over $\mathbb{F}_{p^d}$. Let H be the Hasse-Witt matrix of $\mathcal{C}$ and let $H_\pi = HH^{(p)} \cdots H^{(p^{d-1})}$, where the notation $H^{(q)}$ means element-wise raising to the power q. Let $\kappa(t)$ be the characteristic polynomial of the matrix H_π and let $\chi(t)$ be the characteristic polynomial of the Frobenius endomorphism of the Jacobian of $\mathcal{C}$. Then*

$$\chi(t) \equiv (-1)^g t^g \kappa(t) \mod p.$$

This result provides a quick method to compute the characteristic polynomial of the Frobenius endomorphism and hence the group order of the Jacobian of $\mathcal{C}$ modulo p, when p is not too large. Combined with a Schoof-like algorithm and / or a baby-step / giant-step algorithm, it can lead to a full point-counting algorithm, in particular for genus 2 curves, as was demonstrated in [13, 21].

The obvious solution consists in expanding the product $f^{(p-1)/2}$. Using balanced multiplications, and taking all products modulo X^{gp} this can be done in $O(\mathsf{M}(gp))$ base field operations, whence a time complexity within $O(\mathsf{M}(p))$, if g is kept constant. In what follows, regarding the dependence in p only, we show how to obtain a complexity of $O(\mathsf{M}(\sqrt{p})\log(p))$ base field operations, using the results of the previous sections.

We will make the assumption that the constant term of f is not zero. Note that if it is zero, the problem is actually simpler: writing $f = Xf_1$, the coefficient of degree $ip-j$ in $f^{(p-1)/2}$ is the coefficient of degree $ip-j-(p-1)/2$ in $f_1^{(p-1)/2}$. Hence we can work with a polynomial of degree $2g$ instead of $2g + 1$ and the required degrees are slightly less.

Furthermore, for technical reasons, we assume that $g < p$. This is not a true restriction since for $g \geq p$, all the coefficients of $f^{(p-1)/2}$ up to degree $g(p-1)$ are needed to fill in the matrix H.

Introduction of a linear recurrent sequence. In [11], Flajolet and Salvy already treat the question of computing a selected coefficient in a high power of some given polynomial, as an answer to a SIGSAM challenge. The key point of their approach is that $h = f^{(p-1)/2}$ satisfies the following first-order linear differential equation

$$fh' - \frac{p-1}{2}f'h = 0.$$

From this, we deduce that the coefficients of h satisfy a linear recurrence of order $2g + 1$, with coefficients that are rational functions of degree 1.

Explicitly, let us denote by h_k the coefficient of degree k of the polynomial h, and for convenience, set $h_k = 0$ for $k < 0$. Similarly, the coefficient of degree k of f is denoted by f_k. From the above differential equation, for all k in $\mathbb{Z}$, we deduce that

$$\sum_{i=0}^{2g+1} \left(k + 1 - \frac{(p+1)i}{2} \right) f_i\, h_{k+1-i} = 0.$$

We set $U_k = [h_{k-2g}, h_{k-2g+1}, \ldots, h_k]^t$, and let $A(k)$ be the $(2g+1) \times (2g+1)$ companion matrix:

$$A(k) = \begin{bmatrix} 0 & & & 1 & 0 & \cdots & & 0 \\ 0 & & & 0 & 1 & \cdots & & 0 \\ & \vdots & & & \vdots & \vdots & \ddots & \vdots \\ & & & & & & \ddots & \\ 0 & & & 0 & 0 & \ddots & & 1 \\ \frac{f_{2g+1}((2g+1)(p-1)/2-(k-2g-1))}{f_0 k} & \cdots & \cdots & \cdots & & & & \frac{f_1((p-1)/2-k+1)}{f_0 k} \end{bmatrix}.$$

The initial vector $U_0 = [0, \ldots, 0, f_0^{(p-1)/2}]^t$ can be computed using binary powering techniques in $O(\log(p))$ base field operations; then for $k \geq 0$, we have $U_{k+1} = A(k+1)U_k$. Thus, to answer our specific question, it suffices to note that the vector U_{ip-j} gives the coefficients h_{ip-j} for $j = 1, \ldots, g$ that form the ith row of the Hasse-Witt matrix of $\mathcal{C}$.

Yet, Theorem 1 cannot be directly applied to this sequence, because $A(k)$ has entries that are rational functions, not polynomials. Though the algorithm could be adapted to handle the case of rational functions, we rather use the very specific form of the matrix $A(k)$, so only a small modification is necessary. Let us define a new sequence V_k by the relation

$$V_k = f_0^k k!\, U_k.$$

Then, this sequence is linearly generated and we have $V_{k+1} = B(k+1)V_k$, where

$$B(k) = f_0 k A(k).$$

Therefore, the entries of the matrix $B(k)$ are polynomials of degree at most 1. Note also that the denominators $f_0^k k!$ satisfy the recurrence relation

$$f_0^{k+1}(k+1)! = (f_0(k+1)) \cdot (f_0^k k!).$$

Thus, we will compute separately, first $V_{p-1}, V_{2p-1}, \ldots V_{gp-1}$ and then the denominators $f_0^{p-1}(p-1)!, \ldots, f_0^{gp-1}(gp-1)!$.

To this effect, we proceed iteratively. Let us for instance detail the computation of the sequence $V_{p-1}, V_{2p-1}, \ldots V_{gp-1}$. Knowing V_0, we compute V_{p-1} using Theorem 1. Then we shift all entries of B by p, so another application of Theorem 1 yields V_{2p-1}. Iterating g times, we obtain $V_{p-1}, V_{2p-1}, \ldots V_{gp-1}$ as requested; the same techniques are used to compute $f_0^{p-1}(p-1)!, \ldots, f_0^{gp-1}(gp-1)!$. Then the vectors $U_{p-1}, U_{2p-1}, \ldots U_{gp-1}$ are deduced from

$$U_k = \frac{1}{f_0^k k!} V_k.$$

Lifting to characteristic zero. A difficulty arises from the fact that the characteristic is too small compared to the degrees we are aiming to, so $p!$ is zero in $\mathbb{F}_{p^d}$. The workaround is to do computations in the unramified extension K of $\mathbb{Q}_p$ of degree d, whose residue class field is $\mathbb{F}_{p^d}$. The ring of integers of K will be denoted by O_K; any element of O_K can be reduced modulo p to give an element of $\mathbb{F}_{p^d}$. On the other hand, K has characteristic 0, so p is invertible in K.

We consider an arbitrary lift of f to $O_K[X]$. The reformulation in terms of linear recurrent sequence made in the above paragraph can be performed over K; the coefficients of $f^{(p-1)/2}$ are computed as elements of K and then projected back onto $\mathbb{F}_{p^d}$. This is possible, as they all belong to O_K.

Using the iteration described above, we separately compute the values in K of the vectors V_{ip-1} and the denominators $f_0^{ip-1}(ip-1)!$, for $i = 1, \ldots, g$. To this effect, we apply g times the result given in Theorem 1; this requires to perform

$$O\left(g^{\omega+1}\sqrt{p} + g^3 \mathsf{M}(\sqrt{p})\log(p)\right),$$

operations in K and to store $O(g^2\sqrt{p})$ elements of K.

Computing at fixed precision. Of course, we do not want to compute in the field K at arbitrary precision: for our purposes, it suffices to truncate all computations modulo a suitable power of p. To evaluate the required precision of the computation, we need to check when the algorithm operates a division by p.

To compute the vectors V_{ip-1} and the denominators $f_0^{ip-1}(ip-1)!$, for $i = 1, \ldots, g$, we use Theorem 1. This requires that all integers up to $2\lceil\sqrt{p}\rceil + 1$ are invertible, which holds as soon as $p \geq 11$.

Then, for all $i = 1, \ldots, g$, to deduce U_{ip-1} from V_{ip-1}, we need to divide by $f_0^{ip-1}(ip-1)!$. The element f_0 is a unit in O_K, so the only problem comes from the factorial term. With our assumption that $g < p$, we have $i < p$ and then the p-adic valuation of $(ip-1)!$ is exactly $i-1$. Therefore the worst case is $i = g$, for which we have to divide by p^{g-1}. Hence computing the vectors V_{ip-1} modulo p^g is enough to know the vectors U_{ip-1} modulo p, and then to deduce the Hasse-Witt matrix.

Overall complexity. Storing an element of $O_K/p^g O_K$ requires $O(dg \log(p))$ bits, and multiplying two such elements can be done with $O(\mathsf{M}(dg \log(p)))$ bit-operations. From the results of Section 3, we then deduce the following theorem on the complexity of computing the Hasse-Witt matrix.

Theorem 3 *Let p a prime, $d \geq 1$ and $\mathcal{C}$ a hyperelliptic curve defined over $\mathbb{F}_{p^d}$ by the equation $y^2 = f(x)$, with f of degree $2g+1$. Then, assuming $g < p$, one can compute the Hasse-Witt matrix of $\mathcal{C}$ with a complexity of*

$$O\left(\left(g^{\omega+1}\sqrt{p} + g^3 \mathsf{M}(\sqrt{p})\log(p)\right) \mathsf{M}(dg\log(p))\right)$$

bit-operations and $O\left(dg^3 \sqrt{p} \log(p)\right)$ storage.

The matrix H by itself gives some information on the curve $\mathcal{C}$, for instance H is invertible if and only if the Jacobian of $\mathcal{C}$ is ordinary [33, Corollary 2.3]. However, as stated in Theorem 2, the matrix H_π and in particular its characteristic polynomial $\chi(t)$ tell much more and are required if the final goal is point-counting. Thus, we finally concentrate on the cost of computing the characteristic polynomial of H_π.

The matrix H_π is the "norm" of H and as such can be computed with a binary powering algorithm. For simplicity, we assume that d is a power of 2, then denoting

$$H_{\pi,i} = HH^{(p)} \cdots H^{\left(p^{2^i-1}\right)}.$$

we have

$$H_{\pi,i+1} = H_{\pi,i} \cdot (H_{\pi,i})^{\left(p^{2^i}\right)}.$$

Hence the computation of $H_{\pi,i+1}$ from $H_{\pi,i}$ costs one matrix multiplication and 2^i matrix conjugations. A matrix conjugation consists in raising all the entries to the power p, therefore it costs $O(g^2 \log(p))$ operations in $\mathbb{F}_{p^d}$. The matrix we need to compute is $H_\pi = H_{\pi,\log_2(d)}$. Hence the cost of computing H_π is

$$O\left(dg^2 \log(p) + g^\omega \log(d)\right)$$

operations in $\mathbb{F}_{p^d}$. The general case where d is not a power of 2 is handled by adjusting the recursive step according to the binary expansion of d and yields the same complexity up to a constant factor.

The cost of the characteristic polynomial computation is bounded by the cost of a matrix multiplication [19] and is therefore negligible compared to the other costs.

If we are interested only in the complexity in p and d, i.e. if we assume that the genus is fixed, we get a time complexity for computing $\chi(t) \mod p$ in

$$O\left((\mathsf{M}(\sqrt{p}) + d) \mathsf{M}(d\log(p)) \log(p)\right).$$

Case of large genus. In case of large genus, the algorithm of Theorem 1 is asymptotically not the fastest. In this paragraph, we assume that the function M is essentially linear and we do not take into account the logarithmic factors; adding appropriate epsilons in the exponents would yield a rigorous analysis. The cost in bit-operations of Theorem 3 is at least $g^4\sqrt{p}d$ whereas the cost of the naive algorithm is linear in gpd. If $g > p^{1/6}$, then $g^4\sqrt{p} > gp$, and therefore the naive algorithm is faster.

5   Point-Counting Numerical Example

We have implemented our algorithm using Shoup's NTL C++ library [26]. NTL does not provide any arithmetic of local fields or rings, but allows to work in finite extensions of rings of the form $\mathbb{Z}/p^g\mathbb{Z}$, as long as no division by p occur; the divisions by p are well isolated in the algorithm, so we could handle them separately. Furthermore, NTL multiplies polynomials defined over this kind of structure using an asymptotically fast FFT-based algorithm.

To illustrate that our method can be used as a tool in point-counting algorithms, we have computed the Zeta function of a (randomly chosen) genus 2 curve defined over $\mathbb{F}_{p^3}$, with $p = 2^{32} - 5$. Such a Jacobian has therefore about 2^{192} elements and should be suitable for cryptographic use if the group order has a large prime factor. Note that previous computations were limited to p of order 2^{23} [21].

The characteristic polynomial χ of the Frobenius endomorphism was computed modulo p in 3 hours and 41 minutes, using 1 GB of memory, on an AMD Athlon MP 2200+. Then we used the Schoof-like algorithms of [13] and [14] to compute χ modulo $128 \times 9 \times 5 \times 7$, and finally we used the modified baby-step / giant-step algorithm of [21] to finish the computation. These other parts were implemented in Magma [5] and were performed in about 15 days of computation on an Alpha EV67 at 667 MHz. We stress that this computation was meant as an illustration of the possible use of our method, so little time was spent optimizing our code. In particular, the Schoof-like part and the final baby-step / giant-step computations are done using a generic code that is not optimized for extension fields.

Numerical data. The irreducible polynomial $P(t)$ that was used to define $\mathbb{F}_{p^3}$ as $\mathbb{F}_p[t]/(P(t))$ is

$$t^3 + 1346614179t^2 + 3515519304t + 3426487663.$$

The curve $\mathcal{C}$ has equation $y^2 = f(x)$ where f is given by

$$\begin{aligned}
f(x) = x^5 &+ (2697017539t^2 + 1482222818t + 3214703725)x^3 + \\
&(676673546t^2 + 3607548185t + 1833957986)x^2 + \\
&(1596634951t^2 + 3203023469t + 2440208439)x + \\
&2994361233t^2 + 3327339023t + 862341251.
\end{aligned}$$

Then the polynomial characteristic $\chi(T)$ of the Frobenius endomorphism is given by $T^4 - s_1 T^3 + s_2 T^2 - p^3 s_1 T + p^6$, where

$$s_1 = 332906835893875, \quad s_2 = 1420112352156389846167187570235.$$

The group order of the Jacobian is then

$$627710169154160539591778508077182588386018946581362599397709 = 3^3 \times 13 \times 67 \times 639679 \times 4172680687275363708100101723442360254559339531391.$$

This number has a large prime factor of size 2^{158}, therefore that curve is cryptographically secure.

Measure of the complexity in p. To check the practical asymptotic behaviour of our algorithm, we ran our implementation on a genus 2 curve defined over $\mathbb{F}_{p^3}$ with $p = 2^{34} - 41$. We performed only the Cartier-Manin step, and not the full point-counting algorithm. As the characteristic is about 4 times larger than in the previous example, a complexity linear in $\sqrt{p}$ means a runtime multiplied by about 2. On the same computer, the runtime is 8 hours and 48 minutes. Hence the ratio of the runtimes is about 2.39. The defect of linearity can be explained by taking into account the logarithmic factors. Assuming that $\mathsf{M}(n)$ is $O(n \log(n) \log(\log(n)))$, and neglecting the multi-logarithmic factors, the complexity announced in Theorem 3 is in $O(\sqrt{p}(\log(p))^3)$. With this estimate, the expected ratio between the runtimes becomes about 2.40, that is very close to the measure. This validates our analysis.

6   Conclusion

In this paper, we have presented an improvement of an algorithm by Chudnovsky and Chudnovsky to compute selected terms in a linear sequence with polynomial coefficients. This algorithm is then applied to the computation of the Cartier-Manin operator of hyperelliptic curves, thus leading to improvements in the point-counting problems that occur in cryptography.

This strategy extends readily to curves of the form $y^r = f(x)$ with $r > 2$, for which the Hasse-Witt matrix has a similar form. For more general curves, Mike Zieve pointed to us the work of Stöhr and Voloch [28] that gives formulas that still fit in our context in some cases.

Finally, Mike Zieve pointed out to us the work of Wan [32] that relates Niederreiter's polynomial factorization algorithm to the computation of the Cartier-Manin operator of some variety. The link with our work is not immediate, as that variety has dimension zero. Nevertheless, this remains intriguing, especially if we think of Pollard-Strassen's integer factoring algorithm as a particular case of Chudnovsky and Chudnovsky's algorithm.

Acknowledgements

Part of our numerical experiments were done with the computer of the Medicis center [1]. We thank Takakazu Satoh and Mike Zieve for their interest in our work and their insightful comments and also Bruno Salvy for his numerous helpful suggestions.

References

1. Medicis. http://www.medicis.polytechnique.fr/.
2. A. V. Aho, K. Steiglitz, and J. D. Ullman. Evaluating polynomials at fixed sets of points. *SIAM J. Comput.*, 4(4):533–539, 1975.
3. D. Bailey and C. Paar. Optimal extension fields for fast arithmetic in public-key algorithms. In *Advances in Cryptology – CRYPTO '98*, volume 1462 of *LNCS*, pages 472–485. Springer–Verlag, 1998.
4. A. Borodin and R. T. Moenck. Fast modular transforms. *Comput. System Sci.*, 8(3):366–386, 1974.
5. W. Bosma, J. Cannon, and C. Playoust. The Magma algebra system. I. The user language. *J. Symb. Comp.*, 24(3-4):235–265, 1997. See also http://www.maths.usyd.edu.au:8000/u/magma/.
6. A. Bostan, G. Lecerf, and É. Schost. Tellegen's principle into practice. In *Proceedings of ISSAC'03*, pages 37–44. ACM Press, 2003.
7. P. Bürgisser, M. Clausen, and M. A. Shokrollahi. *Algebraic complexity theory*, volume 315 of *Grundlehren Math. Wiss.* Springer–Verlag, 1997.
8. D. G. Cantor and E. Kaltofen. On fast multiplication of polynomials over arbitrary algebras. *Acta Informatica*, 28(7):693–701, 1991.
9. P. Cartier. Une nouvelle opération sur les formes différentielles. *C. R. Acad. Sci. Paris*, 244:426–428, 1957.
10. D. V. Chudnovsky and G. V. Chudnovsky. Approximations and complex multiplication according to Ramanujan. In *Ramanujan revisited (Urbana-Champaign, Ill., 1987)*, pages 375–472. Academic Press, Boston, MA, 1988.
11. P. Flajolet and B. Salvy. The SIGSAM challenges: Symbolic asymptotics in practice. *SIGSAM Bull.*, 31(4):36–47, 1997.
12. P. Gaudry and N. Gürel. Counting points in medium characteristic using Kedlaya's algorithm. To appear in *Experiment. Math.*
13. P. Gaudry and R. Harley. Counting points on hyperelliptic curves over finite fields. In *ANTS-IV*, volume 1838 of *LNCS*, pages 313–332. Springer–Verlag, 2000.
14. P. Gaudry and É. Schost. Cardinality of a genus 2 hyperelliptic curve over GF($5 \cdot 10^{24} + 41$). e-mail to the NMBRTHRY mailing list. Sept. 2002.
15. G. Hanrot, M. Quercia, and P. Zimmermann. The middle product algorithm, I. Speeding up the division and square root of power series. Preprint.
16. H. Hasse and E. Witt. Zyklische unverzweigte Erweiterungskörper vom primzahlgrade p über einem algebraischen Funktionenkörper der Charakteristik p. *Monatsch. Math. Phys.*, 43:477–492, 1936.
17. E. Kaltofen, R. M. Corless, and D. J. Jeffrey. Challenges of symbolic computation: my favorite open problems. *J. Symb. Comp.*, 29(6):891–919, 2000.
18. K. Kedlaya. Countimg points on hyperelliptic curves using Monsky-Washnitzer. *J. Ramanujan Math. Soc.*, 16:323–338, 2001.

19. W. Keller-Gehrig. Fast algorithms for the characteristic polynomial. *Theor. Comput. Sci.*, 36(2-3):309–317, 1985.
20. J. I. Manin. The Hasse-Witt matrix of an algebraic curve. *Trans. Amer. Math. Soc.*, 45:245–264, 1965.
21. K. Matsuo, J. Chao, and S. Tsujii. An improved baby step giant step algorithm for point counting of hyperelliptic curves over finite fields. In *ANTS-V*, volume 2369 of *LNCS*, pages 461–474. Springer–Verlag, 2002.
22. R. T. Moenck and A. Borodin. Fast modular transforms via division. *Thirteenth Annual IEEE Symposium on Switching and Automata Theory (Univ. Maryland, College Park, Md., 1972)*, pages 90–96, 1972.
23. J. M. Pollard. Theorems on factorization and primality testing. *Proc. Cambridge Philos. Soc.*, 76:521–528, 1974.
24. A. Schönhage. Schnelle Multiplikation von Polynomen über Körpern der Charakteristik 2. *Acta Informatica*, 7:395–398, 1977.
25. A. Schönhage and V. Strassen. Schnelle Multiplikation großer Zahlen. *Computing*, 7:281–292, 1971.
26. V. Shoup. NTL: A library for doing number theory. http://www.shoup.net.
27. V. Shoup. A fast deterministic algorithm for factoring polynomials over finite fields of small characteristic. In *Proceedings of ISSAC'91*, pages 14–21. ACM Press, 1991.
28. K.-O. Stöhr and J. Voloch. A formula for the Cartier operator on plane algebraic curves. *J. Reine Angew. Math.*, 377:49–64, 1987.
29. V. Strassen. Gaussian elimination is not optimal. *Numer. Math.*, 13:354–356, 1969.
30. V. Strassen. Einige Resultate über Berechnungskomplexität. *Jber. Deutsch. Math.-Verein.*, 78(1):1–8, 1976/77.
31. J. von zur Gathen and J. Gerhard. *Modern computer algebra*. Cambridge University Press, 1999.
32. D. Wan. Computing zeta functions over finite fields. *Contemp. Math.*, 225:131–141, 1999.
33. N. Yui. On the Jacobian varietes of hyperelliptic curves over fields of characteristic $p > 2$. *J. Algebra*, 52:378–410, 1978.

Mutual Irreducibility of Certain Polynomials [*]

Michael Dewar and Daniel Panario

School of Mathematics and Statistics, Carleton University
Ottawa, K1S 5B6, Canada
E-mail: mdewar@magma.ca, daniel@math.carleton.ca

Abstract. In a recent paper, Tsaban and Vishne [4] introduce linear transformation shift registers (TSRs) which generate sequences by an entire word with each iteration. The authors recently [1] proved that over $\mathbb{F}_2$, irreducible TSRs occur in pairs. Now the results are generalized and extended for arbitrary finite fields. This aids in the search for irreducible TSRs.

1 Introduction

Linear feedback shift registers (LFSRs) play an important role in engineering for the implementation of sequences over finite fields; see Golomb [3]. LFSRs generate a single bit with each iteration. Recently, Tsaban and Vishne [4] introduced linear transformation shift registers (TSRs) which produce an entire word with each iteration by utilizing word-oriented operations. A linear transformation of order m (the word size) is combined with a LFSR of order n to create the TSR over the finite field $\mathbb{F}_q$. For certain choices, the resulting TSR has primitive characteristic polynomial, and hence maximal order $q^{mn} - 1$. Finding irreducible TSRs is the first step towards finding primitive TSRs.

In a recent paper [1], the authors note the existence of pairs of characteristic polynomials of LFSRs that have the same irreducibility behavior for all possible transformations that form TSRs over $\mathbb{F}_2$. All LFSRs with characteristic polynomial divisible by $x + 1$ exhibit this property. A simple formula allows the computation of the pair of such a LFSR. The individual factors of the LFSRs, along with the order n, determine the pair.

This phenomenon is now generalized and expanded over $\mathbb{F}_q$. A much richer pairing and $(q-1)$-tupling pattern emerges. Several nuances appear which are masked in the $\mathbb{F}_2$ case. Every LFSR is a member of a $(q-1)$-tuple and those with linear factors are also paired in other, more interesting ways. Explicit formulas are given for the calculation of pairs and $(q-1)$-tuples and the development as n grows is explored. Finally, the role of individual divisors of the LFSRs is examined. Several examples of this pattern are provided. Computationally, this allows for time savings in the search for irreducible TSRs. However, the applications of TSRs and their pairs are not the focus of this article. Please refer to [4] and [1] for such explanations.

---

[*] The first author was supported by an NSERC Undergraduate Student Research Award. The second author was supported by NSERC under grant number 238757.

G. Mullen, A. Poli and H. Stichtenoth (Eds.): Fq7 2003, LNCS 2948, pp. 59–68, 2003.

2   Preliminaries

A sequence $s_0, s_1, \ldots,$ of elements from $\mathbb{F}_q$ satisfies a linear feedback shift register (LFSR) $S = \langle a_0, a_1, \ldots, a_{n-1} \rangle, a_0 \neq 0,$ of order n if

$$s_{n+t} = a_0 s_t + a_1 s_{t+1} + \cdots + a_{n-1} s_{n+t-1}$$

for all $t = 0, 1, \ldots.$ We associate S with the polynomial $f_S(x) = a_{n-1} x^{n-1} + \cdots + a_0.$ Unless otherwise specified, f_S is assumed to always have degree strictly less than n in this paper. These LFSRs generate a single new element with each iteration.

Tsaban and Vishne describe a *transformation shift register* (TSR) which generates an entire word v_n of m elements of $\mathbb{F}_q$ with each iteration. For the n words $v_0, v_1, \ldots, v_{n-1}$ and for linear transformation T of order $m,$

$$v_n = T(a_0 v_0 + a_1 v_1 + \cdots + a_{n-1} v_{n-1}).$$

Scalar multiplication and vector addition are used in this computation. We now cite some important results about TSRs.

Proposition 1. *(3.1 in [4]) Let T be a linear transformation of $\mathbb{F}_q^m$ with characteristic polynomial $f_T(x),$ and $S = \langle a_0, a_1, \ldots, a_{n-1} \rangle \in \mathbb{F}_2^n, a_0 \neq 0.$ Then the characteristic polynomial of the TSR $\langle T, S \rangle$ is*

$$f_{\langle T, S \rangle}(x) = f_S(x)^m f_T \left(\frac{x^n}{f_S(x)} \right).$$

The TSR has a maximal period of $q^{mn} - 1$ if, and only if, its characteristic polynomial is primitive.

Proposition 2. *(4.1 in [4]) Let $\mathbb{F}_{q^m}$ be the splitting field of $f_T(x).$ Let α be a root of $f_T(x)$ in $\mathbb{F}_{q^m}.$ Then $f_{\langle T, S \rangle}(x)$ is irreducible over $\mathbb{F}_q$ if, and only if, $x^n - \alpha f_S(x)$ is irreducible over $\mathbb{F}_{q^m}.$*

In an earlier paper [1], the authors investigated *pairs* of LFSRs. The LFSRs S_1 and S_2 form a pair whenever they form irreducible TSRs with exactly the same linear transformations $f_T(x).$ The corresponding polynomials f_{S_1} and f_{S_2} are also said to be a pair. It was further shown that over $\mathbb{F}_2,$ the polynomial $f_{S_1}(x)$ is paired with

$$f_{S_2}(x) := (x+1)^n f_{S_1} \left(\frac{x}{x+1} \right).$$

Furthermore, as the order of the LFSR increases to $n' = n + t,$ we have $(x+1)^i f_{S_1}(x)$ is paired with $(x+1)^{t-i} f_{S_2}(x).$ Finally, if

$$f_{S_1}(x) = (x+1)^{a_0} p_1^{a_1}(x) \cdots p_r^{a_r}(x),$$

then

$$f_{S_2}(x) = (x+1)^{n - a_0 - a_1 \deg p_1 - \cdots - a_r \deg p_r} q_1^{a_1}(x) \cdots q_r^{a_r}(x)$$

where $q_i(x) = (x+1)^{\deg p_i} p_i\left(\frac{x}{x+1}\right)$. In this paper we generalize these results to $\mathbb{F}_q$.

Finally, we also require linear fractional transformations as described by Fitzgerald [2]. For

$$M = \begin{bmatrix} a & b \\ c & d \end{bmatrix} \in GL(2, q),$$

we define

$$M \cdot x = \frac{ax + b}{cx + d}$$

$$M \cdot f(x) = (cx + d)^k f(M \cdot x) \text{ where } k = \deg f.$$

We note that this is an inverse action since $N \cdot (M \cdot f) = (MN) \cdot f$.

3 Pairing of LFSRs over $\mathbb{F}_q$

Pairing of LFSRs is found by computing TSRs for many LFSRs and linear transformations, and by checking for irreducibility and primitivity. Examples of this kind of test are given in Tables 1, 2, and 3. Irreducible linear transformations of orders $m = 2$ over $\mathbb{F}_5$ are listed in Table 1. The coefficients of the characteristic polynomials are listed with the least significant coefficient on the left. Hence, the eighth entry represents $4 + 3x + x^2$. Table 2 lists all LFSRs of order $n = 2$ over $\mathbb{F}_5$ in a similar fashion, and indicates which linear transformations produce irreducible ('I') TSRs. If the resulting TSR is also primitive, a 'P' replaces the 'I'. The numbers at the top of each column refer to the linear transformations in Table 1. Table 3 is similar, though it lists all LFSRs of order $n = 3$.

Note that many rows have non-blank entries (I or P) in the same coordinates. For example, rows 5, 12, 16, and 17 of Table 2 and rows 5, 8, 18, 19, 33, 51, 70, and 96 of Table 3. This phenomenon is mathematically justified in the following subsections.

3.1 Pairing of LFSRs of Fixed Order

Let f_S be the polynomial associated with the LFSR S. The next theorem shows that every f_S is paired with each polynomial in a set of order $q - 1$ (called a trivial pairing) as well as being paired with one polynomial for each distinct linear factor of f_S (called a non-trivial pairing). A polynomial trivially paired with one non-trivially paired with f_S is itself non-trivially paired with f_S.

Theorem 1. *Over $\mathbb{F}_q$, the polynomial $f_S(x)$ with degree $< n$ is paired with $(cx + d)^n f_S\left(\frac{x}{cx+d}\right)$, where $d \in \mathbb{F}_q^*$ and $c = 0$ (a trivial pair) or $1/c$ is a root of $f_S(x)$ (a non-trivial pair).*

Table 1. Characteristic Polynomials of Irreducible Transformations of Degree $m = 2$ over $\mathbb{F}_5$

1	201	6	421
2	301	7	331
3	111	8	431
4	211	9	141
5	321	10	241

Table 2. Irreducibility/Primitivity of TSRs for $n = 2$ over $\mathbb{F}_5$

	1	x	1	2	3	4	5	6	7	8	9	10
1	1	0	I	I			P	P		P		P
2	2	0	I	I			P	P		P		P
3	3	0	I	I			P	P		P		P
4	4	0	I	I			P	P		P		P
5	1	1		I	I	P	P	I		I		
6	2	1	I		I		P	I			I	P
7	3	1	I		I	P			P	I	I	
8	4	1	I					I	P	I	I	P
9	1	2	I					I	P	I	I	P
10	2	2	I		I	P			P	I	I	
11	3	2	I		I		P	I			I	P
12	4	2	I	I	P	P	I		I			
13	1	3	I					I	P	I	I	P
14	2	3	I		I	P			P	I	I	
15	3	3	I		I		P	I			I	P
16	4	3	I	I	P	P	I		I			
17	1	4	I	I	P	P	I		I			
18	2	4	I		I		P	I			I	P
19	3	4	I		I	P			P	I	I	
20	4	4	I					I	P	I	I	P

Table 3. Irreducibility/Primitivity of TSRs for $n = 3$ over $\mathbb{F}_5$

#	1	x	x^2	1	2	3	4	5	6	7	8	9	10
1	1	0	0				I	I	I	I	I	I	I
2	2	0	0				I	I	I	I	I	I	I
3	3	0	0				I	I	I	I	I	I	I
4	4	0	0				I	I	I	I	I	I	I
5	1	1	0	P					I	P			
6	2	1	0	P				P			I		
7	3	1	0	P				P			I		
8	4	1	0	P					I	P			
9	1	2	0		P		P					I	
10	2	2	0		P	I							P
11	3	2	0		P	I							P
12	4	2	0		P		P					I	
13	1	3	0		P	I							P
14	2	3	0		P		P					I	
15	3	3	0		P		P					I	
16	4	3	0		P	I							P
17	1	4	0	P				P			I		
18	2	4	0	P					I	P			
19	3	4	0	P					I	P			
20	4	4	0	P				P			I		
21	1	0	1	P			P						P
22	2	0	1		P	I						I	
23	3	0	1	P					I		I		
24	4	0	1		P			P		P			
25	1	1	1					I	P				P
26	2	1	1	P	P	I	P		I	P	I		P
27	3	1	1	P				P					P
28	4	1	1		P			P				I	
29	1	2	1	P					P			I	
30	2	2	1		P				P		P		
31	3	2	1	P	P	I	P	P	P	I	P		I
32	4	2	1					P			P		I
33	1	3	1	P					I	P			
34	2	3	1		P				P		P		
35	3	3	1	P	P	I		P		P	I	I	P
36	4	3	1				I		P				P
37	1	4	1						P	P			I
38	2	4	1	P	P			P	P	I	I	I	P
39	3	4	1	P				P					P
40	4	4	1	P	I								P
41	1	0	2	P	I							I	
42	2	0	2	P			P			P			
43	3	0	2	P					P				P
44	4	0	2	P							I		I
45	1	1	2	P	P			P	P	I	I	I	P
46	2	1	2	P	I								P
47	3	1	2				P	P			I		
48	4	1	2	P					P				P
49	1	2	2	P			P			P			
50	2	2	2			I			P				P
51	3	2	2	P						I	P		
52	4	2	2	P	P	I		P		P	I	I	P
53	1	3	2	P			P			P			
54	2	3	2					P		P			I
55	3	3	2	P					P			I	
56	4	3	2	P	P	I	P	P	P	I	P		I
57	1	4	2	P	P	I	P		I	P	I		P
58	2	4	2	P			P						I
59	3	4	2						I	P			P
60	4	4	2	P					P				P
61	1	0	3	P					I		I		
62	2	0	3	P					P				P
63	3	0	3	P				P				P	
64	4	0	3	P	I								I
65	1	1	3	P					P				P
66	2	1	3				P	P			I		
67	3	1	3	P	I								P
68	4	1	3	P	P			P	P	I	I	I	P
69	1	2	3	P	P	I		P		P	I	I	P
70	2	2	3	P					I	P			
71	3	2	3				I		P				P
72	4	2	3	P					P		P		
73	1	3	3	P	P	I	P	P	P	I	P		I
74	2	3	3	P					P			I	
75	3	3	3					P			P		I
76	4	3	3	P					P		P		
77	1	4	3	P					P				P
78	2	4	3						I	P			P
79	3	4	3	P			P						I
80	4	4	3	P	P	I	P		I	P	I		P
81	1	0	4	P					P		P		
82	2	0	4	P					I		I		
83	3	0	4	P	I								I
84	4	0	4	P					P				P
85	1	1	4	P			P						I
86	2	1	4	P					P				P
87	3	1	4	P	P	I	P		I	P	I		P
88	4	1	4						I	P			P
89	1	2	4						P		P	I	
90	2	2	4	P	P	I	P	P	P	I	P		I
91	3	2	4	P					P		P		
92	4	2	4	P					P			I	
93	1	3	4				I		P				P
94	2	3	4	P	P	I		P		P	I	I	P
95	3	3	4	P					P		P		
96	4	3	4	P						I	P		
97	1	4	4	P	I								P
98	2	4	4	P			P						P
99	3	4	4	P	P			P	P	I	I	I	P
100	4	4	4				P	P			I		

Proof. Let $\alpha \in \mathbb{F}_{q^m}$ be a root of an irreducible polynomial f_T of degree m, and suppose $x^n - \alpha f_S(x)$ is reducible over $\mathbb{F}_{q^m}$ with root $\beta \in \mathbb{F}_{q^{mk}}$, $k < n$. For $M = \begin{bmatrix} 1 & 0 \\ c & d \end{bmatrix}$, $d \neq 0$, the inverse of M exists. Define $\delta = M^{-1} \cdot \beta$. Then $\beta = M \cdot \delta$ and we have

$$\beta^n - \alpha f_S(\beta) = (M \cdot \delta)^n - \alpha f_S(M \cdot \delta) = 0.$$

Therefore,

$$(c\delta + d)^n \left(\frac{\delta}{c\delta + d} \right)^n - (c\delta + d)^n \alpha f_S \left(\frac{\delta}{c\delta + d} \right) = 0.$$

Hence $x^n - \alpha (cx + d)^n f_S \left(\frac{x}{cx+d} \right)$ is reducible.

Similarly, suppose $x^n - \alpha(cx + d)^n f_S \left(\frac{x}{cx+d} \right)$ has a root $\gamma \in \mathbb{F}_{q^{mk}}$, $k < n$. Define $\mu = M \cdot \gamma$. Then since γ cannot be $-d/c$, we have the both equalities:

$$\gamma^n - \alpha(c\gamma + d)^n f_S \left(\frac{\gamma}{c\gamma + d} \right) = 0,$$

$$\left(\frac{\gamma}{c\gamma + d} \right)^n - \alpha f_S \left(\frac{\gamma}{c\gamma + d} \right) = \mu^n - \alpha f_S(\mu) = 0.$$

Thus $x^n - \alpha f_S(x)$ is irreducible iff $x^n - \alpha(cx + d)^n f_S \left(\frac{x}{cx+d} \right)$ is irreducible.

It remains to show that $(cx+d)^n f_S \left(\frac{x}{cx+d} \right)$ corresponds to a legitimate LFSR of order n by showing that zero is not a root and it has degree strictly less than n. The first fact follows easily from $f_S(0) \neq 0$ and $d \neq 0$ since

$$(cx + d)^n f_S \left(\frac{x}{cx + d} \right) = \sum_{i=0}^{n} a_i x^i (cx + d)^{n-i}$$

$$(c0 + d)^n f_S \left(\frac{0}{c0 + d} \right) = a_0 d^n \neq 0.$$

The second fact follows from

$$(cx + d)^n f_S \left(\frac{x}{cx + d} \right) = x^n (a_{n-1} c + \cdots + a_1 c^{n-1} + a_0 c^n)$$

$$+ \{\text{some polynomial of degree} < n\}$$

$$= x^n c^n f \left(\frac{1}{c} \right) + \{\text{some polynomial of degree} < n\}.$$

The x^n term disappears whenever c is as specified.   ∎

The trivial case $c = 0$ always occurs. Every LFSR f_S is paired with every member of the $(q-1)$-tuple $\{d^n f_S(\frac{x}{d}) : d \in \mathbb{F}_q^*\}$. Since

$$f_{\langle T, S \rangle}(x) = f_S(x)^m f_T\left(\frac{x^n}{f_S(x)}\right)$$

is the characteristic polynomial of a TSR involving f_S, the family of TSRs of the trivial $(q-1)$-tuple is found by direct substitution

$$f'_{\langle T, S' \rangle}(x) = d^{mn} f_{\langle T, S \rangle}\left(\frac{x}{d}\right).$$

More generally, the family of TSRs of all pairs of f_S is also found by substitution

$$f''_{\langle T, S'' \rangle}(x) = (cx + d)^{mn} f_{\langle T, S \rangle}\left(\frac{x}{cx + d}\right)$$
$$= M \cdot f_{\langle T, S \rangle}(x),$$

where c and d are as specified in Theorem 1.

An example of a trivial $(q-1)$-tuple is lines 33, 51, 70, and 96 of Table 3. The thirty-third LFSR of Table 3 is $f_{S_1} = 1 + 3x + x^2$. For $n = 3$ and $c = 0$, $f_{S_d} = d^3 + 3d^2 x + dx^2$. Hence,

$$f_{S_2} = 3 + 2x + 2x^2$$
$$f_{S_3} = 2 + 2x + 3x^2$$
$$f_{S_4} = 4 + 3x + 4x^2,$$

and these are the LFSRs in rows 51, 70, and 96. A non-trivial pair of the fifty-first LFSR of Table 3 is

$$f_{S_5} = (3x + 2)^3 f_{S_2}\left(\frac{x}{3x + 2}\right)$$
$$= 3(3x + 2)^3 + 2(3x + 2)^2 x + 2(3x + 2)x^2$$
$$= 4 + x$$

which is the eighth row of that table. We already noted these are paired in the comment before Subsection 3.1. The other pairs may be found in a similar fashion.

All of these examples of paired f_S have the extra property that $f_{\langle T, S_1 \rangle}$ is primitive if and only if $f_{\langle T, S_2 \rangle}$ is primitive. This does not hold in general. For an example, refer to lines 5 and 23 from Table V of [1].

3.2   Pairing of LFSRs as the Order Varies

The pairs of f_{S_1} for order n determine the pairs of larger order $n+t$. The previous theorem tells us that for $f_{S_2} = (cx + d)^n f_{S_1} \left(\frac{x}{cx+d} \right)$,

$$x^{n+t} - \alpha f_{S_1}(x) \text{ is irreducible,}$$

$$\Leftrightarrow x^{n+t} - \alpha(cx + d)^{n+t} f_{S_1} \left(\frac{x}{cx + d} \right) \text{ is irreducible,}$$

$$\Leftrightarrow x^{n+t} - \alpha(cx + d)^t f_{S_2}(x) \text{ is irreducible.}$$

That is, we have the following theorem.

Theorem 2. *The LFSR $f_{S_1}(x)$ is paired with $(cx + d)^t f_{S_2}(x)$ for LFSR order $n' = n + t$.*

For example, since for $n = 3$, $f_{S_2} = 3 + 2x + 2x^2 \in \mathbb{F}_5[x]$ has non-trivial pair

$$f_{S_5} = (3x + 2)^3 f_{S_2} \left(\frac{x}{3x + 2} \right) = 4 + x,$$

Theorem 2 implies that for $n = 5$, f_{S_2} has pair

$$(3x + 2)^2 f_{S_5} = (3x + 2)^2 (4 + x) = 1 + 2x + 3x^2 + 4x^3.$$

3.3   Factors of the Polynomial Pairs

Pairs of LFSRs can be determined by examining the individual factors of the polynomials $f_{S_1}(x) = p_1^{a_1}(x) \cdots p_r^{a_r}(x)$, where the p_i are irreducible over $\mathbb{F}_q$. Quite simply, the pair of $f_{S_1}(x)$ is

$$f_{S_2}(x) = (cx + d)^n p_1^{a_1} \left(\frac{x}{cx + d} \right) \cdots p_r^{a_r} \left(\frac{x}{cx + d} \right)$$

$$= (cx + d)^{n - a_1 \deg p_1 - \cdots - a_r \deg p_r} q_1^{a_1}(x) \cdots q_r^{a_r}(x)$$

where $q_i(x) = (cx + d)^{\deg p_i} p_i \left(\frac{x}{cx+d} \right)$ is the *pair* of p_i.

That is, f_{S_1} may be factored, the individual divisors replaced with their pairs and then an appropriate power of $(cx + d)$ inserted. This yields f_{S_2}.

In particular, the linear polynomial $p(x) = x - 1/c$ has pair

$$q(x) = (cx + d) \left(\frac{x}{cx + d} - \frac{1}{c} \right) = -\frac{d}{c}.$$

Hence, the degree of the pair of the composite $f_S(x)$ having $p(x) = x - 1/c$ as a factor will have degree less than n. All other (monic) irreducible factors p_i will have pair q_i of the same degree.

For example, consider the polynomial

$$f_S(x) = x^6 + 6x^4 + x^3 + 4x^2 + 2x + 6$$
$$= (x^3 + 6x^2 + 5x + 3)(x^2 + 2)(x + 1) \text{ over } \mathbb{F}_7.$$

For $c = 6$ and $n = 7$, it has pairs for all $1 \leq d \leq 6$,

$$f_{S_d}(x) = (6x + d)(4x^3 + 5dx^2 + 3d^2x + 3d^3)(3x^2 + 3dx + 2d^2)(d)$$
$$= 2dx^6 + 6d^2x^5 + 2d^3x^4 + 4d^4x^3 + 6d^5x^2 + 2d^6x + 6d^7.$$

Whence

$$f_{S_1}(x) = 2x^6 + 6x^5 + 2x^4 + 4x^3 + 6x^2 + 2x + 6$$
$$f_{S_2}(x) = 4x^6 + 3x^5 + 2x^4 + x^3 + 3x^2 + 2x + 5$$
$$f_{S_3}(x) = 6x^6 + 5x^5 + 5x^4 + 2x^3 + 2x^2 + 2x + 4$$
$$f_{S_4}(x) = x^6 + 5x^5 + 2x^4 + 2x^3 + 5x^2 + 2x + 3$$
$$f_{S_5}(x) = 3x^6 + 3x^5 + 5x^4 + x^3 + 4x^2 + 2x + 2$$
$$f_{S_6}(x) = 5x^6 + 6x^5 + 5x^4 + 4x^3 + x^2 + 2x + 1$$

form a $(q-1)$-tuple which are trivially related to each other, but individually, each of them is a non-trivial pair of f_S.

4   Conclusion

Further examination of tables similar to those provided here may yield new interesting results. Perhaps different linear fractional transformations, say for $b \neq 0$, of the LFSRs lead to other connections.

In his paper, Fitzgerald [2] counts the number of polynomials which remain invariant under the action of the matrix M. His techniques are not immediately applicable in our case because $(cx + d)^n f_{S_1}\left(\frac{x}{cx+d}\right) \neq M \cdot f_{S_1}$ as $n > \deg f_{S_1}$. As well, when considering the characteristic polynomial $f_{\langle T,S \rangle}$ of the TSR and its pair $f'_{\langle T,S \rangle} = M \cdot f_{\langle T,S \rangle}$, not all transformations M are applied to all polynomials f of degree mn. M is only applied to those polynomials which represent a TSR. That is, an application of Fitzgerald's methods here would count extraneous invariant polynomials. An example of such an invariant polynomial is $f_S = x^4 - 1$ for $q = 5$, $n = 5$. Tables (similar to 2 and 3) containing examples of invariant polynomials were not provided as they would be too large or too simple. For instance, if $1/c$ is a root of invariant f_S, then

$$f_S(x) = (cx + 1)^n f_S\left(\frac{x}{cx + 1}\right).$$

Plugging in $1/c$ shows that $1/(2c)$ is also a root (unless p, the characteristic, equals 2). Plugging in $1/(2c)$ shows that $1/(3c)$ is a root and so on. Thus f_S

has roots $1/(kc)$ for $1 \leq k < p$. In particular, $\deg f_S \geq p - 1$. An explicit characterization of these invariant f_S would be helpful.

Nevertheless, transformation shift registers may be found efficiently by targeting LFSRs with linear factors (and hence non-trivial pairs). Multiple irreducible TSRs are thus found for the computational price of one.

Acknowledgement. The authors wish to thank the referee for a careful reading of the manuscript and many helpful comments which improved the clarity of this paper.

References

1. M. DEWAR AND D. PANARIO. Linear Transformation Shift Registers. *IEEE Trans. Inform. Theory*, 49:2047–2052, 2003.
2. R. FITZGERALD. Irreducible Polynomials Over Finite Fields that are Invariant Under Linear Fractional Transformations. *Preprint*.
3. S. GOLOMB. *Shift-Register Sequences*. Aegean Park Press, 1982.
4. B. TSABAN AND U. VISHNE. Efficient Linear Feedback Shift Registers with Maximal Period. *Finite Fields Appl.*, 8:256–267, 2002.

Lattice Profile and Linear Complexity Profile of Pseudorandom Number Sequences

Gerhard Dorfer

Department of Algebra and Computational Mathematics, Vienna University of
Technology, Wiedner Hauptstr. 8–10/118, A-1040 Vienna, Austria
`g.dorfer@tuwien.ac.at`

Abstract. The relationship between two concepts measuring structural
properties of pseudorandom numbers, namely the linear complexity profile and the lattice profile, is investigated. In particular an explicit formula expressing the lattice profile in terms of the linear complexity profile
(and vice versa) can be provided once the interrelation is known in certain points. Moreover an intrinsic characterization of lattice profiles is
established.

1 Introduction and Basic Facts

Pseudorandom numbers (PRN) generated by linear congruences as introduced
by Lehmer [8], though still very popular and widely used, comprise severe defi-
ciencies that make them improper in many applications as for instance in quasi-
Monte Carlo methods (cf. [12]). One particularly undesirable feature of these
PRN is their coarse lattice structure. Marsaglia [9] proposed a lattice test for
arbitrary nonlinear congruential generators modulo a prime. This test was investigated and enhanced by several authors, e. g., [5, 15–17].

Recently, in joint work with A. Winterhof [2, 3], we extended a generalized
version of Marsaglia's lattice test for sequences over finite fields to segments of
sequences (η_n) over an arbitrary field $\mathbb{K}$: For given $s \geq 0$ and $N \geq 2$ we say that
(η_n) passes the *s-dimensional N-lattice test* if the vectors $\{\underline{\eta}_n - \underline{\eta}_0 \mid 1 \leq n \leq N - s\}$ span $\mathbb{K}^s$, where

$$\underline{\eta}_n = (\eta_n, \eta_{n+1}, \ldots, \eta_{n+s-1}), \quad 0 \leq n \leq N - s.$$

If (η_n) passes the s-dimensional N-lattice test then it passes all s'-dimensional
N-lattice tests for $s' \leq s$ and if (η_n) fails the s-dimensional N-lattice test then
it fails all s'-dimensional N-lattice tests for $s' \geq s$. The greatest s such that (η_n)
satisfies the s-dimensional N-lattice test is called the *Nth lattice level* of (η_n)
(or the *lattice profile of (η_n) at N*) and is denoted by $S((\eta_n), N)$. Additionally
we define $S((\eta_n), 0) = S((\eta_n), 1) := 0$. The sequence $(S((\eta_n), N))_{N \geq 0}$ is called
the *lattice profile* of (η_n).

Another quality measure appraising the intrinsic structure of PRN is given
by the linear complexity: The *Nth linear complexity* $L((\eta_n), N)$, $N \geq 1$, is the
least order L of a linear recurrence relation over $\mathbb{K}$

$$\eta_{n+L} = \alpha_0 \eta_n + \alpha_1 \eta_{n+1} + \ldots + \alpha_{L-1} \eta_{n+L-1}, \quad 0 \leq n \leq N - L - 1,$$

G. Mullen, A. Poli and H. Stichtenoth (Eds.): Fq7 2003, LNCS 2948, pp. 69–78, 2003.
© Springer-Verlag Berlin Heidelberg 2003

which is satisfied by the first N terms of (η_n) (with the additional conventions that $L((\eta_n), N) = 0$ if the first N terms of (η_n) are all 0 and $L((\eta_n), N) = N$ if the first $N-1$ terms are 0 and the Nth term of (η_n) is nonzero). Moreover we define $L((\eta_n), 0) := 0$. The *linear complexity profile* is the sequence $(L((\eta_n), N))_{N \geq 0}$ and the *linear complexity* $L(\eta_n)$ is defined as

$$L(\eta_n) = \sup_{N \geq 2} L((\eta_n), N).$$

The linear complexity and the linear complexity profile are important cryptographic characteristics of sequences (see e. g. [1, 11, 19]). A low linear complexity of a generator has turned out to be undesirable for more traditional applications in Monte Carlo methods as well (see e. g. [6, 12–14]).

In the following we will use a more compact notation and mostly write $L(N)$ and $S(N)$ instead of $L((\eta_n), N)$ and $S((\eta_n), N)$, respectively, when it is not necessary to stress the role of a particular sequence (η_n) and merely the properties of S and L as functions in N are of interest.

In the remaining part of Sect. 1 we will recall some results on the linear complexity and the lattice profile which will be used later on.

The following proposition (cf. [7, Theorem 6.7.4], [10], or [18]) describes the step-growth of the linear complexity profile.

Proposition 1. (i) *If $L(N) > N/2$ then*

$$L(N + 1) = L(N).$$

(ii) *If $L(N) \leq N/2$, then*

$$L(N + 1) = L(N) \qquad or \qquad L(N + 1) = N + 1 - L(N).$$

Next we list some basic properties of the lattice profile ([2, Proposition 4]).

Proposition 2. (i) $S(N) \leq S(N + 1) \leq S(N) + 1$.
(ii) $S(N) \leq N/2$.

As the main result of [2] the following relation between lattice profile and linear complexity profile for arbitrary sequences is proved.

Theorem 1. *We have either*

$$S(N) = \min\{L(N), N + 1 - L(N)\}$$

or

$$S(N) = \min\{L(N), N + 1 - L(N)\} - 1.$$

In case $L(N) = N + 1 - L(N)$, i. e., $L(N) = (N + 1)/2$, there is a definite value for $S(N)$, namely

$$S(N) = \min\{L(N), N + 1 - L(N)\} - 1 = (N - 1)/2.$$

Furthermore in [2] an example is given which shows that all four possibilities $S(N) = L(N) - 1, L(N), N - L(N), N + 1 - L(N)$ in Theorem 1 occur.

2 Relationship Between Lattice Profile and Linear Complexity Profile

From the point of view of applications Theorem 1 can be interpreted as saying that the linear complexity profile and the lattice profile provide essentially equivalent quality measures for the intrinsic structure of a sequence.

In this paper we continue with the objective to decide which of the four values in Theorem 1 is assumed by $S(N)$ and we try to find out what is the dynamic behind that governs the relationship between $S(N)$ and $L(N)$.

A partial result in this direction is [2, Proposition 5,6 and 10].

Proposition 3. *If* $L := L(N) \leq N/2$ *and*

$$\eta_{n+L} = \alpha_0 \eta_n + \alpha_1 \eta_{n+1} + \ldots + \alpha_{L-1} \eta_{n+L-1}, \quad 0 \leq n \leq N - L - 1, \quad (1)$$

is the linear recurrence relation of least order satisfied by the first N terms of (η_n), *then* $L(N) - 1 \leq S(N) \leq L(N)$ *and*

$$S(N) = L(N) - 1$$

if and only if

$$\alpha_0 + \alpha_1 + \ldots + \alpha_{L-1} = 1.$$

Remark 1. A sequence (η_n) satisfies a recurrence relation (1) with $\alpha_0 + \alpha_1 + \ldots + \alpha_{L-1} = 1$ if and only if every additively shifted sequence $(\eta_n + \alpha)$, $\alpha \in \mathbb{K}$, satisfies the same recurrence.

Next we prove some preparatory results.

Lemma 1. *If* $S(N - 1) = S(N)$ *then*

$$L(N) \leq S(N - 1) + 1.$$

Proof. Put $s := S(N)$. We consider the following matrix:

$$A = \begin{pmatrix} \eta_1 - \eta_0 & \eta_2 - \eta_1 & \cdots & \eta_{N-s-1} - \eta_{N-s-2} \\ \eta_2 - \eta_1 & \cdots & & \eta_{N-s} - \eta_{N-s-1} \\ \vdots & & & \vdots \\ \eta_s - \eta_{s-1} & \cdots & & \eta_{N-2} - \eta_{N-3} \\ \eta_{s+1} - \eta_s & \cdots & \eta_{N-2} - \eta_{N-3} & \eta_{N-1} - \eta_{N-2} \end{pmatrix}.$$

The assumption $S(N-1) = s$ means that the rank of the first s rows of A equals s, thus the first s rows of A are linearly independent. Since $s = S(N)$ the rank of A is also s. Thus the last row of A is a linear combination of the first s rows and there exist $\alpha_0, \ldots, \alpha_{s-1} \in \mathbb{K}$ such that

$$\eta_{s+1+n} - \eta_{s+n} = \alpha_0(\eta_{1+n} - \eta_n) + \ldots + \alpha_{s-1}(\eta_{s+n} - \eta_{s-1+n})$$

for $n = 0, 1, \ldots, N - s - 2$. Rearranging the last equation to

$$\eta_{s+1+n} = -\alpha_0 \eta_n + (\alpha_0 - \alpha_1)\eta_{n+1} + \ldots + (\alpha_{s-2} - \alpha_{s-1})\eta_{s-1+n} + (\alpha_{s-1} + 1)\eta_{s+n},$$

we see that $L(N) \leq s + 1 = S(N - 1) + 1$. $\square$

Chiefly we will use the last result in the following form.

Lemma 2. *If $L(N) > (N+1)/2$ then*

$$S(N) = S(N-1) + 1.$$

Proof. $L(N) > (N+1)/2$ implies $L(N) > S(N-1) + 1$ (due to $S(N-1) \leq (N-1)/2$, Proposition 2 (ii)). By Lemma 1 we obtain $S(N) > S(N-1)$ and Proposition 2 (i) yields $S(N) = S(N-1) + 1$. □

Lemma 3. *Let N be even. If $S(N-1) = N/2 - 1$ and $S(N-2) < N/2 - 1$ then $S(N) = N/2$.*

Proof. We consider the matrix

$$A = \begin{pmatrix} \eta_1 - \eta_0 & \eta_2 - \eta_1 \cdots \eta_{N/2} - \eta_{N/2-1} \\ \eta_2 - \eta_1 & \cdots & \vdots \\ \vdots & & \vdots \\ \eta_{N/2} - \eta_{N/2-1} & \cdots & \cdots \eta_{N-1} - \eta_{N-2} \end{pmatrix}.$$

Let A' denote the matrix consisting of the first $N/2 - 1$ rows of A and let A'' consist of the first $N/2 - 1$ columns of A'. Our assumptions can be interpreted as follows:

(i) $S(N-1) = N/2 - 1$ means that the rows of A' are linearly independent, i.e., the rank of A' is $N/2 - 1$.

(ii) $S(N-2) < N/2 - 1$ implies $\det A'' = 0$, hence, say the i-th row of A'', $1 \leq i \leq N/2 - 1$, is a linear combination of the remaining rows of A''.

Performing the corresponding row-transformation with A we obtain

$$\det A = \det \begin{pmatrix} \eta_1 - \eta_0 & \eta_2 - \eta_1 \cdots \eta_{N/2} - \eta_{N/2-1} \\ \vdots & \vdots \\ 0 & \cdots & 0 \quad \alpha \\ \vdots & \vdots \\ \eta_{N/2} - \eta_{N/2-1} & \cdots \cdots \eta_{N-1} - \eta_{N-2} \end{pmatrix} \quad (\leftarrow i\text{-th row})$$

$$= \pm\alpha \cdot \det \begin{pmatrix} \eta_1 - \eta_0 & \cdots & \eta_{N/2-1} - \eta_{N/2-2} \\ \vdots & & \vdots \\ \eta_{i-1} - \eta_{i-2} & \cdots \eta_{N/2+i-3} - \eta_{N/2+i-4} \\ \eta_{i+1} - \eta_i & \cdots \eta_{N/2+i-1} - \eta_{N/2+i-2} \\ \vdots & & \vdots \\ \eta_{N/2} - \eta_{N/2-1} \cdots & \eta_{N-2} - \eta_{N-3} \end{pmatrix}.$$

Due to (i) we have $\alpha \neq 0$. The transpose of the matrix B appearing in the last step arises from A' by deleting the i-th column which is linearly dependent from the remaining columns (see (ii) and note that A'' is symmetric). Thus the rank of B is equal to the rank of A' which is $N/2 - 1$ and hence $\det B \neq 0$. Consequently $\det A \neq 0$ and $S(N) = N/2$. □

Lemma 4. *If $L(N) = N/2$ and $S(N) = N/2 - 1$ then $S(N - 2) = N/2 - 1$.*

Proof. If $L(N) = N/2$ then $L(N - 1) = N/2$ (Proposition 1) and $S(N - 1) = N/2 - 1$ (second part of Theorem 1).

Assuming $S(N - 2) < N/2 - 1$ Lemma 3 yields $S(N) = N/2$, a contradiction. Hence $S(N - 2) \geq N/2 - 1$ and since $S(N - 2) \leq S(N - 1) = N/2 - 1$ we infer $S(N - 2) = N/2 - 1$. □

Now we are in a position to give a full answer to the question to which extent the linear complexity profile determines the lattice profile.

Theorem 2. *Assume $L(N_1) = N_1/2$, $L(N_2) = N_2/2$, $N_1 < N_2$ and there is no $\bar{N}$ with $N_1 < \bar{N} < N_2$ and $L(\bar{N}) = \bar{N}/2$. Then $S(N)$ is completely determined by $S(N_1)$ for all N with $N_1 \leq N < N_2$. In particular one of the following two cases occurs:*

1. If $S(N_1) = L(N_1)$ then

$$S(N) = \begin{cases} \min\{L(N), N + 1 - L(N)\}, & N_1 \leq N \leq N_2 - 2 \\ N_2/2 - 1 & N = N_2 - 1 \end{cases},$$

2. If $S(N_1) = L(N_1) - 1$ then

$$S(N) = \min\{L(N), N + 1 - L(N)\} - 1, \quad N_1 \leq N \leq N_2 - 1.$$

Additionally in case 2. we have $S(N_2) = N_2/2$.

Proof. An immediate application of Proposition 1 to our assumptions results in

$$L(N) = \begin{cases} N_1/2, & N_1 \leq N \leq (N_1 + N_2)/2 - 1 \\ N_2/2, & (N_1 + N_2)/2 \leq N \leq N_2 \end{cases}.$$

According to Theorem 1 we have either $S(N_1) = N_1/2$ or $S(N_1) = N_1/2 - 1$.

$S(N_1) = N_1/2$: Since $L(N)$ does not increase for N between N_1 and $(N_1 + N_2)/2 - 1$, the unique (see e. g. [2, Lemma 3]) linear recurrence relation of minimal length satisfied by the first N terms of (η_n) persists for $N \in [N_1, (N_1 + N_2)/2 - 1]$. Due to Proposition 3 also $S(N)$ is constant in this interval. For $N \in [(N_1 + N_2)/2, N_2 - 2]$ Lemma 2 ensures that $S(N)$ increases by 1 in each step. This matches exactly with the provided formula. For $N = N_2 - 1$ we have $L(N) = (N + 1)/2$ and thus $S(N) = (N - 1)/2$ follows by the second part of Theorem 1.

$S(N_1) = N_1/2 - 1$: For $N_1 < N \leq N_2 - 1$ the same arguments as in the first case apply. It remains to prove $S(N_2) = N_2/2$. Since $L(N_2) = N_2/2$ we have either $S(N_2) = N_2/2$ or $S(N_2) = N_2/2 - 1$ (Theorem 1). Assume $S(N_2) = N_2/2 - 1$, then Lemma 4 implies $S(N_2 - 2) = N_2/2 - 1$. However, we have already proved that $S(N_2 - 2) = \min\{N_2 - 2, N_2 - 2 + 1 - N_2/2\} - 1 = N_2/2 - 2$, a contradiction. □

Corollary 1. *Let $(L(N))_{N\geq 0}$ be the linear complexity profile of some sequence and $N_0 = 0 < N_1 < N_2 \ldots$ the sequence of all K with $L(K) = K/2$. Then the lattice profile $(S(N))_{N\geq 0}$ of this sequence can be computed from $(L(N))_{N\geq 0}$ provided the values $S(N_i) \in \{L(N_i), L(N_i)-1\}$ are known for all $i \geq 1$. These values obey the following relation: If $S(N_i) = L(N_i) - 1$ for some i then $S(N_{i+1}) = L(N_{i+1})$.*

Proof. The corollary follows immediately from Theorem 2 for all N with $N \leq N_i$ for some i. If there is a largest K with $L(K) = K/2$ then $L(N) = K/2$ for all $N \geq K$ and thus $S(N) = S(K)$ by Proposition 3.   □

In the following figures the progression of $L(N)$ and $S(N)$ for $N_1 \leq N < N_2$ in case 1. and 2. is demonstrated.

Ad 1.:

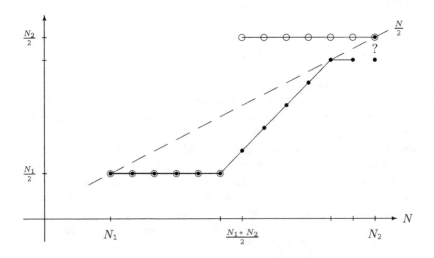

Ad 2.:

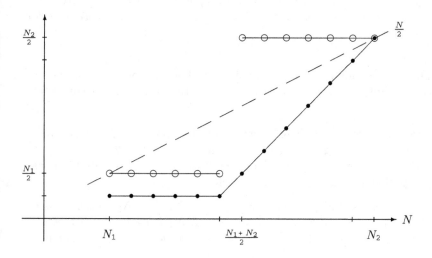

In case 1. we indicated by the question mark that there are two possible values for $S(N_2)$, namely $N_2/2$ or $N_2/2 - 1$. The following examples show that both cases occur.

Example 1. 1. We consider the sequence

$$(\eta_n) = (0, 1, 1, 0, -1, 0, 2, 2).$$

The first 5 terms fulfill the relation $\eta_{n+2} = -\eta_n + \eta_{n+1}$, the whole sequence satisfies $\eta_{n+4} = \eta_n - \eta_{n+2} + \eta_{n+3}$ and we obtain the following values for L and S:

N	2	3	4	5	6	7	8
$L(N)$	2	2	2	2	4	4	4
$S(N)$	1	1	2	2	3	3	3

With $N_1 = 4$ and $N_2 = 8$ we have $S(N_1) = L(N_1)$ and $S(N_2) = L(N_2) - 1$.
2. We consider the sequence

$$(\eta_n) = (0, 1, 1, 0, -1, 0, 1, 1).$$

The only thing changed in comparison to the example before is that the whole sequence now satisfies $\eta_{n+4} = \eta_n - \eta_{n+1}$ and we obtain

N	2	3	4	5	6	7	8
$L(N)$	2	2	2	2	4	4	4
$S(N)$	1	1	2	2	3	3	4

So in this case we have $S(N_1) = L(N_1)$ and $S(N_2) = L(N_2)$.

We point out another consequence of Theorem 2 which will turn out as an important feature of the lattice profile.

Corollary 2. *Let* $(S(N))_{N\geq 0}$ *be the lattice profile of a sequence and* $N_1 < N_2$ *two consecutive numbers* K *with the property* $S(K) = K/2$. *Then we have for all* $N \in [N_1, N_2]$

$$S(N) = \max\{N_1/2, N - N_2/2\}.$$

Proof. The assertion follows directly from Lemma 4 and Theorem 2 if one considers the two possible cases occurring there and how these cases can be connected.
□

In other words Corollary 2 means: when the lattice profile $S(N)$ starts to increase, it steadily increases until it meets the upper bound $N/2$.

Conversely to what we stated in Theorem 2 we can also compute the linear complexity profile in terms of the lattice profile.

Theorem 3. *Suppose* $S(N_1) = N_1/2$, $S(N_2) = N_2/2$, $N_1 < N_2$, *and there is no* $\bar{N}$ *with* $N_1 < \bar{N} < N_2$ *and* $S(\bar{N}) = \bar{N}/2$. *Then* $L(N)$ *is completely determined by* $S(N_2)$ *for all* $N_1 < N \leq N_2$. *In particular, one of the following holds true:*

1. *If* $L(N_2) = S(N_2)$ *then*

$$L(N) = \begin{cases} N_1/2 + 1, & N_1 < N \leq (N_1 + N_2)/2 \\ N_2/2, & (N_1 + N_2)/2 < N \leq N_2 \end{cases},$$

2. *If* $L(N_2) = S(N_2) + 1$ *then*

$$L(N) = \begin{cases} N_1/2, & N_1 < N \leq (N_1 + N_2)/2 \\ N_2/2 + 1, & (N_1 + N_2)/2 < N \leq N_2 \end{cases}.$$

Additionally we have $L(N_1) = N_1/2$ *in the second case. In the first case* $L(N_1)$ *is either* $N_1/2$ *or* $N_1/2 + 1$.

The proof utilizes Theorem 2 and some basic properties of the linear complexity/lattice profile and is omitted here.

3   Intrinsic Description of Lattice Profiles

We have proved several properties of the lattice profile of a sequence. Now we show that some of them characterize lattice profiles among all functions on the nonnegative integers $\mathbb{N}_0$.

Theorem 4. *Let* $S : \mathbb{N}_0 \to \mathbb{N}_0$ *be a function on the nonnegative integers. Then* $S(N)$ *is the lattice profile* $S((\eta_n), N)$ *of some sequence* (η_n) *in the field* $\mathbb{K}$ *if and only if*

(i) $S(N) \leq N/2$,
(ii) $S(N) \leq S(N+1) \leq S(N) + 1$,
(iii) *if* $s = S(N) < S(N+1)$ *then*

$$S(N+1) < \ldots < S(2(N-s)) = N - s.$$

Proof. Proposition 2 and Corollary 2 show that the conditions (i)-(iii) are necessary.

Now we show the sufficiency. (i) implies that $S(0) = S(1) = 0$ and $S(2) \leq 1$. We choose $\eta_1 = \eta_0$ if $S(2) = 0$ and $\eta_1 \neq \eta_0$ if $S(2) = 1$, respectively. Now suppose $\eta_0, \ldots, \eta_{N-1}$ have been chosen and $S((\eta_n), i) = S(i)$ for $0 \leq i \leq N$. If $S(N) = N/2$ then by (i) and (ii) $S(N+1) = N/2$ and whatever η_N is we obtain $S((\eta_n), N+1) = S(N+1)$. Also, if $S(N-1) < S(N) < N/2$ we can choose η_N arbitrarily to result in $S(N+1) = S(N) + 1$ (by (iii)) and $S((\eta_n), N+1) = S((\eta_n), N) + 1$ (by Corollary 2).

Thus we may assume that $S(N-1) = S(N) =: s < N/2$ and we have to show that we can choose η_N such that $S((\eta_n), N+1) = s$ or $S((\eta_n), N+1) = s+1$: If $s = 0$ then $\eta_0 = \ldots = \eta_{N-1}$ and obviously η_N can be chosen properly. For $s > 0$ we consider the $(s+1) \times (N-s)$ matrix

$$A = \begin{pmatrix} \eta_1 - \eta_0 & \eta_2 - \eta_1 & \cdots & \eta_{N-s-1} - \eta_{N-s-2} & \eta_{N-s} - \eta_{N-s-1} \\ \eta_2 - \eta_1 & \cdots & & \vdots & \vdots \\ \vdots & & & & \\ \eta_s - \eta_{s-1} & \cdots & & \eta_{N-2} - \eta_{N-3} & \eta_{N-1} - \eta_{N-2} \\ \eta_{s+1} - \eta_s & \cdots & & \eta_{N-1} - \eta_{N-2} & \eta_N - \eta_{N-1} \end{pmatrix}.$$

Firstly we focus on the first s rows of A and denote this matrix with B. Since $S((\eta_n), N) = s$ the rank of B is s, due to $S((\eta_n), N-1) = s$ the last column of B is a linear combination of the first $N-s-1$ columns. This means that among the first $N-s-1$ columns of B there are s which are linearly independent, say those with indices $i_1, \ldots, i_s$.

Now let C be the matrix consisting of the columns of A with indices $i_1, \ldots, i_s$ and $N-s$, i.e., C is a $(s+1) \times (s+1)$ matrix. We compute $\det C$ by means of expansion by minors with the last column and obtain

$$\det C = (\eta_N - \eta_{N-1})\alpha + \beta,$$

where $\alpha \neq 0$ and α, β only depend on $\eta_0, \ldots, \eta_{N-1}$ and not on η_N. Thus for all but one choice of η_N we have $\det C \neq 0$ so that the rank of A is $s+1$ and consequently $S((\eta_n), N+1) = s+1$.

On the other hand consider the first $N-s-1$ columns of A. $S((\eta_n), N-1) = S((\eta_n), N) = s$ means that the last row of this matrix is a linear combination of the first s rows. So by choosing η_N accordingly, also the last row of A depends on the first S rows of A and for this choice of η_N we have $S((\eta_n), N+1) = s$. This completes the proof.   □

In [4] the results derived here are used to compute counting functions and expected values of the lattice profile at N for sequences in finite fields.

Acknowledgments. Research supported by the Austrian Science Fund FWF under grant No. S 8312.

References

1. Cusick, T. W., Ding, C., Renvall, A.: Stream Ciphers and Number Theory. Amsterdam: Elsevier 1998
2. Dorfer, G., Winterhof, A.: Lattice structure and linear complexity profile of nonlinear pseudorandom number generators. Appl. Algebra Engrg. Comm. Comput. **13**, 499–508 (2003)
3. Dorfer, G., Winterhof, A.: Lattice structure of nonlinear pseudorandom number generators in parts of the period. In: Niederreiter, H. (ed.): Monte Carlo and Quasi-Monte Carlo Methods 2002 (to appear)
4. Dorfer, G., Meidl, W., Winterhof, A.: Counting functions and expected values for the lattice profile at n. Preprint (2003)
5. Eichenauer, J., Grothe, H., Lehn, J.: Marsaglia's lattice test and non-linear congruential pseudo random number generators. Metrika **35**, 241–250 (1988)
6. Eichenauer-Herrmann, J., Herrmann, E., Wegenkittl, S.: A survey of quadratic and inversive congruential pseudorandom numbers. In: Niederreiter, H., et al. (eds.): Monte Carlo and Quasi-Monte Carlo Methods 1996. Lecture Notes in Statistics **127**, pp. 66–97. New York: Springer 1998
7. Jungnickel, D.: Finite Fields: Structure and Arithmetics. Mannheim: Bibliographisches Institut 1993
8. Lehmer, D. H.: Mathematical methods in large-scale computing units. In: Proc. 2nd Sympos. on Large-Scale Digital Calculating Machinery, pp. 141–146. Cambridge, MA: Harvard University Press 1951
9. Marsaglia, G.: The structure of linear congruential sequences. In: Zaremba, S.K. (ed.): Applications of Number Theory to Numerical Analysis, pp. 249–285. New York: Academic Press 1972
10. Massey, J. L.: Shift-register synthesis and BCH decoding, IEEE Trans. Inform. Theory **15**, 122–127 (1969)
11. Menezes, A. J., van Oorschot, P. C., Vanstone, S. A.: Handbook of Applied Cryptography. Boca Raton: CRC Press 1997
12. Niederreiter, H.: Random Number Generation and Quasi-Monte Carlo Methods. Philadelphia: SIAM 1992
13. Niederreiter, H.: New developments in uniform pseudorandom number and vector generation. In: Niederreiter, H., Shiue, P.J.-S. (eds.): Monte Carlo and Quasi-Monte Carlo Methods in Scientific Computing. Lecture Notes in Statistics **106**, pp. 87–120. New York: Springer 1995
14. Niederreiter, H., Shparlinski, I. E.: Recent advances in the theory of nonlinear pseudorandom number generators. In: Fang, K.-T., Hickernell, F.J., Niederreiter, H. (eds.): Monte Carlo and Quasi-Monte Carlo Methods 2000, pp. 86–102. Berlin: Springer 2002
15. Niederreiter, H., Winterhof, A.: On the lattice structure of pseudorandom numbers generated over arbitrary finite fields. Appl. Algebra Engrg. Comm. Comp. **12**, 265–272 (2001)
16. Niederreiter, H., Winterhof, A.: Lattice structure and linear complexity of nonlinear pseudorandom numbers. Appl. Algebra Engrg. Comm. Comp. **13**, 319–326 (2002)
17. Ripley, B.D.: The lattice structure of pseudo-random number generators. Proc. Roy. Soc. London Ser. A **389**, 197–204 (1983)
18. Rueppel, R. A.: Analysis and Design of Stream Ciphers. Berlin: Springer 1986
19. Rueppel, R. A.: Stream ciphers. In: Simmons, G. J. (ed.): Contemporary Cryptology: The Science of Information Integrity, pp. 65–134. New York: IEEE Press 1992

Symplectic Spreads
and Permutation Polynomials

Simeon Ball[*,1] and Michael Zieve[2]

[1] Departament de Matemàtica Aplicada IV, Universitat Politècnica de Catalunya,
Mòdul C3, Campus Nord, 08034 Barcelona, Spain
`simeon@mat.upc.es`
[2] Center for Communications Research, 805 Bunn Drive, Princeton, NJ 08540–1966,
USA
`zieve@idaccr.org`

Abstract. Every symplectic spread of $PG(3, q)$, or equivalently every
ovoid of $Q(4, q)$, is shown to give a certain family of permutation polynomials of $GF(q)$ and conversely. This leads to an algebraic proof of
the existence of the Tits-Lüneburg spread of $W(2^{2h+1})$ and the Ree-Tits
spread of $W(3^{2h+1})$, as well as to a new family of low-degree permutation
polynomials over $GF(3^{2h+1})$.

Let $PG(3, q)$ denote the projective space of three dimensions over $GF(q)$. A
spread of $PG(3, q)$ is a partition of the points of the space into lines. A spread
is called *symplectic* if every line of the spread is totally isotropic with respect
to a fixed non-degenerate alternating form. Explicitly, the points of $PG(3, q)$
are equivalence classes of nonzero vectors (x_0, x_1, x_2, x_3) over $GF(q)$ modulo
multiplication by $GF(q)^*$. Since all non-degenerate alternating forms on $PG(3, q)$
are equivalent (cf. [9, p. 587] or [12, p. 69]), we may use the form

$$((x_0, x_1, x_2, x_3), (y_0, y_1, y_2, y_3)) = x_0 y_3 - x_3 y_0 - x_1 y_2 + y_1 x_2. \qquad (1)$$

Then a symplectic spread is a partition of the points of $PG(3, q)$ into lines such
that $(P, Q) = 0$ for any points P, Q lying on the same line of the spread.

Symplectic spreads are equivalent to other objects. A symplectic spread is
a spread of the generalised quadrangle $W(q)$ (sometimes denoted as $Sp(4, q)$),
whose points are the points of $PG(3, q)$ and whose lines are the totally isotropic
lines with respect to a non-degenerate alternating form. By the Klein corre-
spondence (see for example [4], [12, pp. 189] or [15]), a spread of $W(q)$ gives
an ovoid of the generalised quadrangle $Q(4, q)$ (sometimes denoted $O(5, q)$) and
vice-versa.

Let S be a spread of $PG(3, q)$. There are $q^3 + q^2 + q + 1$ points in $PG(3, q)$,
and each line contains $q+1$ points. Since S is a partition of the points of $PG(3, q)$
into lines, it contains exactly $q^2 + 1$ lines. The group $PGL(4, q)$ acts transitively

---

[*] Supported in part by the Ministerio de Ciencia y Tecnologia, España.

G. Mullen, A. Poli and H. Stichtenoth (Eds.): Fq7 2003, LNCS 2948, pp. 79–88, 2003.
© Springer-Verlag Berlin Heidelberg 2003

on the lines of $PG(3, q)$, so let us assume that S contains the line l_∞, which we define as

$$\langle (0, 0, 0, 1), (0, 0, 1, 0) \rangle.$$

The plane $X_0 = 0$ contains l_∞, so each of the other q^2 lines of the spread contains precisely one of the q^2 points $\{\langle (0, 1, x, y) \rangle \mid x, y \in GF(q)\}$. The plane $X_1 = 0$ also contains l_∞, so the other q^2 lines of the spread are given by two functions f and g such that

$$S = l_\infty \cup \{\langle (0, 1, x, y), (1, 0, f(x, y), g(x, y)) \rangle \mid x, y \in GF(q)\}.$$

The spread condition is satisfied if and only if for each $a \in GF(q)$ the plane $X_1 = aX_0$ is partitioned by the lines of the spread. These planes contain l_∞ and meet the other lines of S in the points

$$\{\langle (1, a, ax + f(x, y), ay + g(x, y)) \rangle \mid x, y \in GF(q)\}.$$

Hence the spread condition is satisfied if and only if

$$(x, y) \mapsto (ax + f(x, y), ay + g(x, y))$$

is a permutation of $GF(q)^2$ for all $a \in GF(q)$.

We are interested here in symplectic spreads. The line l_∞ is totally isotropic with respect to the form (1). The other lines of the spread are totally isotropic with respect to the form (1) if and only if for all x and $y \in GF(q)$

$$((0, 1, x, y), (1, 0, f(x, y), g(x, y)) = -y - f(x, y)$$

is zero. Hence

$$S := l_\infty \cup \{\langle (0, 1, x, y), (1, 0, -y, g(x, y)) \rangle \mid x, y \in GF(q)\}$$

will be a symplectic spread if and only if

$$(x, y) \mapsto (ax - y, ay + g(x, y)) \tag{2}$$

is a permutation of $GF(q)^2$ for all $a \in GF(q)$. Now make the substitution $b = ax - y$ to see that this is equivalent to $x \mapsto a(ax - b) + g(x, ax - b)$ being a permutation of $GF(q)$ for all $a, b \in GF(q)$, which is equivalent to $x \mapsto g(x, ax - b) + a^2 x$ being a permutation of $GF(q)$ for all $a, b \in GF(q)$.

Although merely an observation, this fact seems not to have been noted before, and as we shall see it can be quite useful. So let us formulate this in a theorem.

Theorem 1. *The set of totally isotopic lines*

$$l_\infty \cup \{\langle (0, 1, x, y), (1, 0, -y, g(x, y)) \rangle \mid x, y \in GF(q)\}$$

is a (symplectic) spread if and only if

$$x \mapsto g(x, ax - b) + a^2 x$$

is a permutation of $GF(q)$ for all $a, b \in GF(q)$. ∎

Symplectic spreads of $PG(3,q)$ are rare. All the known examples are given in Table 1 which comes from [11]. In particular, the *regular* spreads are those for which the polynomial $g(x,y)$ has total degree 1. The main result in [3] implies that when q is prime, symplectic spreads of $PG(3,q)$ are regular.

name	$g(x,y)$	q	restrictions
regular	$-nx$	odd	n non-square
Kantor [8]	$-nx^\alpha$	odd	n non-square, $\alpha \mid q$
Thas-Payne [14]	$-nx - (n^{-1}x)^{1/9} - y^{1/3}$	3^h	n non-square, $h > 2$
Penttila-Williams [11]	$-x^9 - y^{81}$	3^5	
Ree-Tits slice [8]	$-x^{2\alpha+3} - y^\alpha$	3^{2h+1}	$\alpha = \sqrt{3q}$
regular	$cx + y$	even	$Tr_{q\to2}(c) = 1$
Tits-Lüneburg [15]	$x^{\alpha+1} + y^\alpha$	2^{2h+1}	$\alpha = \sqrt{2q}$

Table 1. The known examples of symplectic spreads of $PG(3,q)$

Note that from any of the examples in the table we could make many other equivalent symplectic spreads and that the function $g(x,y)$ will not in general have such a nice form. For instance, all the examples in the table give spreads $\mathcal{S}$ that contain the line l

$$\langle(0,1,0,0),(1,0,0,0)\rangle.$$

The linear map τ that switches X_0 and X_3 and switches X_1 and X_2 preserves the form (1) but swiches l_∞ and l. The other $q^2 - 1$ lines in $\mathcal{S}$ are mapped to the lines

$$\{\langle(y,x,1,0),(g(x,y),-y,0,1)\rangle \mid x,y \in GF(q),\ (x,y) \neq (0,0)\}$$

by τ. Writing these lines as the spans of their points on the planes $X_0 = 0$ and $X_1 = 0$, these lines are

$$\{\langle(0,1,u,v),(1,0,-v,\frac{-vx}{y})\rangle \mid x,y \in GF(q),\ (x,y) \neq (0,0)\},$$

where

$$u = \frac{g(x,y)}{xg(x,y) + y^2}$$

and

$$v = \frac{-y}{xg(x,y) + y^2}.$$

(When $y = 0$ we interpret $-vx/y$ to be $1/g(x,0)$.) Now one would have to calculate $-vx/y$ in terms of u and v to deduce the function $g(x,y)$ for the spread $\tau(\mathcal{S})$. For an explicit example of this, consider the Kantor spread $\mathcal{S}$ over

$GF(27)$ with $g(x, y) = -nx^3$, where $n^3 - n = -1$. The function $g(u, v)$ for $\tau(\mathcal{S})$ is

$$nu^{21}v^4 + n^8u^{19}v^{18} + n^2u^{17}v^6 + n^4u^9v^{10} + n^{18}u^5v^{12} + n^{12}u^3.$$

A polynomial h in one variable over $GF(q)$ is called *additive* if $h(x + u) = h(x) + h(u)$ for all $x, u \in GF(q)$. In case $g(x, y) = h_1(x) + h_2(y)$ with h_1 and h_2 being additive polynomials, the symplectic spread corresponds to a translation ovoid of $Q(4, q)$, which in turn comes from a semifield flock of the quadratic cone in $PG(3, q)$. This has been the subject matter of a number of articles, see for example [1], [2] or [10]. The classification of such examples is an open problem whose solution would be of much interest. The partial classification in [2] implies that if there are any further examples over $GF(p^h)$ then $p < 4h^2 - 8h + 2$. Theorem 1 in this case reads: The polynomial $g(x, y) = h_1(x) + h_2(y)$ will give a symplectic spread if and only if $h_1(x) + h_2(ax) + a^2x$ is a permutation polynomial for all $a \in GF(q)$, or equivalently $h_1(x) + h_2(ax) + a^2x$ has no zeros in $GF(q)^*$ for all $a \in GF(q)$.

The two examples where $g(x, y)$ is not of this form are the Tits-Lüneburg spread and the Ree-Tits spread.

Let us first check the Tits-Lüneburg example, where $\alpha = \sqrt{2q}$. In this case

$$g(x, ax - b) + a^2x = x^{\alpha+1} + (ax)^\alpha - b^\alpha + a^2x.$$

So we should have that $x^{\alpha+1} + (ax)^\alpha + a^2x$ is a permutation polynomial for all $a \in GF(q)$, which is easy to see since this polynomial is $(x + a^\alpha)^{\alpha+1} + a^{\alpha+2}$. Note that composing permutation polynomials with permutation polynomials gives permutation polynomials so it is enough to check that $x^{\alpha+1}$ is a permutation polynomial, which it is since $(2^{h+1} + 1, 2^{2h+1} - 1) = 1$.

Now we come to the interesting Ree-Tits slice example, $g(x, y) = -x^{2\alpha+3} - y^\alpha$ where $q = 3^{2h+1}$ and $\alpha = \sqrt{3q}$. This spread was discovered by Kantor [8] as an ovoid of $Q(4, q)$. It is the slice of the Ree-Tits ovoid of $Q(6, q)$. It provides us with an interesting class of permutation polynomials, namely, the polynomials $f_a(x) := b^\alpha - (g(x, ax - b) + a^2x)$,

$$f_a(x) = x^{2\alpha+3} + (ax)^\alpha - a^2x.$$

The polynomial f_a is remarkable in that it is a permutation polynomial over $GF(q)$ whose degree is approximately $\sqrt{q}$. There are only a handful of known permutation polynomials with such a low degree. The bulk of these examples are *exceptional polynomials*, namely polynomials over $GF(q)$ which permute $GF(q^n)$ for infinitely many values n. However, we will show below that f_a is not exceptional, so long as $\alpha > 3$ and $a \neq 0$. There are also some non-exceptional permutation polynomials of degree approximately $\sqrt{q}$ in case q is a square or a power of 2. However, our example is the first for which q is an odd nonsquare.

It follows from [8] and Theorem 1 that f_a is a permutation polynomial. Conversely we now give a direct proof that f_a is a permutation polynomial, which (along with Theorem 1) gives a new proof that the Ree-Tits examples are in fact symplectic spreads. Our proof that f_a is a permutation polynomial uses the method of Hans Dobbertin [5].

Theorem 2. *Let $q = 3^{2h+1}$ and let $\alpha = \sqrt{3q}$. For all $a \in GF(q)$ the polynomial $f_a(x) := x^{2\alpha+3} + (ax)^\alpha - a^2x$ is a permutation polynomial over $GF(q)$.*

Proof. If $f_a(x)$ is a permutation polynomial then so is $\zeta^{2\alpha+3} f_a(x/\zeta)$ for any $\zeta \in GF(q)^*$, and the latter polynomial equals $f_{a\zeta^{\alpha+1}}(x)$. Since $(\alpha+1, q-1) = 2$, it follows that if f_a is a permutation polynomial then so is $f_{a\zeta^2}$ for any $\zeta \in GF(q)^*$. Thus it suffices to verify the theorem for a single nonzero square a, a single nonsquare a, and the value $a = 0$ (in which case the theorem is trivial). Since -1 is a non-square in $GF(3^{2h+1})$ we can assume from now on that $a^2 = 1$.

Suppose that f_a is not a permutation polynomial. Let x, y be distinct elements of $GF(q)$ such that $f_a(x) = f_a(y) = d$. The equations $f_a(x) = d$ and $f_a(x)^\alpha = d^\alpha$ give

$$x^{2\alpha+3} + ax^\alpha - x = d \tag{3}$$

$$x^{6+3\alpha} + ax^3 - x^\alpha = d^\alpha. \tag{4}$$

By viewing these equations as low-degree polynomials in x^α whose coefficients are low-degree polynomials in x, we can solve for x^α as a low-degree rational function in x. Namely, multiplying (3) by $x^{\alpha+3}$ and then subtracting (4) gives

$$ax^{2\alpha+3} - x^{\alpha+4} - ax^3 + x^\alpha = dx^{\alpha+3} - d^\alpha; \tag{5}$$

multiplying (3) by a and subtracting (5) gives

$$x^\alpha(x^4 + dx^3) = ax + da - ax^3 + d^\alpha. \tag{6}$$

This expresses x^α as a low-degree rational function in x, so long as $x \notin \{0, -d\}$. For later use we record this equation in the form $F(x^\alpha, x) = 0$ where

$$F(T, U) := U^4T + dU^3T - aU - da + aU^3 - d^\alpha.$$

Note that x and y are not both in $\{0, -d\}$, for if so then $d = f_a(0) = 0$ so $x = y = 0$, contradiction. Thus, by swapping x and y if necessary, we may assume $x \notin \{0, -d\}$.

Solving for x^α in (6) and substituting into (3) gives a low-degree polynomial satisfied by x:

$$(ax + da - ax^3 + d^\alpha)^2 + a(x + d)(ax + da - ax^3 + d^\alpha) = (x^2 + dx)^3.$$

By expanding this equation we get

$$(d^\alpha a - d^3)x^3 - x^2 + dx - d^2 + d^{2\alpha} = 0. \tag{7}$$

Next we handle the cases $y = 0$ and $y = -d$. If $y = 0$ then $d = f_a(y) = 0$ and (7) implies $x = 0$, contradiction. If $y = -d$ then the analogue of (6) with y in place of x says that $d^\alpha = -ad^3$, so $d^{2\alpha-6} = 1$ and since $(q - 1, 2\alpha - 6) = 2$ that $d^2 = 1$ and hence $a = -1$. Then equation (7) simplifies to $dx(x + d)^2 = 0$. Since $d = 0$ implies $x = 0$ we have $x \in \{0, -d\}$, again a contradiction.

Hence we may assume $y \notin \{0, -d\}$, and moreover we may assume $d \neq 0$ and $d^3 \neq -d^\alpha a$. We can also assume that $d^3 \neq d^\alpha a$. For, if $d^3 = d^\alpha a$ then $d^{2\alpha-6} = 1$ and again since $(q-1, 2\alpha - 6) = 2$ that $d^2 = 1$ and hence $a = 1$. Then equation (7) simplifies to $x(x+d) = 0$, a contradiction.

In particular, equation (7) remains valid if we substitute y for x. Thus x and y are roots of the polynomial

$$\psi(t) := (d^\alpha a - d^3)t^3 - t^2 + dt - d^2 + d^{2\alpha}. \tag{8}$$

We express the roots of $\psi(t)$ in terms of x. Since $\psi(x) = 0$, we know that $t - x$ is a factor of $\psi(t)$: in fact, writing $A := d^\alpha a - d^3$, we have

$$\psi(t)/(t-x) = At^2 + (Ax - 1)t + (Ax^2 + d - x). \tag{9}$$

The discriminant of this quadratic polynomial is

$$\delta := (Ax - 1)^2 - A(Ax^2 + d - x) = 1 - A(x + d).$$

If $\delta = 0$ then $x = -d + 1/A$ and $y = (Ax - 1)/A = -d$ which we have already excluded, so assume from now on that $\delta \neq 0$.

Substituting $d = x^{2\alpha+3} + ax^\alpha - x$ we find that

$$A = -x^{6\alpha+9} + ax^{3\alpha+6} - ax^{3\alpha} - ax^\alpha - x^3$$

and

$$\delta = (x^{4\alpha+6} - ax^{3\alpha+3} + x^{2\alpha} + ax^{\alpha+3} - 1)^2.$$

Thus putting $\sqrt{\delta} = x^{4\alpha+6} - ax^{3\alpha+3} + x^{2\alpha} + ax^{\alpha+3} - 1$, we can write the roots of $\psi(t)/(t-x)$ as

$$y_1 := x - (\sqrt{\delta} + 1)/A = \frac{x^{3\alpha+4} + ax^{2\alpha+1} + x^\alpha + ax}{x^{3\alpha+3} + ax^{2\alpha} + a}$$

and

$$y_2 := x + (\sqrt{\delta} - 1)/A = \frac{x^{3\alpha+7} - ax^{2\alpha+4} - x^{\alpha+3} + x^{\alpha+1} + ax^4 - a}{x^{3\alpha+6} - ax^{2\alpha+3} + x^\alpha + ax^3}.$$

Now one can verify that $F(y_2^\alpha, y_1) = 0$ and $F(y_1^\alpha, y_2) = 0$. But we know that $F(y^\alpha, y) = 0$ and $y \in \{y_1, y_2\}$. Since $y_1 \neq y_2$, this implies $F(T, y) = 0$ has more than one root. But this is a linear polynomial in T, a contradiction. ∎

Recall that a polynomial f over $GF(q)$ is called *exceptional* if it permutes $GF(q^n)$ for infinitely many n. We now show that, except in some special cases, f_a is not exceptional. Our proof relies on the classification of monodromy groups of indecomposable exceptional polynomials, due to Fried, Guralnick, and Saxl [6]. A polynomial is *indecomposable* if it is not the composition of two polynomials of lower degree.

Lemma 1. *When $\alpha > 3$ and $a \neq 0$, $f_a(x)$ is indecomposable.*

Proof. The derivative of f_a is $f_a' = -a^2$, which is a nonzero constant. If $f_a(x) = g(h(x))$ then $-a^2 = f_a'(x) = g'(h(x))h'(x)$, so both g' and h' are nonzero constants. Thus $g(x) = u(x^3) + cx$ and $h(x) = v(x^3) + dx$ for some polynomials u and v and nonzero constants c and d. Since the degree of f_a is not divisible by 9, either g or h has degree not divisible by 3, and hence must have degree 1. Thus f_a is indecomposable. ∎

Theorem 3. *When $\alpha > 3$ and $a \neq 0$, $f_a(x)$ is not exceptional.*

Proof. This follows directly from the preceeding lemma and [6, Theorems 13.6 and 14.1], according to which there is no indecomposable exceptional polynomial of degree $2\alpha + 3$ over a finite field of characteristic 3. ∎

In all the examples in Table 1 the polynomial g is of the form $g(x, y) = h_1(x) + h_2(y)$. In Glynn [7] such a polynomial $g(x, y)$ with this property is called *separable*. Every known example of a symplectic spread of $PG(3, q)$ is equivalent to a symplectic spread with $g(x, y)$ separable. In the examples not only is the polynomial $g(x, y) = h_1(x) + h_2(y)$ separable but $h_2(y) = Cy^\sigma$, where $y \mapsto y^\sigma$ is an automorphism of $GF(q)$. We can classify these examples in the case when q is even using Glynn's Hering classification of inversive planes [7].

Theorem 4. *Let q be even. If $g(x, y) = h_1(x) + Cy^\sigma$ is a separable polynomial that gives a symplectic spread of $PG(3, q)$ then the spread is either a regular spread or a Tits-Lüneburg spread.*

Proof. If $C = 0$ then Theorem 1 implies $h_1(x) + a^2x$ is a permutation polynomial for all $a \in GF(q)$. Let x and y be distinct elements of $GF(q)$, and put $d = (h_1(x) + h_1(y))/(x + y)$. Then $h_1(x) + dx = h_1(y) + dy$, so the polynomial $h_1(x) + dx$ is not a permutation polynomial, a contradiction.

Now assume that $C \neq 0$. Put $z = h_1(x) + Cy^\sigma$ and rewrite this as $y = C^{-1}z^{1/\sigma} - C^{-1}h_1(x)^{1/\sigma}$. Define the function $s(x, z) := C^{-1}z^{1/\sigma} - C^{-1}h_1(x)^{1/\sigma}$. Then $g(x, y) = z$ if and only if $s(x, z) = y$. We have already seen in equation (2) that $g(x, y)$ will give a symplectic spread if and only if

$$(x, y) \mapsto (ax - y, ay + g(x, y))$$

is a permuatation of $GF(q)^2$. This is equivalent to the condition that for all $(x, y) \neq (u, v)$

$$(ax - y, ay + g(x, y)) \neq (au - v, av + g(u, v))$$

for all $a \in GF(q)$. If these pairs were equal then eliminating a this gives the condition that for all $(x, y) \neq (u, v)$

$$(y - v)^2 + (x - u)(g(x, y) - g(u, v)) \neq 0.$$

Now put $z = g(x, y)$ so that $s(x, z) = y$, and put $w = g(u, v)$ so that $s(u, w) = v$. Then we have that $(x, z) \neq (u, w)$

$$(s(x, z) - s(u, w))^2 + (x - u)(z - w) \neq 0.$$

When q is even this is exactly the polynomial condition on such a polynomial $s(x, z)$ that Glynn studies in [7] and that he classifies as coming from either a regular spread or a Tits-Lüneburg spread.   ■

When q is odd we can use Thas' classification of flocks of the quadratic cone in $PG(3, q)$ whose planes are incident with a common point from [13] to prove the following theorem. We realise that for many readers familiar with flocks and semifield flocks the next theorem is immediate, but we include a proof for those readers who may not be.

Theorem 5. *Let q be odd. If $g(x, y) = h_1(x)$ is a separable polynomial that gives a symplectic spread of $PG(3, q)$ then the spread is either a regular spread or a Kantor spread.*

Proof.

Consider the set of q planes of $PG(3, q)$

$$\{X_0 + h_1(x)X_1 + xX_3 = 0 \mid x \in GF(q)\}.$$

We claim that any two of these planes intersect in a line which is disjoint from the degenerate quadric $X_1X_3 = X_2^2$. Indeed take two planes coordinatised by x and y, $x \neq y$. Then the points in their intersection (z_0, z_1, z_2, z_3) satisfy $(h_1(x) - h_1(y))z_1 + (x - y)z_3 = 0$, and the points which also lie on the degenerate quadric satisfy

$$(h_1(x) - h_1(y))z_1^2 + (x - y)z_2^2 = 0.$$

If $z_1 \neq 0$ then $h_1(x) + (z_2/z_1)^2 x$ is not a permutation polynomial, a contradiction. If $z_1 = 0$ then $z_2 = 0$ and $z_0 = -xz_3 = -yz_3$. But $x \neq y$ implies that $z_3 = 0$ and $z_0 = 0$ which is nonsense. We have shown that the set of planes form a flock of the quadratic cone in $PG(3, q)$. Moreover all these planes are incident with $(0, 0, 1, 0)$. By a theorem of Thas [13] this flock is either linear or of Kantor type. In other words, the spread is either regular or Kantor.   ■

In general the permutation polynomial condition from Theorem 1 requires the existence of q^2 permutation polynomials, one for each pair $(a, b) \in GF(q)^2$. If $g(x, y) = h_1(x) + h_2(y)$ and $h_2(y)$ is additive then Theorem 1 simplifies to: The polynomial $g(x, y) = h_1(x) + h_2(y)$ will give a symplectic spread if and only if $f_a(x) := h_1(x) + h_2(ax) + a^2 x$ is a permutation polynomial for all $a \in GF(q)$. This condition only requires the existence of q permutation polynomials. Moreover as we saw in the proof of Theorem 2, if the non-zero terms in h_1 and h_2 have suitable degrees, many of these permutation polynomials may be equivalent.

Let us investigate this further. We define a set of polynomials $\{f_a(x) \mid a \in GF(q)\}$ to be of *class* Δ if there exists a t and d such that

$$f_a(bx) = b^t f_{ab^d}(x)$$

for all $b^{q-1/\Delta} = 1$ and a and $x \in GF(q)$. Now we can lessen the condition in Theorem 1 for $\Delta < q - 1$.

name	q	Δ	$(q-1, \Delta d) + 1$
regular	odd	1	1
Kantor	odd	1	$(q-1, (\alpha-1)/2) + 1$
Thas-Payne	3^h	.	q
Penttila-Williams	3^5	11	23
Ree-Tits slice	3^{2h+1}	1	3
regular	even	1	1
Tits-Lüneburg	2^{2h+1}	1	2

Table 2. The class Δ of the known examples of symplectic spreads of $PG(3, q)$

Theorem 6. *Let the set of q polynomials $\{f_a(x) \mid a \in GF(q)\}$, where $f_a(x) = h_1(x) + h_2(ax) + a^2 x$ and h_2 is additive, be of class Δ. The f_a is a permutation polynomial for all $a \in GF(q)$ if and only if f_a is a permutation polynomial for $a = 0$ and $a = \varepsilon^r$, for all $1 \leq r < (q-1, \Delta d)$, where ε is a fixed primitive element.*

Proof. Write $a = \varepsilon^{n_1(q-1,\Delta d)+n_0}$ where $n_0 < (q-1, \Delta d)$. Now choose b such that $b^d = \varepsilon^{-n_1(q-1,\Delta d)}$. ∎

In Table 2 we have listed the class for the known examples and the quantity $(q-1, \Delta d) + 1$, the number of permutation polynomials that need to be checked in each case. Inspired by this table we used the mathematical package GAP to look at polynomials over $GF(q)$, $q = p^h$, of the form $g(x, y) = Dx^t + Cy^\sigma$ for all σ a power of p and D and C elements of $GF(q)$ where the corresponding set of polynomials $\{f_a(x) \mid a \in GF(q)\}$ is of class Δ with Δ small. An exhaustive search was carried out for $\Delta \leq 23$ and $q \leq 67^2 = 4489$, $\Delta = 2$ and $q < 3^8 = 6561$, $\Delta = 1$ and $q < 3^9 = 19683$. No new examples of symplectic spreads were found.

References

1. Bader, L., Lunardon, G.: On non-hyperelliptic flocks. European J. Combin. **15** (1994) 411–415
2. Ball, S., Blokhuis, A., Lavrauw, M.: On the classification of semifield flocks. Adv. Math. (to appear)
3. Ball, S., Govaerts, P., Storme, L.: On ovoids of $Q(4, q)$ and $Q(6, q)$. Preprint
4. Cameron, P. J.: Projective and Polar Spaces. QMW Maths Notes 13 (1991). Updated version,

   `http:\\www.maths.qmw.ac.uk\~pjc\pps`

5. Dobbertin, H.: Uniformly representable permutation polynomials. In: Helleseth, T., Kumar, P. V., Yang, K. (eds.): Sequences and their Applications. Springer-Verlag, New York (2002) 1–22
6. Fried, M., Guralnick, R., Saxl, J.: Schur covers and Carlitz's conjecture. Israel J. Math. **82** (1993) 157–225
7. Glynn, D. G.: The Hering classification for inversive planes of even order. Simon Stevin **58** (1984) 319–353

8. Kantor, W.: Ovoids and translation planes. Canad. J. Math. **34** (1982) 1195–1207
9. Lang, S.: Algebra. Third Edition, Addison Wesley, Reading (1993)
10. Lavrauw, M.: Scattered subspaces with respect to spreads and eggs in finite projective spaces. Ph. D. thesis, Technical University of Eindhoven, The Netherlands (2001)
11. Penttila, T., Williams, B.: Ovoids of parabolic spaces. Geom. Dedicata **82** (2000) 1–19
12. Taylor, D. E.: The Geometry of the Classical Groups. Sigma Series in Pure Mathematics, Vol. 9. Heldermann Verlag, Berlin (1992)
13. Thas, J. A.: Generalized quadrangles and flocks of cones. European J. Combin. **8** (1987) 441–452
14. Thas, J. A., Payne, S. E.: Spreads and ovoids in finite generalised quadrangles. Geom. Dedicata **52** (1994) 227–253
15. Tits, J.: Ovoides et Groupes de Suzuki. Arch. Math. XIII (1962) 187–198

What Do Random Polynomials over Finite Fields Look like? [*]

Daniel Panario[1]

School of Mathematics and Statistics, Carleton University
Ottawa, K1S 5B6, Canada
E-mail: `daniel@math.carleton.ca`

Abstract. In this paper, we survey old and new results about random univariate polynomials over a finite field $\mathbb{F}_q$. We are interested in three aspects: (1) the decomposition of a random polynomial in terms of its irreducible factors, (2) the usage of random polynomials in algorithms, and (3) the average-case analysis of algorithms that use polynomials over finite fields.

1 Introduction

Let $\mathbb{F}_q$ be a finite field. Along this paper we only consider univariate monic polynomials over $\mathbb{F}_q$. We are interested in three aspects:

1. how is a random polynomial in terms of its irreducible factors?
2. random polynomials in algorithms, and
3. average-case analysis of algorithms that use polynomials over finite fields.

It is well-known (and we will see it later) that a polynomial of degree n is irreducible with probability close to $1/n$. Can we say something more about the behavior of a random polynomial? For example,

- how many irreducible factors should we expect in a random polynomial?
- how often will it be squarefree?
- what is the expected largest (smallest) degree among its irreducible factors? and the second largest one?
- how is the degree distribution among its irreducible factors?
- how often a polynomial is m-smooth (all irreducible factors of degree smaller or equal to m)?
- how often are two polynomials m-smooth and coprime?
- and so on.

Random polynomials over finite fields are used in many algorithms. For example, Rabin [63] proposes a randomized algorithm for finding irreducible polynomials (see also [3]). The index calculus method for computing discrete logarithms in finite fields also takes polynomials at random [7, 16, 57].

---

[*] The author was funded by NSERC grant number 238757.

G. Mullen, A. Poli and H. Stichtenoth (Eds.): Fq7 2003, LNCS 2948, pp. 89–108, 2003.

Moreover, average-case analysis of algorithms that deal with polynomials over finite fields can be obtained by counting polynomials with particular properties. Thus, properties of random polynomials like the ones stated above can be used to explain the behavior of algorithms. Typical areas where these studies can be used are:

- irreducibility tests for polynomials,
- polynomial factorization, and
- discrete logarithm problem.

Flajolet, Gourdon, and Panario [25] give a framework that can be systematically employed to explain the most important features of these algorithms. This framework has two basic components: generating functions to express the properties of interest for the analysis of the algorithm and asymptotic analysis when exact estimations are not possible. In our case, this generic methodology closely relates finite fields and their applications to combinatorics and analytic number theory.

1.1   Outline of the Paper

We present the basic framework in Section 2. First, we introduce the two components of this method: generating functions and asymptotic analysis. Then, we give some simple examples of its usage (number of squarefree polynomials, average number of irreducible factors, and number of irreducible factors of a fixed degree).

The algorithmic applications form the second part of this paper. Irreducibility tests are discussed in Section 3; polynomial factorization algorithms are commented in Section 4; and cryptographic applications are presented in Section 5.

Finally, as a summary of the results, a simplified picture of a random polynomial and a list of open problems are stated in Section 6.

We consider the natural measure of cost, that is, operations in the field $\mathbb{F}_q$ of coefficients of the polynomials. Unless specified otherwise, asymptotic results are considered for n, the degree of the polynomial, tending to infinity. We do not mention here the cost of doing arithmetic in finite fields; see [37] or [41], for example.

This paper is an extended transcription of the author's invited talk at the 7th Finite Fields and their Applications Conference.

2   Basic Framework

We extensively use a methodology that belongs to the realm of "analytic combinatorics", and it has been successfully used in analyzing algorithms; see [27, 64]. Although this framework is more general, we focus only on polynomials over a finite field $\mathbb{F}_q$; see [25] for more details.

2.1   Generating Functions

Let $P(z)$ and $Q(z)$ be the generating functions of polynomials and squarefree polynomials over $\mathbb{F}_q$, respectively. [For simplicity, we consider only monic polynomials.] The coefficient $Q_n = [z^n]Q(z)$ equals the number of monic squarefree polynomials of degree n, and the coefficient $P_n = [z^n]P(z)$ represents the number of monic polynomials of degree n. These generating functions can be obtained by considering an enumerator of a fixed irreducible factor of degree k

$$1 + z^k + z^{2k} + \cdots .$$

This enumerator counts the number of times, $0, 1, 2, \ldots$, that this particular irreducible factor appears in a polynomial. Let I_k be the number of monic irreducible polynomials over $\mathbb{F}_q$ of degree k (we avoid carrying the finite field in this notation since in this paper it will be always $\mathbb{F}_q$). Then, considering the I_k irreducible factors of degree k, and varying on k, we obtain (distributively) the generating function of polynomials over $\mathbb{F}_q$

$$P(z) = \prod_{k \geq 1} \left(1 + z^k + z^{2k} + \cdots\right)^{I_k} = \prod_{k \geq 1} \left(\frac{1}{1 - z^k}\right)^{I_k} .$$

Since P_n is q^n, we have $P(z) = (1 - qz)^{-1}$, and we conclude that

$$P(z) = \prod_{k \geq 1} \left(\frac{1}{1 - z^k}\right)^{I_k} = \frac{1}{1 - qz}. \tag{1}$$

Let $I(z)$ be the generating function of irreducible polynomials, that is, $I(z) = \sum_{k \geq 1} I_k z^k$. The last equation implicitly determines I_n. Indeed, from

$$\frac{1}{1 - qz} = \prod_{k=1}^{\infty} (1 - z^k)^{-I_k},$$

we get

$$\log \frac{1}{1 - qz} = \sum_{k \geq 1} (I_k) \log(1 - z^k)^{-1} = \sum_{k \geq 1} \frac{I(z^k)}{k}.$$

Expanding the logarithm and equating coefficients we get

$$\frac{q^n}{n} = \sum_{k | n} \frac{I_{n/k}}{k}.$$

Finally, Moebius inversion formula gives the classical relation

$$I_n = \frac{1}{n} \sum_{k | n} \mu(k) q^{n/k}.$$

An important consequence for algorithms that use polynomials over finite fields is that

$$I_n = \frac{q^n}{n} + O\left(\frac{q^{n/2}}{n}\right),$$

and hence, a fraction very close to $1/n$ of the polynomials of degree n over $\mathbb{F}_q$ is irreducible.

2.2   Asymptotic Analysis

Generating functions encode exact information in their coefficients. In many cases, the extraction of the coefficients from a generating function is a difficult task. Fortunately, there are powerful methods that allow us to determine the asymptotic form of the coefficients of complicated generating functions directly from their singularities. In particular, it is well-known that the behavior near a dominant positive singularity (one with the smallest modulus) is an important source of coefficient asymptotics.

These methods give conditions under which the asymptotic behavior of the coefficients can be determined using a local asymptotic expansion near a dominant singularity. In other words, these methods give conditions for which the following implication is valid

$$f(z) \sim \sigma(z) \Rightarrow [z^n]f(z) \sim [z^n]\sigma(z),$$

where $f(z)$ is the generating function to be studied and $\sigma(z)$ is its approximation near the singularity.

Most of the generating functions $f(z)$ of interest here are singular at $z = 1/q$ with an isolated singularity of algebraic-logarithmic type. In these cases, we can apply the following result from [26].

Theorem 1. *Let $f(z)$ be a function analytic in a domain*

$$\mathcal{D} = \{z \colon |z| \leq z_1, |Arg(z - 1/q)| > \frac{\pi}{2} - \varepsilon\},$$

where $z_1 > 1/q$ and ε are positive real numbers. Let $k \geq 0$ be any integer, and α a real number with $\alpha \neq 0, -1, -2, \ldots$. If in a neighborhood of $z = 1/q$, $f(z)$ has an expansion of the form

$$f(z) = \frac{1}{(1-qz)^\alpha}\left(\log \frac{1}{1-qz}\right)^k (1 + o(1)), \tag{2}$$

then the coefficients satisfy, asymptotically,

$$[z^n]f(z) = q^n \frac{n^{\alpha-1}}{\Gamma(\alpha)}(\log n)^k (1 + o(1)). \tag{3}$$

We often find generating functions of the form $p(z)f(z)$ in which $p(z)$ is a polynomial and $f(z)$ satisfies the condition of Theorem 1. In such cases, if $h(z) = p(z)f(z)$, then

$$[z^n]h(z) = q^n p(1/q)\frac{n^{\alpha-1}}{\Gamma(\alpha)}(\log n)^k (1 + o(1)). \qquad (4)$$

The translation from Equation (2) to Equation (3) or (4) is achieved by the so-called transfer lemmas that require analytic continuation of $f(z)$ outside its circle of convergence. Such a condition is usually verified by inspection.

However, there are some situations in which generating functions do not satisfy the hypothesis of Theorem 1. For instance, some of the generating functions in Section 5 have a natural boundary at $|z| = 1$ (each point at the unit circle is singular), so analytic continuation is not possible. This is a situation similar to the partition generating function. We use saddle point method in these cases. All asymptotic enumeration methods required in this paper are explained in the excellent presentations by Odlyzko [58] and by Flajolet and Sedgewick [27].

2.3   Examples

We now derive, as a first example, the number of squarefree polynomials over $\mathbb{F}_q$. This result was first proven by Carlitz [11]. Using an enumerator as above, we have

$$Q(z) = \prod_{k \geq 1} \left(1 + z^k\right)^{I_k},$$

an expression for which it is not so easy to extract coefficients. In this case, a much simpler method can be employed. By considering the multiplicity of its irreducible factors, each polynomial f factors as $f = st^2$, where s is squarefree and t is an arbitrary polynomial. We thus have

$$P(z) = Q(z)P(z^2),$$

and hence,

$$Q(z) = \frac{P(z)}{P(z^2)} = \frac{1 - qz^2}{1 - qz}.$$

We immediately have

$$Q_n = q^n - q^{n-1}, \qquad n \geq 2, \qquad (5)$$

with $Q_n = q^n$ for $n = 0, 1$. This means that, for $n \geq 2$, the proportion of squarefree polynomials is $1 - 1/q$, or in other words, for large finite fields $\mathbb{F}_q$ most polynomials are squarefree.

As a second example, let us consider the study of the expected number of irreducible factors of a random polynomial. In this type of questions we need an extension of the method to take care of what we call "parameters" of the problem. This implies in stating a bivariate generating function in the variables z and

u such that z counts polynomials, and u counts the parameter of interest. In our example, let $A(z, u)$ be the bivariate generating function counting polynomials of degree n with k irreducible factors, that is, the coefficient $[z^n u^k]A(z, u)$ represents the number of polynomials of degree n with k irreducible factors. Using the method above, we have

$$A(z, u) = \prod_{k \geq 1}(1 + uz^k + u^2 z^{2k} + \cdots) = \prod_{k \geq 1}(1 - uz^k)^{-I_k}.$$

Averages and standard deviations are obtained by taking successive derivatives of the bivariate generating function with respect to the parameter u, and then setting $u = 1$ (see, for example, [64]):

$$\frac{[z^n]\frac{\partial A(z,u)}{\partial u}\big|_{u=1}}{[z^n]A(z,1)}, \quad \frac{[z^n]\frac{\partial^2 A(z,u)}{\partial u}\big|_{u=1}}{[z^n]A(z,1)} + \frac{[z^n]\frac{\partial A(z,u)}{\partial u}\big|_{u=1}}{[z^n]A(z,1)} - \left(\frac{[z^n]\frac{\partial A(z,u)}{\partial u}\big|_{u=1}}{[z^n]A(z,1)}\right)^2.$$

In our case, differentiating two times, putting $u = 1$ and applying asymptotic analysis give that the average number of irreducible factors is asymptotic to $\log n$ with a standard deviation of $\sqrt{\log n}$. Indeed, much more is known about this parameter since it is one of the most widely studied with respect to polynomials over finite fields.

Theorem 2. *Let Ω_n be a random variable counting the number of irreducible factors of a random polynomial of degree n over $\mathbb{F}_q$, where each factor is counted with its order of multiplicity.*

1. *The mean value of Ω_n is asymptotic to $\log n + O(1)$.*
2. *The variance of Ω_n is asymptotic to $\log n + O(1)$.*
3. *For any two real constants $\lambda < \mu$,*

$$Pr\left\{\log n + \lambda\sqrt{\log n} < \Omega_n < \log n + \mu\sqrt{\log n}\right\} \to \frac{1}{\sqrt{2\pi}}\int_\lambda^\mu e^{-t^2/2}dt.$$

4. *The distribution of Ω_n admits exponential tails.*
5. *A local limit theorem holds.*

Remarks:

1. The average number of irreducible factors of a random polynomial of degree n appears in [6], Ex. 3.6. Then, it also appears in [52], Ex. 4.6.2.5, and with more details in [28, 50, 55].
2. The variance is sketched in [28], and is given with more terms in [50]. The latter also covers the case of *distinct* factors.
3. For any two real constants $\lambda < \mu$, if

$$Pr\left\{\log n + \lambda\sqrt{\log n} < \Omega_n < \log n + \mu\sqrt{\log n}\right\} \to \frac{1}{\sqrt{2\pi}}\int_\lambda^\mu e^{-t^2/2}dt,$$

then it is said that Ω_n satisfies a central limit theorem, or that a Gaussian limit distribution holds. The existence of this limit distribution provides information on the distribution near the mean value. The central limit for Ω_n is in [28], Corollary 1.

4. Exponential tails essentially indicate that large deviations from the mean value are unlikely. In the case of the number of irreducible factors of random polynomials, exponential tails are proven in [29].
5. Local limit theorems basically deal with density functions. They are studied in depth in [4, 5]. For our particular case, the local limit theorem holds as a consequence of the results in [36].

As a final example, let us consider the number of irreducible factors of fixed degree in a random polynomial. The number of linear factors, that is the number of roots, of a random polynomial seems to be first studied by Zsigmondy [68] for the prime field case; Knopfmacher and Knopfmacher [49] present a detailed analysis including variance. The case of polynomials with no roots is interesting when studying the distinct values that a polynomial can take. This is related to permutation polynomials and was studied by Uchiyama [66]; see also [15].

The generating function of polynomials with no linear factors is

$$\prod_{k \geq 2} \left(\frac{1}{1 - z^k} \right)^{I_k} = \frac{1}{1 - qz}(1 - z)^{I_1} = \frac{1}{1 - qz}(1 - z)^q.$$

It is not difficult to extract coefficients from this generating function since it is a convolution of two simple generating functions. If we prefer to use singularity analysis, we obtain that, for large n, the number of polynomials of degree n with no irreducible factors of degree 1 is asymptotic to $q^n(1 - 1/q)^q$. This means that the probability of obtaining one such polynomial tends to $1/e = 0.3678\ldots$ when q grows (computer experiments show that for $q > 11$ this is already a good approximation). In other words, "most" polynomials are reducible and have at least one irreducible factor of degree 1!

In general, the number of irreducible factors of a specified degree d in polynomials of degree n was studied by Williams [67]. A detailed analysis including variance and "distinct irreducibles" case appears in [50].

3   Irreducibility Tests for Polynomials

Rabin [63] presents a probabilistic algorithm for constructing irreducible polynomials over finite fields. The central idea is to take polynomials at random and test them for irreducibility. Since the proportion of irreducible polynomials of degree n (over any finite field) is close to $1/n$, we expect to find an irreducible polynomial after approximately n tries.

In order that this works we need an irreducibility test. Rabin derives an algorithm for testing the irreducibility of a polynomial from the following theorem.

Theorem 3. *Let $p_1, \ldots, p_k$ be the distinct prime divisors of n, and denote $n/p_i = n_i$, for $1 \leq i \leq k$. A polynomial $f \in \mathbb{F}_q[x]$ of degree n is irreducible in $\mathbb{F}_q[x]$ if and only if $\gcd(f, x^{q^{n_i}} - x \bmod f) = 1$, for $1 \leq i \leq k$, and f divides $x^{q^n} - x$.*

Rabin's algorithm simply computes the above gcds one by one. Since $x^{q^{n_i}} - x \in \mathbb{F}_q[x]$ is the product of all monic irreducible polynomials over $\mathbb{F}_q$ of degree dividing n_i (see Theorem 4 below), each gcd tests the existence of irreducible factors in f of several degrees. A polynomial is discarded when one gcd is different from 1. Hence, the analysis of Rabin's algorithm can be carried out by studying the number of polynomials with irreducible factors belonging to a set $\mathcal{T}$ but not belonging to a set $\mathcal{S}$ ($\mathcal{T}$ would be the set of degrees being tested in step i, while $\mathcal{S}$ would be the set of degrees already tested in the previous steps of the algorithm). The corresponding generating function is

$$\frac{1}{1 - qz} \left(\prod_{j \in \mathcal{T}} \left(1 - z^j\right)^{I_j} - \prod_{j \in \mathcal{S} \cup \mathcal{T}} \left(1 - z^j\right)^{I_j} \right).$$

Now, the set of degrees being checked at step i of the algorithm depends on the divisors of $n_1, n_2, \ldots, n_{i-1}$, where $n_j = n/p_j$ is as stated in Theorem 3. Extracting coefficients here is, essentially, an impossible task (the possible cases are when n is prime or product of two primes, see [59]). A delicate asymptotic analysis provides uniform results [59].

Variants for the computation of $x^{q^{n_i}} - x \bmod f$ have been presented in [42] and in [35]. These variants are analyzed in a similar way.

Soon after Rabin gives his algorithm, Ben-Or [3] proposes an algorithm based on the following theorem (for example, see Theorem 3.20 of [53]).

Theorem 4. *For $i \geq 1$, the polynomial $x^{q^i} - x \in \mathbb{F}_q[x]$ is the product of all monic irreducible polynomials in $\mathbb{F}_q[x]$ whose degrees divides i.*

We should point out that both Gauss and Galois suggested using this theorem as a step for factoring polynomials (see [41] and the references therein for a historic account).

Ben-Or's algorithm tests the irreducibility of a polynomial by searching for irreducible factors degree by degree. Since the mean number of irreducible factors of degree k of a random polynomial of degree n approaches $1/k$ as n tends to infinity [50], if the polynomial is reducible, Ben-Or's algorithm quickly discards it. Moreover, since the degrees of the polynomials involved in the gcds are smaller than in Rabin's algorithm, these gcds are less expensive.

In order to analyze this algorithm, we have to study the probability that a random polynomial of degree n contains no irreducible factors of degree up to a certain value m (such polynomials are sometimes called m-rough). Car [10] gives estimates for m-roughness that depend on the Buchstab function for m large with respect to n, say $m > c_1 n \log \log n / \log n$. On the other extreme, Gao and Panario [35] show that for m small with respect to n, say $m < c_2 \log n$, the estimate $e^{-\gamma}/m$ holds, where γ is Euler's constant. The *Buchstab* function is the unique continuous solution of the difference-differential equation

$$\begin{aligned} u\omega(u) &= 1 & 1 \leq u \leq 2, \\ (u\omega(u))' &= \omega(u-1) & u > 2. \end{aligned}$$

It was introduced by Buchstab [9] when studying the analogous problem for integer numbers, that is, numbers with no small prime factors. Much is known about this function. For example, it is known that the Buchstab function quickly tends to $e^{-\gamma} = 0.56416\ldots$ (see Fig. 1).

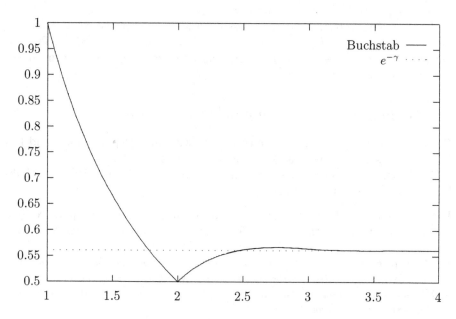

Fig. 1. The relation between the Buchstab function and $e^{-\gamma}$ in the interval $[1, 4]$.

The study of the probability that a random polynomial is m-rough for the complete range $1 \leq m \leq n$, is given by Panario and Richmond [60]. The estimates are in terms of the Buchstab function when $m \to \infty$. When m is fixed singularity analysis is applied.

Theorem 5. *The smallest degree S_n among the irreducible factors of a random polynomial of degree n over $\mathbb{F}_q$ satisfies*

$$Pr(S_n \geq m) = \frac{1}{m} \omega \left(\frac{n}{m} \right) + O \left(\max \left\{ \frac{1}{m^2}, \frac{\log n}{mn} \right\} \right),$$

when m tends to infinity with n.

Using Theorem 5 it is not difficult to prove that the expected smallest degree among the irreducible factors of a random polynomial is asymptotic to $e^{-\gamma} \log n$. More general, the expected rth smallest degree among the irreducible factors of a random polynomial is asymptotic to $e^{-\gamma} \log^r n/r!$.

These studies generalize to the size of the *smallest* components in random decomposable structures [61, 62]. The results include limit distributions and local theorems for the size of the rth smallest component of an object of size n.

Expectation, variance and higher moments of the rth smallest component are also derived. The results apply to several combinatorial structures in the exp-log class for both labelled and unlabelled objects. This class of combinatorial objects includes permutations, polynomials over finite fields, 2-regular graphs, random mappings (functional digraphs), random mappings patterns, arithmetical semi-groups, etc; see [61] for details and references.

Similar generic results for this class of objects but for the *number* of irreducible components have been developed by Flajolet and Soria [28, 29]. The size of the *largest* components in random decomposable structures have been carried out by Gourdon [43, 44].

4   Factorization of Polynomials over Finite Fields

Factoring polynomials is a fundamental task with many applications, and has been largely studied; see Chapter 14 of [37] or [41] for recent surveys.

There exists a general factorization algorithm (that Knuth [52] calls "folklore") that works in three stages:

ERF *elimination of repeated factors* replaces a polynomial by a squarefree one which contains all the irreducible factors of the original polynomial with exponents reduced to 1;

DDF *distinct-degree factorization* splits a squarefree polynomial into a product of polynomials whose irreducible factors have all the same degree;

EDF *equal-degree factorization* factors a polynomial whose irreducible factors have the same degree.

We do not include the algorithms in this paper [37, 41]. We observe that the first stage of this process is normally done by means of the so-called *squarefree factorization* but for simplicity we consider ERF here. In any case, this is not the crucial stage of the algorithm as we will see later.

We only state the main features of the analysis of each stage; for the complete details see the paper by Flajolet, Gourdon and Panario [25].

We start by considering the elimination of repeated factors. By Equation 5 most polynomials are squarefree. If we call "nonsquarefree part" the factor that remains after dividing the polynomial by the result of computing ERF, it is not difficult to prove that the nonsquarefree part is expected to be constant. Indeed, the bivariate generating function counting the total degree of the nonsquarefree part is

$$\prod_{k \geq 1} (1 + z^k + u z^{2k} + u^2 z^{3k} + \cdots) = \prod_{k \geq 1} \left(1 + \frac{z^k}{1 - u^k z^k} \right)^{I_k}.$$

By differentiating with respect to u, setting $u = 1$ and applying singularity analysis, we obtain the mean degree of the non-squarefree part

$$N_q = \sum_{k \geq 1} \frac{k I_k}{q^{2k} - q^k},$$

and we have $N_q \sim 1/q$ when $q \to \infty$. Combining this information with the cost of each step of the algorithm, one shows that ERF accounts, essentially, for one gcd computation.

The distinct-degree factorization (DDF) follows immediately from Theorem 4: check factors of the polynomial f by computing $\gcd(x^{q^i} - x, f)$ for $i = 1, 2, \ldots$; if a gcd is different from 1, remove this factor from f and continue iterating. This procedure factors f into factors that contain one or more irreducible factors of the same degree. We stop this procedure when $i > n/2$: the remaining factor is either 1 or irreducible.

A natural idea is the "early abort" strategy: stop the iteration of DDF when $2i$ exceeds the degree of the remaining factor. The analysis of the early abort strategy requires information on the largest and second largest expected degree of irreducible factors of a random polynomial. Naturally, the Dickman function [17, 18] appears in these estimates since it models the analogous problem for integer numbers. A technical analysis [25, 43] provides an expected largest degree tending to $c_1 n$ where $c_1 = 0.62432\ldots$ is precisely Golomb's constant [23] that models the expected largest length among the cycles of a random permutation. The second largest irreducible factor has expected degree $c_2 n$ where $c_2 = 0.20958\ldots$, where a generalized Dickman function plays an important role. As in the case of the smallest components already commented, all results generalize to other decomposable combinatorial structures [43, 44].

The global saving of the early abort rule is of 36%, and the expected cost of DDF dominates the whole factorization process. This gives a firm justification to a fact that was known from a worst-case perspective.

DDF does not completely factor a polynomial that has different factors of same degree. We have the following theorem [25, 51].

Theorem 6. *1. The probability that DDF yields the complete factorization is asymptotic to*

$$c_q = \prod_{k \geq 1} \left(1 + \frac{I_k}{q^k - 1}\right) (1 - q^{-k})^{I_k},$$

where $c_2 = 0.6656\ldots$, $c_{257} = 0.5618\ldots$, $c_\infty = e^{-\gamma} = 0.5614\ldots$.
2. The number of degree values for which there is more than one irreducible factor in the polynomial produced by DDF has an average that is asymptotic to the constant

$$\sum_{k \geq 1} (1 - q^{-k})^{I_k} \left((1 - q^{-k})^{-I_k} - 1 - \frac{I_k q^{-k}}{1 - q^{-k}}\right).$$

3. The degree of the part of the polynomial that remains to be factored by the EDF algorithm has expectation $\log n + O(1)$, and standard deviation of approximately $\sqrt{n}$.

Finally, the factorization problem is reduced to factoring polynomials that have all their irreducible factors of the same (known) degree (EDF). The fastest algorithms for this task are randomized. Indeed, it is not known a deterministic polynomial time algorithm for this stage of the factorization process. One such algorithm would yield a deterministic polynomial time algorithm for the whole factoring problem, a well-known open problem [2, 22, 31].

The analysis of EDF is an interesting combination of a recursive partitioning problem akin to digital trees (better known as "tries") with estimates on the degree of irreducible factors of random polynomials.

On the other hand, Theorem 6 implies that EDF is executed less than 50% of the times and with a total degree of roughly $\log n$. Since the cost of these algorithms heavily depends on the degree of the polynomial in the input, we can conclude that the cost of the EDF stage is computationally small though it cannot be completely discarded.

Factorization algorithms of the 1990's involve the analysis of irreducible factors whose degrees lie in intervals [42, 47, 65]. These algorithms split the interval $[1, n]$ into parts. For each subinterval, the product of all irreducible factors of the original polynomial whose degree lies in that interval is computed. Using Theorem 4, a gcd computation determines if the polynomial contains irreducible factors in the subinterval. For each subinterval with more than one irreducible factor the standard DDF algorithm is applied to compute the distinct-degree factorization.

The analysis of these algorithms requires information on how irreducible factors are distributed among its parts, given a partition of $[1, n]$. For example, the probability of no interval with more than one irreducible factor, the average number of factors in one subinterval, and so on, provide useful information and the methodology presented here seems amenable to this problem.

When factoring random polynomials, polynomially growing interval sizes seem to be a good option for the partition. An example of polynomially growing interval size is the quadratic partition: $[1, 1], [2, 4], [5, 9], [10, 16], \ldots$. We expect to have a decreasing number of irreducible factors as their degrees increase, and hence, polynomially growing interval size partitions seem to distribute the factors in a more balanced fashion. Preliminary results show that this is the case. Using this information, von zur Gathen and Gerhard [40] provide algorithms for factoring very large degree random polynomials over $\mathbb{F}_2$. For example, their algorithm factors a random polynomial of degree about $250\,000$ in one day of CPU time.

5   Cryptographic Applications

Let g be a generator of the multiplicative group of $\mathbb{F}_q$. For any element $h \in \mathbb{F}_q$, $h \neq 0$, there exists an integer x, $0 \leq x \leq q - 2$, such that $h = g^x$. We call x the *discrete logarithm* of h in the base g.

A fundamental task in cryptography is the *discrete logarithm problem*: find a computationally feasible algorithm to compute the discrete logarithm of h, for any $h \in \mathbb{F}_q$, $h \neq 0$. Indeed, the security of many public-key cryptosystems depends on the assumption that finding discrete logarithms is hard, at least for certain groups. For instance, the security of the following cryptographic applications depends on the current inability to solve the discrete logarithm problem efficiently: Diffie-Hellman key exchange scheme [19], El Gamal's cryptosystem [21], and pseudorandom bit generators [8, 32].

We observe that this problem can be defined over any group. However, in this paper, we restrict ourselves to the case of discrete logarithm problem over $\mathbb{F}_{2^n}$. We view the elements in $\mathbb{F}_{2^n}$ as polynomials of degree n over $\mathbb{F}_2$. We point out that all results in this section apply to $\mathbb{F}_q$, where $q = p^n$, p is a small prime and n is large, but for practical reasons we only consider $p = 2$.

The breakthrough in the computation of discrete logarithms in such groups was the development of the *index calculus method*. Odlyzko [57] provides an excellent account of this problem. The method consists of two parts: a construction of a large database of logarithms, and the computation of individual logarithms. Let S be a set of irreducible polynomials over $\mathbb{F}_p$, where p is the characteristic of $\mathbb{F}_q$.

1) Choose an integer s in $[1, q-1]$ uniformly at random, and form the polynomial $h \equiv g^s \pmod{f}, \deg h < n$. Check if h factors completely into irreducibles over the set S. If not, discard it and iterate. If it does, say $h = \prod_{v \in S} v^{e_v(h)}$, record the congruence

$$s \equiv \sum_{v \in S} e_v(h) \log_g v \pmod{q - 1}.$$

Repeat the above steps until "slightly more" than $\#S$ congruences are obtained. Then solve the system to determine $\log_g v$ for all $v \in S$.

2) Let h^* be the element whose logarithm we want to compute. Choose an integer s in $[1, q - 1]$ uniformly at random and form the polynomial $h \equiv h^* g^s \pmod{f}, \deg h < n$. Check if h factors completely into irreducibles over the set S. If not, discard it and iterate. If it does, say $h = \prod_{v \in S} v^{e_v(h)}$, compute the required discrete logarithm as

$$\log_g h^* \equiv -s + \sum_{v \in S} e_v(h) \log_g v \pmod{q - 1}.$$

There are many variations of this "basic" version; see [57] for details. Normally, S is the set of irreducible polynomials of degree smaller or equal to m, the so-called m-smooth polynomials. Since we repeat the search until we find an m-smooth polynomial, the analysis of the index calculus method requires information on the number of polynomials that are m-smooth. Using the methodology described in this paper, Odlyzko [57] obtains the generating function of the number $N_q(m; n)$ of monic polynomials over $\mathbb{F}_q$ of degree n which are m-smooth:

$$S_m(z) = \sum_{n \geq 0} N_q(m; n) z^n = \prod_{k=1}^{m} \left(\frac{1}{1 - z^k} \right)^{I_k}.$$

For the cryptographical applications, m tends to infinity with n. More precisely, we have $m = \sqrt{n \log n}/\sqrt{2 \log 2}$; see [57]. Hence, singularity analysis does not apply since we have a natural boundary in $|z| = 1$ and analytic continuation is not possible. Odlyzko uses the saddle point method for deriving an asymptotic estimation for the numbers $N_q(m; n)$ as $n \to \infty$, uniformly for m in the range $n^{1/100} \le m \le n^{99/100}$. [Actually, his results hold for $n^\delta \le m \le n^{1-\delta}$, where $\delta > 0$.]

Blake *et al.* propose a variant of the index calculus method over $\mathbb{F}_{2^n}$. This variant is known as the Waterloo algorithm [7, 57]. It improves the running time of the method by introducing a heuristic argument that makes its analysis not rigorous.

The central idea is to change the search of one m-smooth polynomial of degree n by two m-smooth polynomials of degree at most $(n-1)/2$.

0) Set A to 0.
1) If $\deg h(x) \le m$ and $h(x) = \prod_i h_i(x)^{e_i}$, then $\log_g h(x) \equiv \sum_i e_i \log_g h_i(x)$ mod $(q-1)$, and stop.
2) Generate a random integer a; set A to $A + a$ and $h(x)$ to $h(x)g(x)^a$. Apply the extended Euclidean algorithm to $h(x)$ and $f(x)$ to obtain polynomials $t(x)h(x) \equiv r(x) \pmod{f(x)}$, with $\deg r(x), \deg t(x) \le (n-1)/2$.
3) Factor $t(x) = \prod_i p_i(x)^{d_i}$ and $r(x) = \prod_j p_j(x)^{d_j}$. If $\deg p_i(x) \le m$ and $\deg p_j(x) \le m$ for all i, j, then compute the required discrete logarithm as

$$\log_g h(x) = \sum_j d_j \log_g p_j(x) - \sum_i d_i \log_g p_i(x) - A,$$

and stop. Otherwise, return to 2.

The correctness of the method is based on the known fact that if we apply the extended Euclidean algorithm to $h(x)$ and $f(x)$, then there exist two polynomials $r(x)$ and $t(x)$ both of degree smaller or equal to $(n-1)/2$ such that $t(x)h(x) \equiv r(x) \pmod{f(x)}$; see [7].

It is not difficult to check that the polynomials $r(x), t(x)$ of degree $\le (n-1)/2$ can be taken as *relatively prime* polynomials. Then, for the analysis of the running time of the algorithm, we have to estimate the probability that two random monic polynomials of degree $\le (n-1)/2$ are relatively prime, and that they decompose into irreducible polynomials with degree $\le m$.

Let $P(m; (n-1)/2)$ be the probability that a polynomial of degree $(n-1)/2$ is m-smooth, and $P(m; (n-1)/2, (n-1)/2)$ be the probability that a pair of polynomials each of degree $(n-1)/2$ is m-smooth. Blake *et al.* approximate this probability by the probability that each polynomial $r(x)$ and $t(x)$ has degree about $(n-1)/2$ and that each polynomial factors into irreducibles of degree $\le m$ *independently* of the other. They show experimental data validating this heuristic argument.

Drmota and Panario [20] provide a rigorous proof of this heuristic. Let $N_q(m; n_1, n_2)$ denote the number of coprime pairs of monic polynomials f and g over $\mathbb{F}_q$ of degrees n_1 and n_2, respectively, which are m-smooth. The generating function $F_m(z, w)$ of interest here is

$$F_m(z, w) = \sum_{n_1, n_2 \ge 0} N_q(m; n_1, n_2)\, z^{n_1} w^{n_2}.$$

Considering an enumerator of an irreducible factor of degree k that counts the presence of the factor in one of the polynomials f or g but not in both, we have

$$1 + z^k + z^{2k} + \cdots + w^k + w^{2k} + \cdots .$$

Varying on the possible I_k irreducible factors of degree k we have

$$F_m(z, w) = \prod_{k=1}^{m} \left(1 + z^k + z^{2k} + \cdots + w^k + w^{2k} + \cdots\right)^{I_k}$$

$$= \prod_{k=1}^{m} \left(\frac{1 - z^k w^k}{(1 - z^k)(1 - w^k)}\right)^{I_k} = \frac{S_m(z)S_m(w)}{S_m(zw)}.$$

Now a bivariate saddle point argument similar to Odlyzo's provides the following theorem (see [20]).

Theorem 7. *Let $\delta > 0$ be given. Then we have, uniformly for $m, n_1, n_2 \to \infty$ with $n_1^\delta \leq m \leq n_1^{1-\delta}$ and $n_2^\delta \leq m \leq n_2^{1-\delta}$,*

$$N_q(m; n_1, n_2) \sim \left(1 - \frac{1}{q}\right) N_q(m; n_1) N_q(m; n_2).$$

In other words, Blake *et al.* approximation is correct in asymptotic terms, and we provide a precise estimation for this relation. The results generalize to provide estimates for the probability that two random monic polynomials of degree *at most* $(n-1)/2$ are relatively prime and m-smooth.

The basic index calculus method works with a set S formed for m-smooth polynomials. Another possibility is to use a non-smooth factor base. Garefalakis and Panario [38, 39] propose a different factor base: all irreducible polynomials with degree in an interval between m_2 and m_1. The required generating function is a simple generalization of the m-smooth generating function, and again, saddle point method is needed for the asymptotic approximation.

The theoretical estimation is as good as the one for the standard base. As in the smooth case, there is no much freedom for the upper limit m_1 of the interval: m_1 behaves exactly as m in the basic version. However, the lower limit m_2 remains a free variable which can be chosen (almost) at will; see the details in [39].

The running time of the algorithm is again dominated by the first stage. This is when a tradeoff takes place regarding the size of the factor base: large $\#S$ means small number of repetitions (until a useful congruence is found), but many such congruences are needed for the system to be solvable.

In practical terms, there is a tradeoff associated to m_2. The influence of the parameter m_2 is considered experimentally [38], however more computational studies are needed to draw some conclusions about the best m_2 and this variant.

The generalized factor base also applies to the Waterloo and the Coppersmith [16] variants. Moreover, Odlyzko [57] describes other variants. These algorithms use clever algebraic manipulations to compute polynomials of "low" degree, which are subsequently factored. The computation of these polynomials is completely independent of the factor base. Therefore, a different factor base like the one proposed is "compatible" with all the variants.

Finally, we should mention that the analysis of Coppersmith's algorithm [16] and the related Adleman's function field sieve algorithm [1] are still open problems. Here the main problem to study is the distribution of very sparse irreducible polynomials; see [33].

6   Conclusions

We surveyed on a methodology for counting properties of random polynomials over finite fields. This general framework is based on generating functions and asymptotic analysis. It does not only allow the study of properties of random polynomials but also provides precise average-case analysis of polynomial algorithms. Moreover, we show the relation between properties of random polynomials over finite fields and properties of random decomposable combinatorial structures.

A simplified picture of a random polynomial over a finite field is as follows:

- it is irreducible with probability tending to 0 as $n \to \infty$;
- it contains $\log n$ number of irreducible factors (concentrated);
- it has $c_k n$ expected kth largest degree irreducible factor ($c_1 = 0.62432\ldots$ and $c_2 = 0.20958\ldots$);
- it has $e^{-\gamma} \log n$ and $e^{-\gamma} \log^2 n/2$ expected first and second smallest degree irreducible factors (not concentrated).

There are other polynomial problems where this methodology has not been fully employed yet. One example is finding roots of a polynomial. This problem can be considered as a variation of factoring polynomials, but there are methods especially tailored for this task (see the references in [37, 41]). The methodology presented here together with results similar to the ones in [25] should provide analysis for these algorithms.

Other important problems are polynomial gcd computations, and fast algorithms for polynomial multiplication (a la Karatsuba) and for "repeated squaring" (also called "binary powering" or "square and multiply") methods. For the gcd problem, some results are in [30, 48, 54]. The analysis of repeated squaring algorithm seems to follow from the recent work of Grabner et al. [46] for the analysis of the similar problem of computing linear combinations of points in an elliptic curve. Karatsuba's algorithm has been less studied than the other problems. However, the analysis of mergesort algorithm in Flajolet and Golin [24] may serve as a starting point for the study of Karatsuba's algorithm since both algorithms have a similar recursive structure.

There are other algebraic related problems that are even less understood from an average-case perspective. There has been no analysis of the polynomial factorization algorithms based on linear algebra due to Berlekamp [6] and

Niederreiter [56]. It is not clear how to use the methodology of this paper to analyze those algorithms.

Solving sparse linear systems of equations over finite fields is another fundamental problem that has not been deeply studied from this perspective. This is an important task, for example, for the index calculus method [57].

Finally, it would be interesting to have a similar methodology to the one presented here for studying properties of random polynomials in several variables. This problem has been studied by Carlitz [12, 13], Cohen [14] and Hayes [45]. The only average-case analysis of an algorithm for factoring bivariate polynomials that we know is due to Gao and Lauder [34]. However, there seems to be no generic methodology for analyzing algorithms for the factorization, irreducibility test, and so on, of polynomials over finite fields in several variables.

Acknowledgment. The author would like to thank his co-authors. It has been a privilege to work with them.

References

1. L. ADLEMAN. The function field sieve. In *Proc. 1st ANTS Symp.*, vol. 877 of *Lecture Notes in Computer Science*, 1994, 108–121.
2. E. BACH, J. VON ZUR GATHEN AND H.W. LENSTRA JR. Factoring Polynomials over Special Finite Fields. *Finite Fields and Their Applications*, 7:5–28, 2001.
3. M. BEN-OR. Probabilistic algorithms in finite fields. In *Proc. 22nd IEEE Symp. Foundations Computer Science*, 394–398, 1981.
4. E. BENDER. Central and local limit theorems applied to asymptotic enumeration. *J. Combin. Theory*, Ser. A, 15:91–111, 1973.
5. E. BENDER AND B. RICHMOND. Central and local limit theorems applied to asymptotic enumeration II: multivariate generating functions. *J. Combin. Theory,* Ser. A, 34:255–265, 1983.
6. E.R. BERLEKAMP. *Algebraic Coding Theory*. McGraw Hill, New York NY, 1968.
7. I.F. BLAKE, R. FUJI-HARA, R.C. MULLIN AND S.A. VANSTONE. Computing discrete logarithms in finite fields of characteristic two. *SIAM J. Alg. Disc. Meth.*, 5:276–285, 1984.
8. M. BLUM AND S. MICALI. How to generate cryptographically strong sequences of pseudorandom bits. *SIAM J. Comput.*, 13:850–864, 1984.
9. A.A. BUCHSTAB. Asymptotic estimates of a general number theoretic function. *Mat. Sbornik*, 44:1239–1246, 1937.
10. M. CAR. Théorèmes de densité dans $\mathbb{F}_q[x]$. *Acta Arith.*, 48:145–165, 1987.
11. L. CARLITZ. The arithmetic of polynomials in a Galois field. *Amer. J. Math.*, 54:39–50, 1932.
12. L. CARLITZ. The distribution of irreducible polynomials in several indeterminates. *Illinois J. Math.*, 7:371–375, 1963.
13. L. CARLITZ. The distribution of irreducible polynomials in several indeterminates II. *Canad. J. Math.*, 17:261–266, 1965.
14. S.D. COHEN. The distribution of irreducible polynomials in several indeterminates over a finite field. *Proc. Edinburgh Math. Soc.*, 16:1–17, 1968.
15. S.D. COHEN. The values of a polynomial over a finite field. *Glasgow Math. J.*, 14:205–208, 1973.

16. D. COPPERSMITH. Fast evaluation of logarithms in fields of characteristic two. *IEEE Trans. Info. Theory*, 30:587–594, 1984.
17. N. DE BRUIJN. On the number of positive integers $\leq x$ and free of prime factors $> y$. *Indag. Math*, 13:2–12, 1951,.
18. K. DICKMAN On the frequency of numbers containing prime factors of a certain relative magnitude. *Ark. Mat. Astr. Fys.*, 22:1–14, 1930.
19. W. DIFFIE AND M. HELLMAN. New directions in cryptography. *IEEE Trans. Inform. Theory*, 22:644–654, 1976.
20. M. DRMOTA AND D. PANARIO. A rigorous proof of the Waterloo algorithm for the discrete logarithm problem. *Designs, Codes and Cryptography*, 26:229-241, 2002.
21. T. ELGAMAL. A public key cryptosystem and a signature scheme based on discrete logarithms. *IEEE Trans. Info. Theory*, 31:469–472, 1985.
22. S. A. EVDOKIMOV. Factorization of polynomials over finite fields in subexponential time under GRH. In *Proc. 1st ANTS Symp.*, vol. 877 of *Lecture Notes in Computer Science*, 1994, 209–219.
23. S.R. FINCH. *Mathematical Constants*. Encyclopedia of Mathematics and its Applications vol. 94, Cambridge University Press, 2003.
24. P. FLAJOLET AND M. GOLIN. Mellin transform and asymptotics: the mergesort recurrence. *Acta Inf.*, 31:673–696, 1994.
25. P. FLAJOLET, X. GOURDON AND D. PANARIO. The complete analysis of a polynomial factorization algorithm over finite fields. *J. of Algorithms*, 40:37–81, 2001.
26. P. FLAJOLET AND A. ODLYZKO. Singularity analysis of generating functions. *SIAM J. of Disc. Math.*, 2:216-240, 1990.
27. P. FLAJOLET AND R. SEDGEWICK. *Analytic Combinatorics*. In preparation; see http://algo.inria.fr/flajolet/Publications/books.html
28. P. FLAJOLET AND M. SORIA. Gaussian limiting distributions for the number of components in combinatorial structures. *J. of Combin. Theory*, Ser. A, 53:165–182, 1990.
29. P. FLAJOLET AND M. SORIA. General combinatorial schemas: Gaussian limiting distributions and exponential tails. *Discrete Math.*, 114:159–180, 1993.
30. C. FRIESEN AND D. HENSLEY. The statistics of continued fractions for polynomials over a finite field. *Proc. Amer. Math. Soc.*, 124:2661–2673, 1996.
31. S. GAO. On the deterministic complexity of polynomial factoring. *Journal of Symbolic Computation*, 31:19–36, 2001.
32. S. GAO, J. VON ZUR GATHEN, AND D. PANARIO. Gauss periods: orders and cryptographical applications. *Math. Comp.*, 67:343–352, 1998.
33. S. GAO, J. HOWELL, AND D. PANARIO. Irreducible polynomials of given forms. In R.C. Mullin and G.L. Mullen, editors, *Finite Fields: Theory, Applications, and Algorithms (Fourth International Conference on Finite Fields: Theory, Applications, and Algorithms)*, vol. 225 of *Contemporary Mathematics*, American Mathematical Society, 43–54, 1999.
34. S. GAO AND A. LAUDER. Hensel lifting and polynomial factorisation. *Math. Comp.*, 71:1663–1676, 2002.
35. S. GAO AND D. PANARIO. Tests and constructions of irreducible polynomials over finite fields. In F. Cucker and M. Shub, editors, *Foundations of Computational Mathematics*, Springer Verlag, 346–361, 1997.
36. Z. GAO AND B. RICHMOND. Central and local limit theorems applied to asymptotic enumeration IV: multivariate generating functions. *J. of Comput. Appl. Math.*, 41:177–186, 1992.
37. J. VON ZUR GATHEN AND J. GERHARD. *Modern Computer Algebra*. Cambridge University Press, 1999.

38. T. GAREFALAKIS AND D. PANARIO. The index calculus method using non-smooth polynomials. *Mathematics of Computation*, 70:1253–1264, 2001.

39. T. GAREFALAKIS AND D. PANARIO. Polynomials over finite fields free from large and small degree irreducible factors. *Journal of Algorithms*, 44:98-120, 2002.

40. J. VON ZUR GATHEN AND J. GERHARD. Polynomial factorization over $\mathbb{F}_2$. *Math. Comp.*, 71:1677-1698, 2002.

41. J. VON ZUR GATHEN AND D. PANARIO. A survey on factoring polynomials over finite fields. *Journal of Symbolic Computation*, 31:3–17, 2001.

42. J. VON ZUR GATHEN AND V. SHOUP. Computing Frobenius maps and factoring polynomials. *Comput complexity*, 2:187–224, 1992.

43. X. GOURDON. *Combinatoire, algorithmique et géométrie des polynômes*. PhD thesis, École Polytechnique, 1996.

44. X. GOURDON. Largest component in random combinatorial structures. *Discrete Math.*, 180:185–209, 1998.

45. D.R. HAYES. The distribution of irreducibles in $\mathbb{F}_q[x]$. *Trans. American Math. Soc.*, 117:101–127, 1965.

46. P. GRABNER, C.HEUBERGER, H. PRODINGER AND J. THUSWALDNER. Efficient linear combinations in elliptic curve cryptography. Preprint, 2003.

47. E. KALTOFEN AND V. SHOUP. Subquadratic-time factoring of polynomials over finite fields. In *Proc. 27th ACM Symp. Theory of Computing*, 398–406, 1995.

48. J. KNOPFMACHER AND A. KNOPFMACHER. The exact length of the Euclidean algorithm in $F_q[X]$. *Mathematika*, 35:297-304, 1988.

49. A. KNOPFMACHER AND J. KNOPFMACHER. Counting polynomials with a given number of zeros in a finite field. *Lin. and Multilin. Alg.*, 26:287–292, 1990.

50. J. KNOPFMACHER AND A. KNOPFMACHER. Counting irreducible factors of polynomials over a finite field. *SIAM J. on Disc. Math.*, 112:103–118, 1993.

51. A. KNOPFMACHER AND R. WARLIMONT. Distinct degree factorizations for polynomials over a finite field. *Trans. Amer. Math. Soc.*, 37:2235–2243, 1995.

52. D.E. KNUTH. *The Art of Computer Programming, vol.2: Seminumerical Algorithms*. Addison-Wesley, Reading MA, 3rd edition, 1997.

53. R. LIDL AND H. NIEDERREITER. *Introduction to Finite Fields and Their Applications*. Encyclopedia of Mathematics and its Applications vol. 20, 2nd edition, Cambridge University Press, 1994.

54. K. MA AND J. VON ZUR GATHEN. Analysis of Euclidean algorithms for polynomials over finite fields. *J. of Symb. Comp.*, 9:429-455, 1990.

55. M. MIGNOTTE AND J.L. NICOLAS. Statistiques sur $\mathbb{F}_q[x]$. *Ann. de l'Inst. Henri Poincaré*, 19:113–121, 1983.

56. H. NIEDERREITER. Factoring polynomials over finite fields using differential equations and normal bases. *Math. Comp.*, 62:819–830, 1994.

57. A. ODLYZKO. Discrete logarithms and their cryptographic significance. In *Advances in Cryptology, Proc. of Eurocrypt 1984*, vol. 209 of *Lecture Notes in Computer Science*, Springer-Verlag, 224–314, 1985.

58. A. ODLYZKO. Asymptotic enumeration methods. In *Handbook of Combinatorics*, R. Graham, M. Grötschel, and L. Lovász, Eds., vol. 2. Elsevier, 1063–1229, 1995.

59. D. PANARIO, B. PITTEL, B. RICHMOND AND A. VIOLA. Analysis of Rabin's irreducibility test for polynomials over finite fields. *Random Struct. Alg.*, 19: 525–551, 2001.

60. D. PANARIO AND B. RICHMOND. Analysis of Ben-Or's polynomial irreducibility test. *Random Struct. Alg.*, 13:439–456, 1998.

61. D. PANARIO AND B. RICHMOND. Smallest components in decomposable structures: exp-log class. *Algorithmica*, 29:205–226, 2001.

62. D. PANARIO AND B. RICHMOND. Exact largest and smallest size of components in decomposable structures, *Algorithmica*, 31:413–432, 2001.
63. M.O. RABIN. Probabilistic algorithms in finite fields. *SIAM J. Comp.*, 9:273–280, 1980.
64. R. SEDGEWICK AND P. FLAJOLET. *An Introduction to the Analysis of Algorithms.* Addison Wesley, 1996.
65. V. SHOUP. A new polynomial factorization algorithm and its implementation. *J. Symb. Comp.*, 20:363–397, 1996.
66. S. UCHIYAMA. Note on the mean value of $v(f)$ II. *Proc. Japan Acad.*, 31:321–323, 1955.
67. K.S. WILLIAMS. Polynomials with irreducible factors of specified degree. *Canad. Math. Bull.*, 12:221–223, 1969.
68. K. ZSIGMONDY. Über die Anzahl derjenigen ganzen ganzzahligen Functionen $n$ten Grades von x, welche in Bezug auf einen gegebenen Primzahlmodul eine vorgeschriebene Anzahl von Wurzeln besitzen. *Sitzungsber. Wien Abt II*, 103:135–144, 1894.

Combinatorics of the Two-Variable Zeta Function

Iwan M. Duursma [*]

Department of Mathematics, University of Illinois at Urbana-Champaign,
Urbana IL 61801, USA, duursma@math.uiuc.edu

Abstract. We consider the rank polynomial of a matroid and some
well-known applications to graphs and linear codes. We compare rank
polynomials with two-variable zeta functions for algebraic curves. This
leads us to normalize the rank polynomial and to extend it to a rational
rank function. As applications to linear codes we mention: A formulation
of Greene's theorem similar to an identity for zeta functions of curves
first found by Deninger, the definition of a class of generating functions
for support weight enumerators, and a relation for algebraic-geometric
codes between the matroid of a code and the two-variable zeta function
of a curve.

1 Introduction

Matroids were introduced by Whitney to generalize the abstract properties of
linear dependence. In the next section we will give the general definition of a
matroid. For a special case, let $E = \{1, 2, \dots, n\}$ be a finite set that labels a list
$S = (x_1, x_2, \dots, x_n)$ of vectors in a vector space V. The matroid on E is defined
as the collection of all subsets of E of full rank. Various problems involving S
can be solved in terms of the matroid without reference to other properties of S.
We are mainly interested in those problems that can be solved in terms of the
rank polynomial of the matroid. The rank polynomial enumerates the number of
subsets of E of given size and rank. Such problems include among others graph
coloring problems, where S is the set of columns in the vertex-edge incidence
matrix of a graph, weight distribution problems, where S is the set of columns
of a generator matrix of a linear code, and decision problems for combinatorial
games. They are the subject of Sections 2-4. These sections can be seen as an
introduction to matroids: we give the basic definitions and collect basic results.
We also mention some problems that can not be solved in terms of the rank
polynomial.

In Sections 5-8, we study zeta functions for algebraic curves and their connection with matroids. Section 5 deals with the problem of enumerating divisors of
given degree and dimension on a given curve over a finite field. Its formal solution
is given by the two-variable zeta function of Pellikaan. VanderGeer and Schoof
reformulated this zeta function for number fields. Section 6 describes Deninger's

---

[*] This work supported by NSF Grant DMS-0099761.

transformation that relates the two zeta functions. In Section 7, we define a two-variable zeta function for linear codes that describes the Hamming weight distribution of a linear code over its base field and over all finite extensions of the base field. It is similar to the Pellikaan two-variable zeta function of an algebraic curve. In Section 8, we rewrite the rank polynomial of a matroid as a normalized rank function. It is similar to the vanderGeer-Schoof two-variable zeta function in its version for curves. Under Deninger's transformation, the normalized rank function of a matroid becomes the two-variable zeta function of a matroid. The normalized rank function of a matroid does not reveal the parameters of the matroid (its size and its rank, or for a code its length and its dimension). For a given choice of parameters it gives the same information as the rank polynomial. The uniform matroid of rank k on $E = \{1, 2, \dots, n\}$ has as subsets of full rank all subsets of size at most k. For all n and k the normalized rank function is

$$\frac{(1 - xy)}{(1 - x)(1 - y)} = \cdots + x^2 + x + 1 + y + y^2 + \cdots .$$

The zeta function of the uniform matroid is

$$\frac{1}{(1 - T)(1 - uT)}.$$

Sections 9-12 describe some applications of the connection between zeta functions and matroids. In Section 9, we show that for linear codes Deninger's transformation is equivalent to Greene's theorem. Section 10 describes a decomposition of the zeta function. The decomposition is used in Section 11 to describe a class of generating functions for support weight enumerators. In Section 12, we relate the matroid of an algebraic-geometric code to the two-variable zeta function of a curve. We recover a lower bound of Munuera for the weight hierarchy of an algebraic geometric code.

Reference texts for matroids are [41], [28]. The latter book does not discuss rank polynomials. A good survey on the rank polynomial and its applications is [7]. The books [6], [16] on graph theory have excellent chapters on the rank polynomial. Reference texts for algebraic curves over finite fields are [24], [33], and for coding theory [38], [23].

2 Matroids

A matroid $M = (E, \mathcal{I})$ consists of a finite set E and a collection $\mathcal{I}$ of subsets of E called *independent sets* such that

(I1) $\emptyset \in \mathcal{I}$.

(I2) If $I_1 \in \mathcal{I}$ and $I_2 \subset I_1$, then $I_2 \in \mathcal{I}$.

(I3) if $I_1, I_2 \in \mathcal{I}$ such that $|I_2| < |I_1|$ then there exists $e \in I_1 - I_2$ such that $I_2 \cup \{e\} \in \mathcal{I}$.

For vectors $e_1, e_2, \ldots, e_n$ in some vector space V we obtain a matroid on the set $E = \{1, 2 \ldots, n\}$ by taking for $\mathcal{I}$ the collection of all subsets of E that correspond to independent vectors. In particular, for a given matrix, we can define the matroid of its columns. The matroid of a graph is defined as the matroid of its vertex-edge incidence matrix. The matroid of a linear code is defined as the matroid of its generator matrix. Both cases will be treated in more detail in the next two sections. On the other hand not all matroids are *representable* by vectors in some vector space.

The axiom (I3) guarantees that all maximal independent subsets $I \subset A$ of a given subset $A \subset E$ have the same size, so that we can talk about the rank $r(A)$ of a subset A. A maximal independent subset $I \subset E$ is called a *basis*. Clearly, a matroid on E is determined by the collection $\mathcal{B}$ of all bases. A collection $\mathcal{B}$ of subsets of E defines the bases for a matroid if and only if [28, Corollary 1.2.5]

(B1) $\mathcal{B}$ is non-empty.
(B2) if $B_1, B_2 \in \mathcal{B}$ such that $e \in B_1 - B_2$ then there exists $e' \in B_2 - B_1$ such that $(B1 - e) \cup \{e'\} \in \mathcal{B}$.

Under the same conditions, for bases B_1, B_2 and for $e \in B_1 - B_2$, we can use axiom (I3) to extend $\{e\}$ to a basis of the form $(B_2 - e') \cup e$. Combination of this simple result with (B2) shows that $\mathcal{B}^* = \{E - B : B \in \mathcal{B}\}$ is the collection of bases for a matroid $M^* = (E, \mathcal{I}^*)$ called the *dual matroid* of M.

The non-Pappus matroid is defined on $E = \{1, 2, 3, 4, 5, 6, 7, 8, 9\}$ with bases all 3-sets except the eight 3-sets of collinear points in Figure 1. For some time it was thought to be the smallest possible example of a nonrepresentable matroid till Vamos gave an example on eight points. The Vamos matroid is defined on $E = \{1, 2, 3, 4, 5, 6, 7, 8\}$ with bases all 4-sets except $\{1, 2, 3, 4\}$, $\{1, 2, 5, 6\}$, $\{1, 2, 7, 8\}$, $\{3, 4, 5, 6\}$, $\{3, 4, 7, 8\}$.

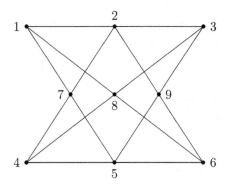

Figure 1: Non-Pappus matroid.

A subset $A \subset E$ is called *closed* if $r(A \cup x) = r(A) + 1$ for all $x \in E - A$. A *hyperplane* H is a maximal proper closed set. It has rank $r(H) = r(E) - 1$.

Closed sets are also called flats and all flats can be obtained as intersection of hyperplanes [28, Theorem 1.7.8]. A matroid on E is determined by the collection $\mathcal{H}$ of all hyperplanes. A collection $\mathcal{H}$ of subsets of E defines the hyperplanes for a matroid if and only if [28, Proposition 2.1.18]

(H1) $E \notin \mathcal{H}$.
(H1) No proper subset of a hyperplane is a hyperplane.
(H2) For distinct $H_1, H_2 \in \mathcal{H}$ and $x \notin H_1 \cup H_2$ there exists $H_3 \in \mathcal{H}$ that contains $H_1 \cap H_2$ and x.

For a subset $A \subset E$ we define the *degree* $|A|$ and the *rank* $r(A)$ of the subset, where

$$r(A) = \max\{|I| : I \subset A, I \in \mathcal{I}\}.$$

We also consider the *corank* and the *nullity* of a subset.

$$\begin{array}{ll} r(E) - r(A) & \text{(corank)}, \\ |A| - r(A) & \text{(nullity)}. \end{array}$$

In general, the corank and nullity are nonnegative integers. They satisfy the duality, for $\bar{A} = E - A$ [28, Proposition 2.1.9],

$$r(E) - r(A) = |\bar{A}| - r^*(\bar{A}).$$

The rank polynomial (or Whitney polynomial, or corank-nullity polynomial) is defined as

$$W(x,y) = \sum_{A \subset E} x^{r(E)-r(A)} y^{|A|-r(A)}.$$

So that the dual matroid M^* of M has $W^*(x,y) = W(y,x)$. In this paper we are particularly interested in properties of a matroid that can be studied through its rank polynomial. We give two theorems by Edmonds that show that the natural packing and covering problem for a matroid can be decided from its rank polynomial [41, Section 8.4].

Theorem 1 (Packing problem). *[15] A matroid M on the set E has k disjoint bases if and only if, for all subsets $A \subset E$,*

$$k \cdot (r(E) - r(A)) \leq |\bar{A}|.$$

Theorem 2 (Covering problem). *[15] A matroid M on the set E can be covered by k independent sets if and only if, for all subsets $A \subset E$,*

$$k \cdot r(A) \geq |A|.$$

The reference [7] discusses in detail various other applications of the rank polynomial. We include an example from [7] of two matroids with the same rank polynomial (Figure 2). They differ in many other aspects, for example they have different lattice of flats. The matroid on the left has three hyperplanes of size 2 (namely $\{2,5\}, \{3,5\}, \{4,5\}$), whereas the matroid on the right has two hyperplanes of size 2 ($\{1,4\}, \{3,5\}$).

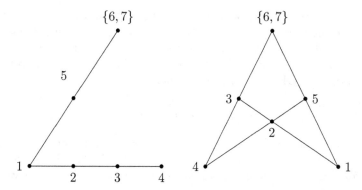

Figure 2: Non-isomorphic matroids with the same rank polynomial.

3 Graphs

Through work of Tutte and others there have been important contributions from graph theory to matroid theory and vice versa. In the previous section we defined the matroid of a graph as the matroid of its vertex-edge incidence matrix D. For a directed graph the entry $D_{v,e}$ takes value -1 if the edge e originates in v, $+1$ if it arrives at v, and value 0 if it does neither or both. Edges of an undirected graph can be signed arbitrarily with no effect on the matroid.

An equivalent definition of this matroid can be given in terms of *circuits* or minimal dependent sets. Thus $C \subset E$ is a circuit if

$$r(C) < |C| \quad \text{and for all } z \in C : r(C - z) = |C - z|.$$

In dual terms, for $\bar{C} = E - C$,

$$r^*(\bar{C}) < r^*(E), \quad \text{and for all } z \notin \bar{C} : r^*(\bar{C} \cup z) = r^*(E).$$

In other words, a circuit is the complement of a hyperplane in the dual matroid. And a matroid on E is determined by the collection $\mathcal{C}$ of its circuits. A collection $\mathcal{C}$ of subsets of E defines the circuits for a matroid if and only if [28, Corollary 1.1.5]

(C1) $\emptyset \notin \mathcal{C}$.
(C1) No proper subset of a circuit is a circuit.
(C2) For distinct $C_1, C_2 \in \mathcal{C}$ and $z \in C_1 \cap C_2$ there exists $C_3 \in \mathcal{C}$ contained in $C_1 \cup C_2 - z$.

The matroid defined by the circuits of a graph is called the *cycle matroid* of the graph. The dual matroid is called the *cocycle matroid* of the graph. A connected planar graph G and its dual G^* can be given orientations such that $D^* D^T = 0$. The matroids defined with D and D^*, respectively, are thus in duality and correspond to the *cycle matroid* and the *cocycle matroid* of the graph G.

The incidence matrix D represents the boundary operator $\partial : C_1 \rightarrow C_0$ from the space of 1-chains, defined on edges, to the space of 0-chains, defined on vertices. The boundary operator maps the edge $e = (v_-, v_+)$ to $\partial e = v_+ - v_-$. Let ∂C_1 be the image of C_1 in C_0. For a subset A of the edge set E, we consider the restriction $\partial|A : C_1(A) \rightarrow \partial C_1$. The complex

$$\cdots \rightarrow 0 \rightarrow C_1(A) \xrightarrow{\partial|A} \partial C_1 \rightarrow 0 \rightarrow \cdots$$

has nontrivial homology groups of dimensions $h_1(A) = \dim \text{Ker } \partial|A = |A| -$ rank $\partial|A$ and $h_0(A) = \dim \text{Coker } \partial|A = \text{rank } \partial - \text{rank } \partial|A$. Thus the rank polynomial for a graph gives information about the homology of the various boundary operators defined on subgraphs.

We give two problems on graphs that have a solution in terms of the rank polynomial of the corresponding matroid. A flow mod m in a connected graph is the assignment x of a nonzero integer residue mod m to each edge such that $Dx = 0$. The number of flows mod m (of nowhere zero 1-cycles) on a connected graph G is [7, Proposition 6.3.4.]

$$(-1)^{|E|-|V|+1} W(-1, -m)$$

An m-coloring of a connected graph is the assignment y of an integer residue mod m to each vertex such that $D^t y$ is nonzero on each edge. The number of m-colorings (of nowhere zero 1-coboundaries) on a connected graph is [7, Proposition 6.3.1.]

$$m(-1)^{|V|-1} W(-m, -1)$$

The following game on matroids is described in [27] as a variation of Shannon's switching game for graphs [41, Section 19.4]. Obviously, no basis is contained in a hyperplane. Therefore every basis has non-trivial intersection with every circuit in the dual matroid. Players B and C take turns picking an element from the underlying set $E = \{1, 2, \ldots, n\}$ of the matroid. Player B wins if he conquers a basis for the matroid. Player C wins if he conquers a circuit for the dual matroid. The game has precisely one winner. The following are equivalent [27], [15].

(1) Player C plays first and player B can win against all possible strategies of C.
(2) The matroid M has two disjoint bases.

The second (global) condition can be verified with the (local) conditions of Edmonds packing theorem in the previous section. And the existence of a winning strategy for B is revealed by the rank polynomial of the matroid.

The matroid of a graph describes certain properties of the edges in the graph, but in general does not determine the spectrum of a graph or properties related to the spectrum. A trivial example is given by the two graphs in Figure 3. They have the same matroid, but have different spectrum (the spectrum of Chung's

Laplacian [8] is $(0, 1/2, 3/2, 2)$ and $(0, 1, 1, 2)$, respectively). Clearly they have different diameter. The two graphs in Figure 4 also have the same matroid (in fact with the given orientation their incidence matrices have the same row space), but have different spectrum (the spectrum of Chung's Laplacian is $(0, 2/3, 4/3, 2)$ and $(0, 1, 1, 2)$, respectively).

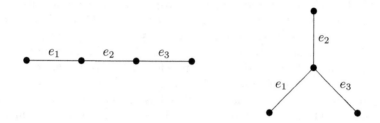

Figure 3: Non-isomorphic graphs with the same matroid.

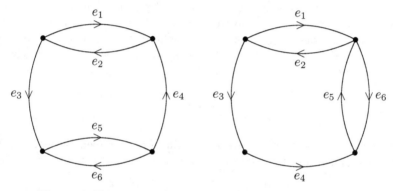

Figure 4: Non-isomorphic graphs with the same matroid.

The number of cycles (or closed backtrackless tail-less paths) of given length differs for the two graphs. The Ihara zeta function is a generating function for the number of cycles and using Bass's generalization it can be computed as [5], [19],

$$Z(X, u) = \frac{(1 - u^2)^\chi}{\det(1 - Au + Qu^2)},$$

where $\chi = |V| - |E|$ is the Euler characteristic, A is the adjacency matrix, and Q is a diagonal matrix with entries the vertex degrees minus one.

4   Linear Codes

A finite sequence of n points in projective space, not all contained in one hyperplane, becomes a matroid with the natural definitions for the degree and rank of a subsequence of points. The matroid definition of a hyperplane as a maximal proper closed subset of $\{1, 2, \ldots, n\}$ applies. But we also have the collection of hyperplanes of the projective space. The properties of the first type of hyperplanes are described by the lattice of flats of the matroid. The rank polynomial of the matroid in general does not give full information about the number of such hyperplanes. On the other hand the number of hyperplanes in projective space over a finite field depends only on the size of the field and the dimension of the projective space. The theorem by Greene below shows that the rank polynomial of a matroid of n points gives the number of projective hyperplanes that contain precisely a of the n points, for any given $a = 0, 1, \ldots, n$. This makes the rank polynomial an important invariant for linear codes.

A linear code is a subspace of the space of all n letter words over a finite field $\mathbf{F}$ of q elements. After choosing generators for the subspace, the code can be described as the row space of a matrix G of full rank called the *generator matrix*. The matroid $M(C)$ of a linear code C is defined as the matroid associated to its generator matrix. The matroid $M(C^*)$ of the dual code equals the dual matroid $M(C)^*$ of the code.

Assuming that the code has no zero columns in its generator matrix the columns of the matrix define points in projective space. The generator matrix defines a natural $q - 1$ to 1 map from nonzero codewords to the hyperplanes of the projective space. The Hamming weight distribution of the code gives the number of codewords of given Hamming weight. Equivalently, it gives the number of hyperplanes that contain a given number of points. Let

$$A(x, y) = \sum_{i=0}^{n} A_i x^{n-i} y^i$$

be the Hamming weight enumerator for the linear code. Greene's theorem relates the weight enumerator of a code to the rank polynomial of its matroid [17].

$$\frac{A(x, y)}{(x - y)^k y^{n-k}} = W\left(\frac{qy}{x - y}, \frac{x - y}{y}\right)$$

In particular, for the number A_n of words with n nonzero coordinates we find

$$A_n = (-1)^k W(-q, -1).$$

This is similar to the expression for the number of nowhere zero 1-coboundaries in the graph coloring problem. For a linear code over the field of two elements, we have the useful evaluation [31], [20], [7] (with incorrect sign)

$$W(-2, -2) = (-1)^n (-2)^{\dim C \cap C^\perp}.$$

There are other properties of a linear code that can not be obtained from the matroid structure alone. Examples are coset weight enumerators and the covering radius. The following two codes over the field of five elements have the same matroid.

$$C_1 = \begin{pmatrix} 1 & 0 & 0 & | & 0 & 2 & 4 \\ 0 & 1 & 0 & | & 4 & 0 & 2 \\ 0 & 0 & 1 & | & 2 & 4 & 0 \end{pmatrix} \quad C_2 = \begin{pmatrix} 1 & 0 & 0 & | & 0 & 3 & 3 \\ 0 & 1 & 0 & | & 3 & 0 & 3 \\ 0 & 0 & 1 & | & 3 & 3 & 0 \end{pmatrix}$$

In each case, a subset of $\{1, 2, 3, 4, 5, 6\}$ is independent if it is of size at most three and different from $\{2, 3, 4\}, \{1, 3, 5\}, \{1, 2, 6\}$. The two codes are not equivalent. Any six letter word is at Hamming distance at most 2 from the second code. But the word 111000 is at distance 3 from the first code.

Binary codes are completely determined by their matroid. For a code is determined by its Plücker coordinates and the Plücker coordinate corresponding to a full minor is 1 if the columns in the minor are a basis for the matroid and 0 otherwise. On the other hand, inequivalent binary codes can have the same rank polynomial. The following two codes are the smallest pair that we could find.

$$C_1' = \begin{pmatrix} 1\,0\,0\,0 & 1\,0\,0\,0 & 0\,1\,1\,0\,1 \\ 0\,1\,0\,0 & 0\,1\,0\,0 & 1\,1\,1\,0\,0 \\ 0\,0\,1\,0 & 0\,0\,1\,0 & 1\,0\,1\,1\,1 \\ 0\,0\,0\,1 & 0\,0\,0\,1 & 0\,0\,0\,1\,1 \end{pmatrix}$$

$$C_2' = \begin{pmatrix} 1\,0\,0\,0 & 1\,0\,0\,0 & 0\,1\,1\,0\,1 \\ 0\,1\,0\,0 & 0\,1\,0\,0 & 1\,1\,1\,1\,1 \\ 0\,0\,1\,0 & 0\,0\,1\,0 & 1\,0\,1\,0\,1 \\ 0\,0\,0\,1 & 0\,0\,0\,1 & 0\,0\,0\,1\,1 \end{pmatrix}$$

5 Special Divisors on Curves

Let $X/\mathbf{F}$ be an algebraic curve (projective, non-singular, absolutely irreducible) over a finite field $\mathbf{F}$ of q elements. Let g denote the genus of the curve and let K denote a canonical divisor. Let $h = |Pic_0(X)|$ be the number of distinct divisor classes on X of given degree. Riemann-Roch gives, for the dimension $l(D)$ of the linear space $L(D) = \{f \in \mathbf{F}(X) : (f) + D \geq 0\} \cup \{0\}$,

$$l(D) - l(K - D) = \deg D + 1 - g.$$

A divisor E on X is special if both $l(E) > 0$ and $l(K - E) > 0$. On the set of special divisors $\mathcal{E}$ of a curve we have the natural degree map $|E| = \deg E$. We define the rank of a special divisor as the number of independent linear conditions that it imposes on the canonical linear system.

$$r(E) = l(K) - l(K - E)$$

For E a finite sum of points this definition agrees with the rank of the set of points embedded in projective space with the canonical embedding. Although this definition of degree and rank does not make the set of special divisors a matroid, we find that several notions that were introduced for matroids, such as corank, nullity and rank polynomial, are very useful in describing properties of special divisors. For the corank and nullity of a special divisor we find

$$r(K) - r(E) = l(K - E) - 1, \quad |E| - r(E) = l(E) - 1.$$

In particular corank and nullity are in duality if we define the complement of E in K to be the special divisor $K - E$. It is easily verified that for $P_1 + P_2 + \cdots + P_n \sim$

K, the definition of degree and rank define a matroid on the set of special divisors $0 \le E \le P_1 + P_2 + \cdots + P_n$ (see also Section 12). The matroid has rank polynomial

$$W(x,y) = \sum_{E} x^{l(K-E)-1} y^{l(E)-1}$$

The matroid is the union of two disjoint bases. The easy directions in Theorems 1 and 2 (with $k = 2$) then imply

$$2(l(K-E)-1) \le \deg(K-E) \quad \text{and} \quad 2(l(K)-l(K-E)) \ge \deg E,$$

respectively. Thus Clifford's theorem arises as a special case of the packing and covering theorems for matroids. For an arbitrary special divisor E, and for a sufficiently large base field $\mathbf{F}$, we can always find $P_1, P_2, \ldots, P_n$ such that $0 \le E \le P_1 + P_2 + \cdots + P_n \sim K$.

Let $h^0(D) = \dim H^0(X, \mathcal{O}(D))$ and $h^1(D) = \dim H^1(X, \mathcal{O}(D))$ be the cohomological dimensions for the line bundle associated to the divisor D. To describe the dimensions for all divisors, we write the two-variable vanderGeer-Schoof zeta function [37, Section 8] as a rank function

$$W^{GS}(x,y) = \sum_{[D]} x^{h^0(D)} y^{h^1(D)}.$$

The summation is over divisor classes $[D]$. The decomposition in [37, Section 8] yields

$$W^{GS}(x,y) = \sum_{i=0}^{2g-2} \sum_{\deg[D]=i} x^{h^0} y^{h^1} + \sum_{i>2g-2} h x^{i+1-g} + \sum_{i<0} h y^{-i-1+g},$$

$$= \sum_{i=0}^{2g-2} \sum_{\deg[D]=i} x^{h^0} y^{h^1} + h \frac{x^g}{1-x} + h \frac{y^g}{1-y}. \tag{1}$$

A different decomposition with a finite term that sums over special divisors only is presented in Section 10.

6   Zeta Functions for Curves

We make the connection between rank polynomials and zeta functions. Let a_i be the number of effective divisors of degree i on X. With Riemann-Roch,

$$a_i = \begin{cases} 0, & \text{for } i < 0. \\ h\,(q^{i+1-g}-1)/(q-1), & \text{for } i > 2g-2. \end{cases}$$

Thus, for $i \notin \{0, 1, \ldots, 2g\}$,

$$a_i - (q+1)a_{i-1} + q a_{i-2} = 0. \tag{2}$$

Define the Hasse-Weil zeta function as the generating function

$$Z(T) = \sum_{i \geq 0} a_i T^i.$$

Let $p_i = a_i - (q+1)a_{i-1} + qa_{i-2}$. With (2), the zeta function can be written as the rational function

$$Z(T) = \frac{P(T)}{(1-T)(1-qT)}, \qquad P(T) = p_0 + p_1 T + \cdots + p_{2g} T^{2g}.$$

Pellikaan [30] defines a two-variable zeta function as the power series

$$Z(T, u) = \sum_{[D]} \frac{u^{l(D)} - 1}{u - 1} T^{\deg D},$$

and gives a decomposition

$$(u-1)Z(T, u)$$

$$= \sum_{i=0}^{2g-2} \sum_{\deg[D]=i} u^{l(D)} T^{\deg D} + \sum_{i>2g-2} hu^{i+1-g} T^i - \sum_{i \geq 0} hT^i. \quad (3)$$

The summation is over divisor classes $[D]$. For a base field $\mathbf{F}$ of q elements and for $u = q$, it agrees with the Hasse-Weil zeta function: $Z(T, q) = Z(T)$. The vanderGeer-Schoof zeta function gives a generalization to number fields. In the version for curves it is defined as

$$\zeta^{GS}(s, t) = \sum_{[D]} q^{sh^0(D)+th^1(D)}.$$

We use it in the form

$$W^{GS}(x, y) = \sum_{[D]} x^{h^0} y^{h^1}.$$

So that $\zeta^{GS}(s, t) = W^{GS}(q^s, q^t)$. Deninger [10, Proposition 2.1] establishes the relation

$$(u-1)T^{1-g}Z(T, u) = W^{GS}(uT, T^{-1}). \quad (4)$$

The short proof uses that the two sides agree termwise for the decompositions (1) and (3). For later use, we write the left side as

$$(u-1)T^{1-g}Z(T, u)$$

$$= \sum_{i=0}^{2g-2} \sum_{\deg[D]=i} (u^{h^0} - 1)T^{h^0-h^1} + \sum_{i>2g-2} h(u^{i+1-g} - 1)T^{i+1-g}. \quad (5)$$

Example 1. The hyperelliptic curve

$$y^2 + y = \frac{x^3 + x}{x^2 + x + 1}$$

over the field of two elements is of genus two with special divisor classes: the zero class, the classes of the five rational points, and the canonical class (Figure 5). It has two-variable zeta function

$$Z(T, u) = \frac{1 + (4 - u)T + (9 - 3u)T^2 + (4u - u^2)T^3 + u^2 T^4}{(1 - T)(1 - uT)}$$

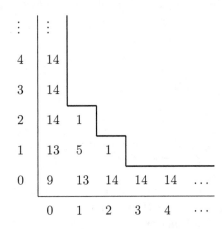

$\vdots$	$\vdots$					
4	14					
3	14					
2	14	1				
1	13	5	1			
0	9	13	14	14	14	$\cdots$
	0	1	2	3	4	$\cdots$

Figure 5: Number of divisor classes with given h^0, h^1 (Example 1).

7   Zeta Functions for Linear Codes

In the previous section we saw that the rank polynomial is defined naturally for a curve and that it is related to the zeta function of the curve. The zeta function of a curve is the generating function for the number of effective divisors. In this section we define the zeta function of a linear code as the generating function for the normalized binomial moments of the code.

Let C be a linear code of length n over the finite field $\mathbf{F}$ of q elements. For a subset $S \subset \{1, 2, \ldots, n\}$ let C_S be the subcode of C of words with support on S. Let k_S denote the dimension of C_S.

$$k_S = \begin{cases} 0, & \text{for } 0 \leq |S| < d. \\ k - (n - |S|), & \text{for } n - d^\perp < |S| \leq n. \end{cases}$$

The number of nonzero words in C_S counted up to multiplication by scalars is given by $(q^{k_S} - 1)/(q - 1)$. Let

$$B_i^1 = \sum_{|S|=i} \frac{q^{k_S} - 1}{q - 1}.$$

The B_i^1 are called the *binomial moments* of the code. As in [34], we have

$$B_i^1 = \begin{cases} 0, & \text{for } 0 \le i < d. \\ \binom{n}{i}(q^{i+k-n} - 1)/(q-1), & \text{for } n - d^\perp < i \le n. \end{cases}$$

Let $b_i = B_{d+i}^1/\binom{n}{d+i}$ be the *normalized binomial moment*. Extend the definition of b_i to all $i \in Z$, by setting

$$b_i = \begin{cases} 0, & \text{for } i < 0. \\ (q^{i+d+k-n} - 1)/(q-1), & \text{for } i > n - d - d^\perp. \end{cases}$$

Define the zeta function as the generating function

$$Z(T) = \sum_{i \ge 0} b_i T^i. \tag{6}$$

For $i \notin \{0, 1, \dots, n - d - d^\perp + 2\}$,

$$b_i - (q+1)b_{i-1} + qb_{i-2} = 0. \tag{7}$$

Let $p_i = b_i - (q+1)b_{i-1} + qb_{i-2}$. With (7), the zeta function can be written as the rational function

$$Z(T) = \frac{P(T)}{(1-T)(1-qT)}, \quad \text{where}$$

$$P(T) = p_0 + p_1 T + \cdots + p_{n+2-d-d^\perp} T^{n+2-d-d^\perp}.$$

For the one-variable zeta function,

$$(q-1)T^{k+d-n} Z(T)$$

$$= \sum_{i=0}^{n} \frac{1}{\binom{n}{i}} \sum_{|S|=i} (q^{ks} - 1)T^{i+k-n} + \sum_{i>n} (q^{i+k-n} - 1)T^{i+k-n}.$$

Define the two-variable zeta function for linear codes by replacing q with the variable u. Then

$$Z(T, u) = \frac{P(T, u)}{(1-T)(1-uT)}$$

is a rational function with $\deg_T P(T, u) = n + 2 - d - d^\perp$, and

$$(u-1)T^{k+d-n} Z(T, u)$$

$$= \sum_{i=0}^{n} \frac{1}{\binom{n}{i}} \sum_{|S|=i} (u^{ks} - 1)T^{i+k-n} + \sum_{i>n} (u^{i+k-n} - 1)T^{i+k-n}. \tag{8}$$

The above form is convenient as a definition, but obviously it can be sharpened to get a smaller finite term.

$$(u-1)T^{k+d-n}Z(T,u)$$

$$= \sum_{i=0}^{n+2-d-d^{\perp}} \frac{1}{\binom{n}{i}} \sum_{|S|=i} (u^{k_S}-1)T^{i+k-n} + \sum_{i>n+2-d-d^{\perp}} (u^{i+k-n}-1)T^{i+k-n}. \quad (9)$$

Example 2. The formally self-dual binary code

$$C = \begin{pmatrix} 1\,0\,0\,0 & 1\,1\,0\,0 \\ 0\,1\,0\,0 & 0\,1\,1\,0 \\ 0\,0\,1\,0 & 0\,0\,1\,1 \\ 0\,0\,0\,1 & 1\,0\,0\,1 \end{pmatrix}$$

has as non-generic column sets: $|S| = 3$ and $k_S = 1$ (4 times), $|S| = 4$ and $k_S = 1$ (25 times), and $|S| = 5$ and $k_S = 2$ (4 times) (See Figure 6). The two-variable zeta function is

$$Z(T,u) = \frac{1}{14} \frac{1 + (4-u)T + (9-3u)T^2 + (4u-u^2)T^3 + u^2T^4}{(1-T)(1-uT)}$$

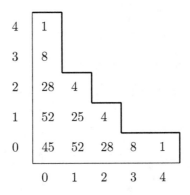

Figure 6: Number of coordinate subsets with given corank, nullity (Example 2).

8   Normalized Rank Functions

For an algebraic curve we have the zeta function $Z(T,u)$ and the rank function $W^{GS}(x,y)$. Equation (4) provides the transformation

$$(u-1)T^{1-g}Z(T,u) = W^{GS}(uT,T^{-1}) \quad (10)$$

For a linear code of length n and dimension k we will show that

$$(u-1)T^{k+d-n}Z(T,u) = W_n^+(uT,T^{-1}), \quad (11)$$

for

$$W_n^+(x,y) = W_n(x,y) + \frac{x^{k+1}}{1-x} + \frac{y^{n-k+1}}{1-y},$$

and

$$W_n(x,y) = \sum_{i=0}^{n} \frac{1}{\binom{n}{i}} \sum_{A \in G, |A|=i} x^{r(G)-r(A)} y^{|A|-r(A)}.$$

Thus Equation (10) for curves holds for linear codes if we normalize the rank polynomial of the linear code and add infinite tails. We call W_n the *normalized rank polynomial* of a matroid and W_n^+ the *normalized rank function* of a matroid. For a general matroid, the length n and dimension k of a code should be replaced with the size n and the rank k of the matroid. The rank polynomial

$$W(x,y) = \sum_A x^{r(G)-r(A)} y^{|A|-r(A)}$$

for a linear code is defined with the matroid of the columns G of a generator matrix. For a subset $S \subset G$ with complement A, let $h^0 = r(G) - r(A)$ and $h^1 = |A| - r(A)$. We rewrite the left side of (11) starting from the expression in (8). With $k_S = \dim(C_S) = r(G) - r(A)$ and $|A| - r(A) = |A| - r(G) + r(G) - r(A) = n - |S| - k + k_S$,

$$(u-1)T^{k+d-n} Z(T,u)$$

$$= \sum_{i=0}^{n} \frac{1}{\binom{n}{i}} \sum_{|S|=i} (u^{h^0} - 1) T^{h^0 - h^1} + \sum_{i>n} (u^{i+k-n} - 1) T^{i+k-n}. \quad (12)$$

Thus

$$(u-1)T^{k+d-n} Z(T,u)$$

$$= \sum_{i=0}^{n} \frac{1}{\binom{n}{i}} \sum_{|S|=i} (uT)^{h^0} (T^{-1})^{h^1} + \sum_{i>n} (uT)^{i+k-n} - \sum_{i \geq 0} T^{i+k-n},$$

$$= W_n(uT, T^{-1}) + \frac{(uT)^{k+1}}{1-uT} - \frac{T^{k-n}}{1-T},$$

$$= W_n^+(uT, T).$$

Equation (12) generalizes the definition of the zeta function $Z(T,u)$ in (8) to matroids. A priori the zeta function has an interpretation only for representable matroids, as generating function for weight distributions. With this definition, (11) holds more generally for matroids. Note that to the minimum distance d of a code corresponds more generally the size of the smallest cocircuit d of a matroid. In dual terms, $n - d$ is the size of the largest hyperplane in a matroid. The genus g in (10) has as corresponding matroid parameter the maximal corank of a cocircuit, or by duality the maximal nullity of a hyperplane.

$$g = \max\{r^*(E) - r^*(C) : C \in \mathcal{C}^*\} = (n-k) - (d-1).$$
$$g = \max\{|H| - r(H) : H \in \mathcal{H}\} = (n-d) - (k-1).$$

Standard terminology refers to circuits of size one as loops and to cocircuits of size one as bridges or isthmuses. A linear code is without bridges if and only if $d > 1$ and without loops if and only if $d^{\perp} > 1$.

Theorem 3. *Let M be a matroid without bridges of size n and rank k with normalized rank polynomial $W_n(x, y)$. The average normalized rank polynomial for the n restricted matroids of size $n - 1$ is $W_n(x, y) - y^{n-k}$. The normalized rank function $W_n^{+}(x, y)$ of a matroid equals the average normalized rank function of its restricted matroids of size $n - 1$.*

Proof. The second claim is immediate from the first. The matroids restricted to $n - 1$ elements have the same rank as the original matroid. The contribution of a subset A with $|A| < n$ is the same for the original and for the averaged rank polynomial. The unique set A of size $|A| = n$ contributes y^{n-k} to the original rank polynomial and 0 to the averaged rank polynomial.

A uniform matroid of size n and rank k has

$$W_n(x, y) = x^k + \cdots + x + 1 + y + \cdots + y^{n-k},$$

$$W_n^{+}(x, y) = \frac{1 - xy}{(1 - x)(1 - y)},$$

$$Z(T, u) = \frac{-T^{-1}}{(1 - uT)(1 - T^{-1})} = \frac{1}{(1 - T)(1 - uT)}.$$

9   Greene's Theorem

The zeta function of a code was defined in Section 7 in terms of its binomial moments, that is to say in terms of its weight enumerator. The significant finite term in the rank function is defined in terms of the rank polynomial of the code. Thus (11) yields a relation between the weight enumerator of a code and its rank polynomial. We show that the relation is Greene's theorem.

First we relate the weight enumerator of a code to its binomial moments. The proof is short and well-known. Let $B^1(x, y) = \sum_{i=0}^{n} B_i^1 x^{n-i} y^i$ be an enumerator for the binomial moments. For a weight enumerator $A(x, y) = x^n + (q - 1)A^1(x, y)$, we claim that $B^1(x, y) = A^1(x + y, y)$. The support of a codeword of weight i contributes one to B_j^1 for each subset S of size j that contains the support. So that $B_j^1 = \sum_i \binom{n-i}{n-j} A_i^1$, and $B^1(x, y) = A^1(x + y, y)$. Let $B(x, y) = (x + y)^n + (q - 1)B^1(x, y) = A(x + y, y)$. Let

$$A_n(x, y) = \sum_{i=0}^{n} \frac{1}{\binom{n}{i}} A_i x^{n-i} y^i. \tag{13}$$

Define $A_n^1(x, y), B_n(x, y), B_n^1(x, y)$ in the same way. From the definition of the zeta function,

$$(q - 1)Z(T)T^d \equiv B_n^1(1, T) \pmod{T^{n+1}}.$$

With (11),

$$W_n^+(qT, T^{-1})T^{n-k} \equiv B_n^1(1, T) \pmod{T^{n+1}}. \tag{14}$$

Adding

$$T^{n-k}\left(\frac{-(T^{-1})^{n-k+1}}{1 - T^{-1}}\right) = 1/(1 - T)$$

to both sides gives an equality among polynomials of degree n.

$$W_n(qT, T^{-1})T^{n-k} = B_n(1, T) \tag{15}$$

The normalization on both sides is the same at each coefficient of T^i and the relation holds if we replace W_n and B_n with W and B, respectively. We find

$$B(x, y) = y^{n-k}x^k W(q\frac{y}{x}, \frac{x}{y})). \tag{16}$$

Finally $A(x + y, y) = B(x, y)$ gives Greene's theorem.

$$\frac{A(x, y)}{(x - y)^k y^{n-k}} = W(\frac{qy}{x - y}, \frac{x - y}{y}).$$

The purpose of the above proof is to make the relation between (11) and Greene's theorem explicit. Relation (16), that provides the connection, has a direct and much shorter proof. Let C_i^l be the number of $S \subset \{1, 2, \ldots, n\}$ with $|S| = i$ and $k_S = \dim(C_S) = l$. Then

$$B^1(x, y) = \sum_i \sum_l C_i^l \frac{q^l - 1}{q - 1}x^{n-i}y^i, \qquad B(x, y) = \sum_i \sum_l C_i^l q^l x^{n-i}y^i.$$

On the other hand, let S have complement A in G. As in the previous section, we use the rank polynomial

$$W(x, y) = \sum_A x^{r(G)-r(A)}y^{|A|-r(A)}$$

with $r(G) - r(A) = k_S$ and $|A| - r(A) = n - |S| - k + k_S$.

$$W(x, y) = y^{-k} \sum_i \sum_l C_i^l (xy)^l y^{n-i}.$$

And (16) follows. By using (15) we find a normalized version of Greene's theorem. For $A(x + y, y) = B(x, y)$ homogeneous of degree n,

$$A_n(1, \frac{T}{1 - T}) \equiv B_n(1, T)(1 - T) \pmod{T^{n+1}}.$$

As in [12],

$$A_n(1, t)(1 + t)^{n+1} \equiv W_n(\frac{qt}{1+t}, \frac{1+t}{t})(1 + t)^{k+1}t^{n-k} \pmod{t^{n+1}}.$$

In another form, using (14),

$$A_n^1(1,t)(1+t)^{d+1}t^{-d}$$

$$\equiv W_n^+(\frac{qt}{1+t}, \frac{1+t}{t})(1+t)^{k+d\,|\,1-n}t^{n\ \ k-d} \quad (\mathrm{mod}\ t^{n-d+1}). \quad (17)$$

Theorem 4 ([11]). *Let C be a linear code of length n with minimum distance d > 2. The expression $A_n^1(1,t)(1+t)^{d+1}t^{-d}$ modulo t^{n-d+1} is the same for the code and for the average over its n punctured codes (obtained by restriction of the code to $n-1$ coordinates).*

Proof. Apply Theorem 3 to the right side of (17).

10   Decomposition of Zeta Functions

The decomposition of W^{GS} in (1) and of $Z(T,u)$ in (3) uses as finite term a summation over all divisor classes of degree $0 \leq [D] \leq 2g-2$. This decomposition goes back to Weil [40] and is followed in [24], [33]. Duursma [14, Lemma 5] and vanderGeer-Schoof [37, Section 2] use a different decomposition. Two properties of divisors ensure that $W^{GS}(x,y)$ is a rational function.

(W1) The exponents $h^0(D)$ and $h^1(D)$ are nonnegative, such that for almost all $[D]$ either $h^0 = 0$ or $h^1 = 0$.

(W2) The number of $[D]$ with given difference $h^0 - h^1$ is constant and equal to h.

Starting from these properties, we define as finite term

$$W^*(x,y) = \sum_{h^0 \geq h^1}{}'(x^{h^0}y^{h^1} - x^{h^0-h^1}) + \sum_{h^1 \geq h^0}{}'(x^{h^0}y^{h^1} - y^{h^1-h^0}).$$

As in [37], divisors with $h^0 = h^1$ in $\sum'$ are counted with multiplicity 1/2.

$$W^{GS}(x,y) = W^*(x,y) + \sum_{h^0 \geq h^1}{}'x^{h^0-h^1} + \sum_{h^1 \geq h^0}{}'y^{h^1-h^0},$$

$$= W^*(x,y) + h\frac{1-xy}{(1-x)(1-y)}.$$

The tail in $W^{GS}(x,y)$ assumes that all h divisor classes of a given degree are non-special. The finite term $W^*(x,y)$ contains the necessary corrections for the special divisors. The same decomposition applied to $Z(T,u)$ yields

$$Z(T,u) = \sum_{h^0 \leq h^1}{}' \frac{u^{h^0}-1}{u-1} T^{h^0-h^1+g-1}$$

$$+ \sum_{h^0 \geq h^1}{}' \frac{u^{h^1}-1}{u-1} u^{h^0-h^1} T^{h^0-h^1+g-1} + \sum_{h^0 \geq h^1}{}' \frac{u^{h^0-h^1}-1}{u-1} T^{h^0-h^1+g-1}.$$

After collecting the variables uT, T^{-1}

$$(u-1)T^{1-g}Z(T,u) = \sum_{h^0 \leq h^1}{}' T^{-h^1}\left((uT)^{h^0} - T^{h^0}\right) +$$

$$+ \sum_{h^0 \geq h^1}{}' (uT)^{h^0}\left(T^{-h^1} - (uT)^{-h^1}\right) + \sum_{h^0 \geq h^1}{}' (uT)^{h^0-h^1} - T^{h^0-h^1}$$

$$= W^{GS}(uT, T^{-1}).$$

And we recover (10). As in [14], we define $Z^*(T, u)$ via

$$Z(T,u) = Z^*(T,u) + \frac{hT^g}{(1-T)(1-uT)}. \tag{18}$$

So that $Z^*(T, u)$ is a polynomial with contributions by special divisors only. The decomposition is compatible with the decomposition of $W(x, y)$,

$$(u-1)T^{1-g}Z^*(u,T) = W^*(uT, T^{-1}).$$

Conditions (W1) and (W2) hold for the normalized rank function W_n^+ of a matroid, for $h = 1$. And the above decomposition carries through. For a matroid we find

$$W_n^+(x,y) = W_n^*(x,y) + \frac{1-xy}{(1-x)(1-y)},$$

$$Z(T,u) = Z^*(T,u) + \frac{T^g}{(1-T)(1-uT)},$$

for polynomials $W_n^*(x,y)$ and $Z^*(T,u)$, such that W_n^* is the normalization of the polynomial

$$W^*(x,y) = \sum_{h^0 \geq h^1 > 0}{}' x^{h^0-h^1}\left((xy)^{h^1} - 1\right) + \sum_{h^1 \geq h^0 > 0}{}' y^{h^1-h^0}\left((xy)^{h^0} - 1\right).$$

11   Support Weights

With a linear code is associated a second family of weight enumerators, the so called support weight enumerators. They first appeared in [21] and [18]. They became object of intense study after Wei described an application to cryptography [39]. For an overview of results we refer to the paper by Tsfasman and Vladuts [34]. The papers [13], [3] make a connection with matroids. A linear subspace $D \subset C$ has support

$$\text{Supp}(D) = \{i : \exists v \in D, v_i \neq 0\}.$$

The support weight of D is $|\text{Supp}(D)|$ (also called effective length or generalized Hamming weight). The j-th support weight enumerator is defined as

$$A^j(x,y) = \sum_{D \subset C, \dim D = j} x^{n-|\text{Supp}(D)|} y^{|\text{Supp}(D)|} = \sum_{i=0}^n A_i^j x^{n-i} y^i.$$

For the weight enumerator $A(x, y)$ of the code, we find $A(x, y) = A^0(x, y) + (q - 1)A^1(x, y)$. The support weight enumerators are determined by the rank polynomial of the matroid of the code. But there is no immediate analogue of Greene's theorem and unlike the usual weight enumerators, support weight enumerators can in general not be obtained as the evaluation of the rank polynomial with suitable arguments. But they can be written as a linear combination of such evaluations. For this we need to consider the linear codes $C^{(m)}$ that are generated by the code C but have their coefficients in the extension field $\mathbf{F}_{q^m}$. Klove [22, Lemma 4] gives for the weight enumerator of $C^{(m)}$,

$$A(C^{(m)})(x, y) = \sum_{j=0}^{m} [m]_j A^j(x, y).$$

where $[m]_j = (q^m - 1)(q^m - q) \cdots (q^m - q^{j-1})$. Let $A^{(m)}(x, y)$ be the first support weight enumerator for the code $C^{(m)}$. That is, $A(C^{(m)})(x, y) = x^n + (q^m - 1)A^{(m)}(x, y)$. The above relation becomes

$$[m]_1 A^{(m)}(x, y) = \sum_{j=1}^{m} [m]_j A^j(x, y). \tag{19}$$

The relation appears in [21]. Theorem 3.6 [34] is similar but lacks the factor $[m]_1 = (q^m - 1)$. The relation is invertible so that $A^j(x, y)$ can be expressed as a linear combination of $A^{(1)}(x, y), \dots, A^{(j)}(x, y)$.
Let C_i^l be the number of $S \subset \{1, 2, \dots, n\}$ with $k_S = \dim(C_S) = l$ and $|S| = i$. The Gaussian coefficient

$$\begin{bmatrix} l \\ j \end{bmatrix}_q = \frac{(q^l - 1)(q^l - q) \cdots (q^l - q^{j-1})}{(q^j - 1)(q^j - q) \cdots (q^j - q^{j-1})}$$

gives the number of j-dimensional subspaces in a l-dimensional vector space over $\mathbf{F}_q$. For the code $C^{(m)}$ we define binomial moments $B_i^{(m)}$.

$$B^{(m)}(x, y) = \sum_i \sum_{|S|=i} \begin{bmatrix} k_S \\ 1 \end{bmatrix}_{q^m} x^{n-i} y^i =$$

$$= \sum_i \sum_l C_i^l \begin{bmatrix} l \\ 1 \end{bmatrix}_{q^m} x^{n-i} y^i = \sum_i B_i^{(m)} x^{n-i} y^i.$$

Then $B^{(m)}(x, y) = A^{(m)}(x + y, y)$. For the support weights, define binomial moments B_i^j via

$$B^j(x, y) = \sum_i \sum_{|S|=i} \begin{bmatrix} k_S \\ j \end{bmatrix}_q x^{n-i} y^i =$$

$$= \sum_i \sum_l C_i^l \begin{bmatrix} l \\ j \end{bmatrix}_q x^{n-i} y^i = \sum_i B_i^j x^{n-i} y^i.$$

Then $B^j(x, y) = A^j(x + y, y)$. As in [3], Equation (19) now follows from the relation for Gaussian coefficients

$$[m]_1 \begin{bmatrix} l \\ 1 \end{bmatrix}_{q^m} = q^{ml} - 1 = \sum_{j=1}^{m} [m]_j \begin{bmatrix} l \\ j \end{bmatrix}_q.$$

The left side is the number of nonzero row vectors of length l over $\mathbf{F}_{q^m}$. The right side is the number of nonzero $m \times l$-matrices over $\mathbf{F}_q$ counted by their rank j.

The zeta function $Z(T)$ of a linear code was defined in (6) as the generating function for the normalized binomial moments.

$$[x^{n-i} y^i T^{n-d}] Z(T)(y + xT)^n = \binom{n}{i} b_{i-d} = B_i^1 = B_i^{(1)}.$$

In general, for a code $C^{(m)}$ over the extension field $\mathbf{F}_{q^m}$ we have

$$[T^{n-d}] Z(T, q^m)(y + xT)^n = B^{(m)}(x, y).$$

The decomposition $[m]_1 B^{(m)}(x, y) = \sum_{j=1}^{m} [m]_j B^j(x, y)$ implies the existence of generating functions $Z_j(T)$ with

$$[m]_1 Z(T, q^m) = \sum_{j=1}^{m} [m]_j Z_j(T), \tag{20}$$

$$[T^{n-d}] Z_j(T)(y + xT)^n = B^j(x, y). \tag{21}$$

To describe $Z_j(T)$ we make use of (18). For the polynomial part, let

$$(u - 1) Z^*(T, u) = \sum_{j=1}^{g^\perp} [u]_j Z_j^*(T), \tag{22}$$

where $[u]_j = (u - 1)(u - q) \cdots (u - q^{j-1})$. So that the right side gives the Newton decomposition of the polynomial $(u - 1) Z^*(T, u)$ considered as polynomial of degree $g^\perp$ in u with respect to the sequence $(1, q, q^2, \dots, q^{g^\perp - 1})$. For the rational part, let

$$G_j(T) = \sum_{l \geq 0} \begin{bmatrix} l \\ j \end{bmatrix}_q T^l = \frac{T^{j-1}}{(1 - T)(1 - qT) \cdots (1 - q^j T)} \tag{23}$$

be a generating function for the Gaussian coefficients. Then

$$[m]_1 \frac{1}{(1 - T)(1 - q^m T)} = \sum_{j=1}^{m} [m]_j G_j(T).$$

Adding the two contributions yields

Theorem 5. *The generating function $Z_j(T)$ for the j-th support weights, as defined by (21), can be written as*

$$Z_j(T) = Z_j^*(T) + hT^g G_j(T) \qquad (h = 1, g = n + 1 - k - d)$$

for a polynomial $Z_j^*(T)$ as in (22) and for $G_j(T)$ as in (23). For $j > g^\perp$, $Z_j^*(T)$ vanishes.

Example 3. The first order Reed-Muller code $[16, 5, 8]$ has

$$\begin{aligned} Z^*(T, u) &= \frac{1}{429} + \frac{3}{143}T + \frac{15}{143}T^2 + \frac{5}{13}T^3 + \frac{1}{13}uT^4 \\ &= Z_1^*(T) + (u - 2) Z_2^*(T). \end{aligned}$$

Its dual, the second order Reed-Muller code $[16, 11, 4]$, has

$$\begin{aligned} Z^*(T, u) &= \frac{1}{13} + \frac{5}{13}T + \frac{15}{143}uT^2 + \frac{3}{143}u^2T^3 + \frac{1}{429}u^3T^4 \\ &= Z_1^*(T) + (u - 2) Z_2^*(T) + (u - 2)(u - 4) Z_3^*(T) \\ &\quad + (u - 2)(u - 4)(u - 8) Z_4^*(T). \end{aligned}$$

The $g^\perp$ polynomials $Z_1^*(T), \dots, Z_{g^\perp}^*(T)$ determine $Z(T, u)$ and thus the weight distribution of $C^{(m)}$ for all $m \geq 1$. But some of the $Z_j^*(T)$ may be redundant.

Theorem 6. *The two-variable zeta function $Z(T, u)$ is determined by its values for $u = q, \dots, q^\lambda$, or equivalently by $Z_1^*(T), \dots, Z_\lambda^*(T)$, where*

$$\lambda - \max\{h^0 : [x^{h^0} y^{h^0}]W(x, y) \neq 0\} \leq \min\{g, g^\perp\}.$$

In particular, for a linear code $\lambda = \max\{k_S : |S| = n - k\}$.

Proof. Let λ be minimal such that for all elements in the matroid either the corank or the nullity is at most λ. Then $W(x, y)$ can be obtained from its evaluations on λ distinct hyperbolas $xy = q_1, q_2, \dots, q_\lambda$.

The problem of how many weight distributions determine all remaining weight distributions is posed in [21]. That paper suggests a solution that is one weaker than the proposition. Also, the solution depends on an unproved condition. The condition always holds and a minor modification strengthens the solution by one, so that it agrees with the Theorem. In Example 3, $\lambda = 1$.

The r-th minimum support weight $d_r(C)$ of the code C is defined as

$$d_r(C) = \min\{|\mathrm{Supp}(D)| : \dim D = r\} = \min\{|S| : k_S = r\}.$$

Wei gives a relation for the weight hierarchy of a code and its dual [39].

$$\begin{aligned} \{1, 2, \dots, n\} &= \{d_r(C) : r = 1, \dots, k\} \\ &\cup \{n + 1 - d_r(C^\perp) : r = 1, \dots, n - k\}. \quad (24) \end{aligned}$$

In terms of the rank polynomial of the code

$$d_r(C) = \min\{n - k + r - s : [x^r y^s] W_C(x, y) \neq 0\}. \tag{25}$$

In Figure 7, the minimum support weights correspond to the k horizontal steps in the path of length n from the closed dot to the open dot. The minimum support weights of the dual code correspond to the $n - k$ vertical steps of the same path in the opposite direction. This proves (24).

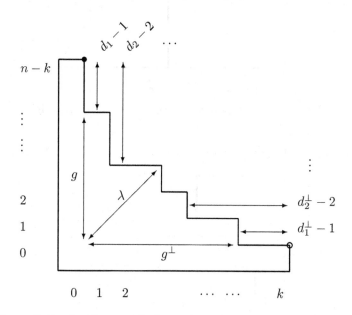

Figure 7: Rank polynomial of a code and its weight hierarchy.

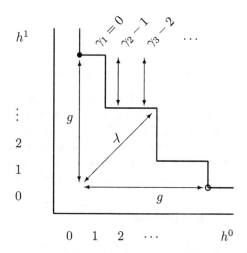

Figure 8: Rank polynomial of a curve and its gonality sequence.

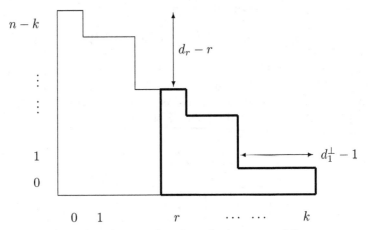

Figure 9: Rank polynomial and $r-$th support weights.

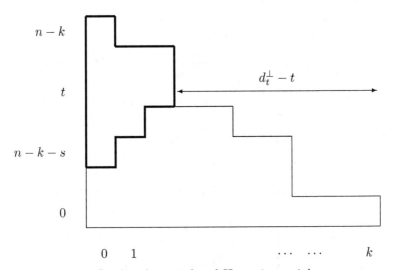

Figure 10: Rank polynomial and Hamming weights up to s.

Figure 9 indicates the part of the rank polynomial that is needed to enumerate the support weights of subspaces of dimension at least r in a code. Figure 10 indicates the part of the rank polynomial that is needed to enumerate words of weight at most s in a code. The part of the rank polynomial that enumerates the words of weight at most s enumerates the support weights of dimension t in the dual code for $k - (d_t^\perp - t) \leq t - (n - k - s)$, or for $d_t(C^*) \geq n - s$.

12   Algebraic-Geometric Codes

Let X be a curve, D a divisor and $\mathcal{P}=(P_1, P_2, \dots, P_n)$ a list of n rational points on the curve. The code $C(D, \mathcal{P})$ is the image of $L(D)$ under evaluation in $\mathcal{P}$.

The rank polynomial of the code is

$$W_C(x, y) = \sum_{E \subset \{1,2,\dots,n\}} x^{h^0} y^{h^1}$$

where h^0 and h^1 are the corank and nullity of a subset, respectively. Define the geometric rank polynomial of the code as

$$W_X(x, y) = \sum_{0 \le E \le P_1 + \cdots + P_n} x^{l(D-E)} y^{i(D-E)}.$$

We establish the relation between W_C and W_X. The dimension of the code and its dual are

$$k = l(D) - l(D - (P_1 + \cdots + P_n)), \quad k^{\perp} = i(D - (P_1 + \cdots + P_n)) - i(D).$$

Let $a = l(D - (P_1 + \cdots + P_n))$ and $a^{\perp} = i(D)$ be the abundance of the code and its dual, respectively [29]. With

$$h^0(E) = k - (l(D) - l(D - E)) = l(D - E) - a,$$
$$h^1(E) = |E| - (l(D) - l(D - E)) = i(D - E) - a^{\perp},$$

we find

$$W_X(x, y) = x^a y^{a^{\perp}} W_C(x, y). \tag{26}$$

The gonality sequence $\{\gamma_t\}$ of a curve is defined with $\gamma_t = \min\{\deg F : l(F) \ge t\}$. Figure 8 shows the relation between the rank function of a curve and its gonality sequence. Combination of (26) with (25) yields Munuera's lower bound for the r-th Hamming distance d_r of a code in terms of the gonality sequence of a curve [25], [42]. With $r - s = h^0 - h^1 = k - |E|$,

$$d_r(C) = \min\{n - |E| : l(D - E) = r + a \text{ and } i(D - E) = s + a^{\perp}\},$$
$$\ge \min\{n - |E| : l(D - E) = r + a\},$$
$$\ge n - \deg D + \gamma_{r+a}.$$

Next, consider a family of codes $C_1 = C, C_2, \dots, C_h$ defined with inequivalent divisors $D_1 = D, D_2, \dots, D_h$ of the same degree $2g - 2 < \deg(D) < n$. All codes have $a = a^{\perp} = 0$.

$$\sum_{i=1}^{h} W_{n,C_i}(x, y) = \sum_{j=0}^{n} \frac{1}{\binom{n}{j}} \sum_{\substack{0 \le E \le P \\ \deg E = j}} \sum_{i=1}^{h} x^{l(D_i - E)} y^{i(D_i - E)},$$

$$= \sum_{j=0}^{n} \sum_{\deg[E]=j} x^{l(D-E)} y^{i(D-E)},$$

$$= \sum_{\deg D - n \le \deg[F] \le \deg D} x^{l(F)} y^{i(F)}.$$

The last summation is over all divisor classes with degree in the given range. The range includes the interval $0 \le \deg[F] \le 2g - 2$. Adding

$$h\frac{x^{k+1}}{1-x} + h\frac{y^{n-k+1}}{1-y} = h\frac{x^{\deg D+2-g}}{1-x} + h\frac{y^{n-\deg D+g}}{1-y}$$

to the equation yields

$$\sum_{i=1}^{h} W_{n,C_i}^{+}(x,y) = W^{GS}(x,y). \tag{27}$$

Theorem 7. *Let $C_1, \dots, C_h$ be codes defined with inequivalent divisors of the same degree $2g - 2 < \deg D < n$. The two-variable zeta function of the h linear codes and the two-variable zeta function of the curve satisfy*

$$\sum_{i=1}^{h} T^{g-g(C_i)} Z_{C_i}(T,u) = Z(T,u)$$

Proof. This follows from (27) with (10) and (11). Note that the Goppa bound for codes gives $g(C_i) = n + 1 - k - d \le g$.

References

1. Alexei Ashikhmin and Alexander Barg. Binomial moments of the distance distribution: bounds and applications. *IEEE Trans. Inform. Theory*, 45(2):438–452, 1999.
2. F. Baldassarri, C. Deninger, and N. Naumann. A motivic version of Pellikaan's two variable zeta function. arXiv:math.AG/0302121.
3. Alexander Barg. The matroid of supports of a linear code. *Appl. Algebra Engrg. Comm. Comput.*, 8(2):165–172, 1997.
4. Alexander Barg. On some polynomials related to weight enumerators of linear codes. *SIAM J. Discrete Math.*, 15(2):155–164 (electronic), 2002.
5. Hyman Bass. The Ihara-Selberg zeta function of a tree lattice. *Internat. J. Math.*, 3(6):717–797, 1992.
6. Béla Bollobás. *Modern graph theory*, volume 184 of *Graduate Texts in Mathematics*. Springer-Verlag, New York, 1998.
7. Thomas Brylawski and James Oxley. The Tutte polynomial and its applications. In *Matroid applications*, volume 40 of *Encyclopedia Math. Appl.*, pages 123–225. Cambridge Univ. Press, Cambridge, 1992.
8. Fan R. K. Chung. *Spectral graph theory*, volume 92 of *CBMS Regional Conference Series in Mathematics*. Published for the Conference Board of the Mathematical Sciences, Washington, DC, 1997.
9. Henry H. Crapo and Gian-Carlo Rota. *On the foundations of combinatorial theory: Combinatorial geometries*. The M.I.T. Press, Cambridge, Mass.-London, preliminary edition, 1970.
10. Christopher Deninger. Two-variable zeta functions and regularized products. arXiv:math.NT/0210269.

11. Iwan Duursma. From weight enumerators to zeta functions. *Discrete Appl. Math.*, 111(1-2):55–73, 2001.
12. Iwan M. Duursma. Results on zeta functions for codes. In proceedings: Fifth Conference on Algebraic Geometry, Number Theory, Coding Theory and Cryptography, University of Tokyo, January 17-19, 2003. arXiv:math.CO/0302172.
13. Iwan M. Duursma. Zeta functions for linear codes. preprint, 1997.
14. Iwan M. Duursma. Weight distributions of geometric Goppa codes. *Trans. Amer. Math. Soc.*, 351(9):3609–3639, 1999.
15. Jack Edmonds. Lehman's switching game and a theorem of Tutte and Nash-Williams. *J. Res. Nat. Bur. Standards Sect. B*, 69B:73–77, 1965.
16. Chris Godsil and Gordon Royle. *Algebraic graph theory*, volume 207 of *Graduate Texts in Mathematics*. Springer-Verlag, New York, 2001.
17. Curtis Greene. Weight enumeration and the geometry of linear codes. *Studies in Appl. Math.*, 55(2):119–128, 1976.
18. Tor Helleseth, Torleiv Kløve, and Johannes Mykkeltveit. The weight distribution of irreducible cyclic codes with block length $n_1((q^l-1)/N)$. *Discrete Math.*, 18(2):179–211, 1977.
19. J. W. Hoffman. Remarks on the zeta function of a graph. LSU Mathematics Electronic Preprint Series 2002-17.
20. François Jaeger. On Tutte polynomials of matroids representable over GF(q). *European J. Combin.*, 10(3):247–255, 1989.
21. Torleiv Kløve. The weight distribution of linear codes over GF(q^l) having generator matrix over GF(q). *Discrete Math.*, 23(2):159–168, 1978.
22. Torleiv Kløve. Support weight distribution of linear codes. *Discrete Math.*, 106/107:311–316, 1992. A collection of contributions in honour of Jack van Lint.
23. F. J. MacWilliams and N. J. A. Sloane. *The theory of error-correcting codes.* North-Holland Publishing Co., Amsterdam, 1977. North-Holland Mathematical Library, Vol. 16.
24. Carlos Moreno. *Algebraic curves over finite fields*, volume 97 of *Cambridge Tracts in Mathematics*. Cambridge University Press, Cambridge, 1991.
25. Carlos Munuera. On the generalized Hamming weights of geometric Goppa codes. *IEEE Trans. Inform. Theory*, 40(6):2092–2099, 1994.
26. Niko Naumann. On the irreducibility of the two variable zeta-function for curves over finite fields. arXiv:math.AG/0209092.
27. James Oxley. What is a matroid? LSU Mathematics Electronic Preprint Series 2002-9.
28. James G. Oxley. *Matroid theory.* Oxford Science Publications. The Clarendon Press Oxford University Press, New York, 1992.
29. Ruud Pellikaan. On the gonality of curves, abundant codes and decoding. In *Coding theory and algebraic geometry (Luminy, 1991)*, volume 1518 of *Lecture Notes in Math.*, pages 132–144. Springer, Berlin, 1992.
30. Ruud Pellikaan. On special divisors and the two variable zeta function of algebraic curves over finite fields. In *Arithmetic, geometry and coding theory (Luminy, 1993)*, pages 175–184. de Gruyter, Berlin, 1996.
31. P. Rosenstiehl and R. C. Read. On the principal edge tripartition of a graph. *Ann. Discrete Math.*, 3:195–226, 1978. Advances in graph theory (Cambridge Combinatorial Conf., Trinity College, Cambridge, 1977).
32. Juriaan Simonis. The effective length of subcodes. *Appl. Algebra Engrg. Comm. Comput.*, 5(6):371–377, 1994.
33. Henning Stichtenoth. *Algebraic function fields and codes.* Universitext. Springer-Verlag, Berlin, 1993.

34. Michael A. Tsfasman and Serge G. Vlăduţ. Geometric approach to higher weights. *IEEE Trans. Inform. Theory*, 41(6, part 1):1564–1588, 1995. Special issue on algebraic geometry codes.
35. W. T. Tutte. A ring in graph theory. *Proc. Cambridge Philos. Soc.*, 43:26–40, 1947.
36. W. T. Tutte. On the algebraic theory of graph colorings. *J. Combinatorial Theory*, 1:15–50, 1966.
37. Gerard van der Geer and René Schoof. Effectivity of Arakelov divisors and the theta divisor of a number field. *Selecta Math. (N.S.)*, 6(4):377–398, 2000.
38. J. H. van Lint. *Introduction to coding theory*, volume 86 of *Graduate Texts in Mathematics*. Springer-Verlag, Berlin, third edition, 1999.
39. Victor K. Wei. Generalized Hamming weights for linear codes. *IEEE Trans. Inform. Theory*, 37(5):1412–1418, 1991.
40. André Weil. *Sur les courbes algébriques et les variétés qui s'en déduisent*. Actualités Sci. Ind., no. 1041 = Publ. Inst. Math. Univ. Strasbourg **7** (1945). Hermann et Cie., Paris, 1948.
41. D. J. A. Welsh. *Matroid theory*. Academic Press [Harcourt Brace Jovanovich Publishers], London, 1976. L. M. S. Monographs, No. 8.
42. Kyeongcheol Yang, P. Vijay Kumar, and Henning Stichtenoth. On the weight hierarchy of geometric Goppa codes. *IEEE Trans. Inform. Theory*, 40(3):913–920, 1994.

Constructions of Mutually Unbiased Bases

Andreas Klappenecker[1] and Martin Rötteler[2]

[1] Department of Computer Science,
Texas A&M University, College Station, TX 77843-3112, USA
klappi@cs.tamu.edu
[2] Department of Combinatorics and Optimization,
University of Waterloo, Waterloo, Ontario, Canada, N2L 3G1
mroetteler@math.uwaterloo.ca

Abstract. Two orthonormal bases B and B' of a d-dimensional complex inner-product space are called mutually unbiased if and only if $|\langle b|b'\rangle|^2 = 1/d$ holds for all $b \in B$ and $b' \in B'$. The size of any set containing pairwise mutually unbiased bases of $\mathbb{C}^d$ cannot exceed $d+1$. If d is a power of a prime, then extremal sets containing $d+1$ mutually unbiased bases are known to exist. We give a simplified proof of this fact based on the estimation of exponential sums. We discuss conjectures and open problems concerning the maximal number of mutually unbiased bases for arbitrary dimensions.

Key words: Quantum cryptography, quantum state estimation, Weil sums, finite fields, Galois rings.

1 Motivation

The notion of mutually unbiased bases emerged in the literature of quantum mechanics in 1960 in the works of Schwinger [18]. Two orthonormal bases B and B' of the vector space $\mathbb{C}^d$ are called *mutually unbiased* if and only if $|\langle b | b' \rangle|^2 = 1/d$ holds for all $b \in B$ and all $b' \in B'$; here $\langle b | b' \rangle$ denotes the standard hermitian inner product on the complex vector space $\mathbb{C}^d$ that is anti-linear in the first argument and linear in the second, $\langle b | b' \rangle = b^\dagger b'$. Schwinger realized that no information can be retrieved when a quantum system which is prepared in a basis state from B' is measured with respect to the basis B. A striking application is the protocol by Bennett and Brassard [5] which exploits this observation to distribute secret keys over a public channel in an information-theoretically secure way (see also [4]).

Any collection of pairwise mutually unbiased bases of $\mathbb{C}^d$ has cardinality $d + 1$ or less, see [3, 11, 13, 15, 22]. Extremal sets attaining this bound are of considerable interest. Ivanović showed that the density matrix of an ensemble of d-dimensional quantum systems can be completely reconstructed from the statistics of measurements with respect to $d + 1$ mutually unbiased bases [14]. Furthermore, he showed that the density matrix cannot be reconstructed from the statistics of fewer measurements.

G. Mullen, A. Poli and H. Stichtenoth (Eds.): Fq7 2003, LNCS 2948, pp. 137–144, 2003.

Let $N(d)$ denote the maximum cardinality of any set containing pairwise mutually unbiased bases of $\mathbb{C}^d$. It is known that $N(d) = d + 1$ holds when d is a prime power, see [3, 14, 22]. We derive a simplified proof of this result, which takes advantage of Weil-type exponential sums. We present two different constructions—both based on Weil sums over finite fields—in the case of odd prime power dimensions. We exploit exponential sums over Galois rings in the case of even prime power dimensions. If the dimension d is not a prime power, then the exact value of $N(d)$ is not known. We discuss lower bounds, conjectures, and open problems in the fourth section.

2   Odd Prime Powers

Let $\mathbb{F}_q$ be a finite field with q elements which has odd characteristic p. Denote the absolute trace from $\mathbb{F}_q$ to the prime field $\mathbb{F}_p$ by $\mathrm{tr}(\,\cdot\,)$. Each nonzero element $x \in \mathbb{F}_q$ defines a non-trivial additive character $\mathbb{F}_q \to \mathbb{C}^\times$ by

$$y \mapsto \omega_p^{\mathrm{tr}(xy)},$$

where $\omega_p = \exp(2\pi i/p)$ is a primitive p-th root of unity. All non-trivial additive characters are of this form.

Lemma 1 (Weil sums). *Let $\mathbb{F}_q$ be a finite field of odd characteristic and χ a non-trivial additive character of $\mathbb{F}_q$. Let $p(X) \in \mathbb{F}_q[X]$ be a polynomial of degree 2. Then*

$$\left| \sum_{x \in \mathbb{F}_q} \chi(p(x)) \right| = \sqrt{q}.$$

We refer to [17, Theorem 5.37] or [7, p. 313] for a proof. We will use this lemma in the following constructions of mutually unbiased bases.

Convention. In the following, we will tacitly assume that the elements of $\mathbb{F}_q$ are listed in some fixed order, and this order will be used whenever an object indexed by elements of $\mathbb{F}_q$ appears.

We begin with a historical curiosity. Schwinger introduced the concept of mutually unbiased bases in 1960. However, he did not construct extremal sets of mutually unbiased bases, except in low dimensions, and no further progress was made during the next twenty years. Alltop constructed in 1980 complex sequences with low correlation for spread spectrum radar and communication applications [1]. It turns out that the sequences given by Alltop provide $p + 1$ mutually unbiased bases in dimension p, for all primes $p \geq 5$. Unfortunately, Alltop was not aware of his contribution to quantum physics, and his work was not noticed until recently. Our first construction generalizes the Alltop sequences to prime power dimensions.

Theorem 1. *Let $\mathbb{F}_q$ be a finite field of characteristic $p \geq 5$. Let B_α denote the set of vectors*

$$B_\alpha = \{b_{\lambda,\alpha} \mid \lambda \in \mathbb{F}_q\}, \qquad b_{\lambda,\alpha} = \frac{1}{\sqrt{q}}\left(\omega_p^{\mathrm{tr}((k+\alpha)^3 + \lambda(k+\alpha))}\right)_{k \in \mathbb{F}_q}.$$

The standard basis and the sets B_α, with $\alpha \in \mathbb{F}_q$, form an extremal set of $q+1$ mutually unbiased bases of the vector space $\mathbb{C}^q$.

Proof. Notice that B_α is an orthonormal basis because

$$\langle b_{\kappa,\alpha} \mid b_{\lambda,\alpha}\rangle = \frac{1}{q}\sum_{k \in \mathbb{F}_q} \omega_p^{\mathrm{tr}((\lambda-\kappa)(k+\alpha))}.$$

Indeed, the right hand side equals 0 when $\kappa \neq \lambda$ because the argument $k + \alpha$ ranges through all values of $\mathbb{F}_q$; and equals 1 when $\kappa = \lambda$.

Note that all components of the sequence $b_{\lambda,\alpha}$ have absolute value $1/\sqrt{q}$, hence the basis B_α and the standard basis are mutually unbiased, for any $\alpha \in \mathbb{F}_q$.

By computing the inner product $|\langle b_{\kappa,\alpha} \mid b_{\lambda,\beta}\rangle|$ for $\alpha \neq \beta$, we see that the terms cubic in k cancel out and, moreover, that the exponent is given by the trace of a quadratic polynomial in k. By Lemma 1 the inner product evaluates to $q^{-1/2}$, hence B_α and B_β are mutually unbiased. $\square$

Remark 1. A remarkable feature of the previous construction is that knowledge of one basis B_α is sufficient because shifting the indices by adding a field element yields the other bases. The construction does not work in characteristic 2 and 3 because in these cases the sets B_α and B_β, with $\alpha \neq \beta$, are not mutually unbiased.

Ivanović gave a fresh impetus to the field in 1981 with his seminal paper [14]. Among other things, he gave explicit constructions of $p+1$ mutually unbiased bases of $\mathbb{C}^p$, for p a prime. His construction was later generalized in the influential paper by Wootters and Fields [22], who gave the first proof of the following theorem. This proof was recently rephrased by Chaturvedi [9], and an alternate proof was given by Bandyopadhyay et al. [3]. We give a particularly short proof by taking advantage of Weil sums.

Theorem 2. *Let $\mathbb{F}_q$ be a finite field with odd characteristic p. Denote by $B_a = \{v_{a,b} \mid b \in \mathbb{F}_q\}$ the set of vectors given by*

$$v_{a,b} = q^{-1/2}\left(\omega_p^{\mathrm{tr}(ax^2 + bx)}\right)_{x \in \mathbb{F}_q}.$$

The standard basis and the sets B_a, with $a \in \mathbb{F}_q$, form an extremal set of $q+1$ mutually unbiased bases of $\mathbb{C}^q$.

Proof. By definition

$$|\langle v_{a,b} \mid v_{c,d}\rangle| = \left|\frac{1}{q}\sum_{x \in \mathbb{F}_q} \omega_p^{\mathrm{tr}((c-a)x^2 + (d-b)x)}\right|. \tag{1}$$

Suppose that $a = c$. The right hand side evaluates to 1 if $b = d$, and to 0 if $b \neq d$. This proves that B_a is an orthonormal basis. The coefficients of the vector $v_{a,b}$ have absolute value $q^{-1/2}$, hence B_a is mutually unbiased with the standard basis. On the other hand, if $a \neq c$, then the right hand side evaluates to $q^{-1/2}$ by Lemma 1, which proves that the bases B_a and B_c are mutually unbiased.   □

Example 1. In dimension 3, this construction yields the bases

$$B_0 = \{v_{0,0}, v_{0,1}, v_{0,2}\} = \{\ 3^{-1/2}(1,1,1),\ 3^{-1/2}(1,\omega_3,\omega_3^2),\ 3^{-1/2}(1,\omega_3^2,\omega_3)\},$$
$$B_1 = \{v_{1,0}, v_{1,1}, v_{1,2}\} = \{\ 3^{-1/2}(1,\omega_3,\omega_3),\ \ 3^{-1/2}(1,\omega_3^2,1),\ \ 3^{-1/2}(1,1,\omega_3^2)\},$$
$$B_2 = \{v_{2,0}, v_{2,1}, v_{2,2}\} = \{\ 3^{-1/2}(1,\omega_3^2,\omega_3^2),\ \ 3^{-1/2}(1,\omega_3,1),\ \ 3^{-1/2}(1,1,\omega_3)\},$$

which form together with the standard basis four mutually unbiased bases.

3   Even Prime Powers

We showed in the last section that extremal sets of $q+1$ mutually unbiased bases exist in dimension q if q is a power of an odd prime. In this section we treat the case when q is a power of two. We cannot use Weil sums because Lemma 1 does not apply in even characteristics. However, it turns out that exponential sums over a finite Galois ring can serve as a substitute.

We recall some elementary facts about finite Galois rings, see [20] for more details. Let $\mathbb{Z}_4$ denote the residue class ring of integers modulo 4. Denote by $\langle 2 \rangle$ the ideal generated by 2 in $\mathbb{Z}_4[x]$. A monic polynomial $h(x) \in \mathbb{Z}_4[x]$ is called *basic primitive* if and only if its image in $\mathbb{Z}_4[x]/\langle 2 \rangle \cong \mathbb{Z}_2[x]$ under the canonical map is a primitive polynomial in $\mathbb{Z}_2[x]$. Let $h(x)$ be a monic basic primitive polynomial of degree n. The ring $\mathrm{GR}(4,n) = \mathbb{Z}_4[x]/\langle h(x) \rangle$ is called the Galois ring of degree n over $\mathbb{Z}_4$.

The construction ensures that $\mathrm{GR}(4,n)$ has 4^n elements. The element $\xi = x + \langle h(x) \rangle$ is of order $2^n - 1$. Any element $r \in \mathrm{GR}(4,n)$ can be uniquely written in the form $r = a + 2b$, where $a, b \in \mathcal{T}_n = \{0, 1, \xi, \dots, \xi^{2^n-2}\}$. This representation in terms of the Teichmüller set $\mathcal{T}_n$ is convenient, since it allows us to characterize the units of $\mathrm{GR}(4,n)$ as the elements $a + 2b$ with $a \neq 0$.

The automorphism $\sigma \colon \mathrm{GR}(4,n) \to \mathrm{GR}(4,n)$ defined by $\sigma(a + 2b) = a^2 + 2b^2$ is called the Frobenius automorphism. This map leaves the elements of the prime ring $\mathbb{Z}_4$ fixed. All automorphisms of $\mathrm{GR}(4,n)$ are of the form σ^k for some integer $k \geq 0$. The trace map $\mathrm{tr} \colon \mathrm{GR}(4,n) \to \mathbb{Z}_4$ is defined by $\mathrm{tr}(x) = \sum_{k=0}^{n-1} \sigma^k(x)$.

Lemma 2. *Keep the notation as above. The exponential sum* $\Gamma \colon \mathrm{GR}(4,n) \to \mathbb{C}$ *defined by* $\Gamma(r) = \sum_{x \in \mathcal{T}_n} \exp(\frac{2\pi i}{4} \mathrm{tr}(rx))$ *satisfies*

$$|\Gamma(r)| = \begin{cases} 0 & \text{if } r \in 2\mathcal{T}_n,\ r \neq 0, \\ 2^n & \text{if } r = 0, \\ \sqrt{2^n} & \text{otherwise.} \end{cases}$$

The above lemma is proved in [8, Lemma 3], see also [23]. This lemma will be crucial in the next construction of mutually unbiased bases.

Theorem 3. *Let $\mathrm{GR}(4, n)$ be a finite Galois ring with Teichmüller set $\mathcal{T}_n$. For $a \in \mathcal{T}_n$, denote by $M_a = \{v_{a,b} \mid b \in \mathcal{T}_n\}$ the set of vectors given by*

$$v_{a,b} = 2^{-n/2} \left(\exp \left(\frac{2\pi i}{4} \mathrm{tr}(a + 2b)x \right) \right)_{x \in \mathcal{T}_n}.$$

The standard basis and the sets M_a, with $a \in \mathcal{T}_n$, form an extremal set of $2^n + 1$ mutually unbiased bases of $\mathbb{C}^{2^n}$.

Proof. By definition,

$$|\langle v_{a,b} | v_{a',b'} \rangle| = \frac{1}{2^n} \left| \sum_{x \in \mathcal{T}_n} \exp \left(\frac{2\pi i}{4} \mathrm{tr} \left((a' - a) + 2(b' - b))x \right) \right|$$

If both vectors belong to the same basis, *i.e.*, when $a = a'$, then Lemma 2 shows that the right hand side evaluates to 0 in case $b \neq b'$, and to 1 in case $b = b'$. This shows that M_a is an orthonormal basis.

If the vectors belong to different bases, *i.e.*, when $a \neq a'$, then Lemma 2 shows that $|\langle v_{a,b} | v_{a',b'} \rangle| = 2^{-n/2}$, hence M_a and $M_{a'}$ are mutually unbiased. The entries of the vectors $v_{a,b}$ have absolute value $2^{-n/2}$, thus the standard basis and M_a are mutually unbiased for all $a \in \mathrm{GR}(4, n)$.   □

Example 2. We illustrate this construction by deriving five mutually unbiased bases in $\mathbb{C}^4$. In this case, the Galois ring $\mathrm{GR}(4, 2) = \mathbb{Z}_4[x]/\langle x^2 + x + 1 \rangle$ with 16 elements is the basis of the construction. The Teichmüller set is given by $\mathcal{T}_2 = \{0, 1, 3\xi + 3, \xi\}$. Recall that an element of $\mathrm{GR}(4, 2)$ can be represented in the form $a + 2b$ with $a, b \in \mathcal{T}_2$. By definition, $\mathrm{tr}(a + 2b) = a + 2b + a^2 + 2b^2$. Computing the basis vectors yields

$$
\begin{aligned}
M_0 &= \{\tfrac{1}{2}(1,\ 1,\ 1,\ 1),\ \tfrac{1}{2}(1, 1, -1, -1),\ \tfrac{1}{2}(1, -1, -1, 1),\ \tfrac{1}{2}(1, -1, 1, -1)\}, \\
M_1 &= \{\tfrac{1}{2}(1, -1, -i, -i),\ \tfrac{1}{2}(1, -1,\ i,\ i),\ \tfrac{1}{2}(1,\ 1,\ i, -i),\ \tfrac{1}{2}(1,\ 1, -i,\ i)\}, \\
M_{3\xi+3} &= \{\tfrac{1}{2}(1, -i, -i, -1),\ \tfrac{1}{2}(1, -i,\ i,\ 1),\ \tfrac{1}{2}(1,\ i,\ i, -1),\ \tfrac{1}{2}(1,\ i, -i,\ 1)\}, \\
M_\xi &= \{\tfrac{1}{2}(1, -i, -i, -1),\ \tfrac{1}{2}(1, -i,\ i,\ 1),\ \tfrac{1}{2}(1,\ i,\ i, -1),\ \tfrac{1}{2}(1,\ i, -i,\ 1)\}.
\end{aligned}
$$

These four bases and the standard basis form an extremal set of five mutually unbiased bases of $\mathbb{C}^4$.

4   Non Prime Powers

In the previous two sections, we established that the number $N(d)$ of mutually unbiased bases in dimension d attains the maximal possible value, $N(d) = d + 1$, when d is a prime power. In contrast, the exact value of $N(d)$ is not known for any dimension d which is divisible by at least two distinct primes, not even in small dimensions such as $d = 6$.

The problem to determine $N(d)$ is similar to the combinatorial problem to determine the number $M(d)$ of mutually orthogonal Latin squares of size $d \times d$. The number $M(d)$ is exactly known for prime powers but not in general when d is divisible by at least two distinct primes, see [6, 16] for more details. Lower bounds on the number of mutually orthogonal Latin squares can be obtained with the help of a lemma by MacNeish. Our next result formulates a similar statement for the number $N(d)$ of mutually unbiased bases.

Lemma 3. *Let $d = p_1^{a_1} \cdots p_r^{a_r}$ be a factorization of d into distinct primes p_i. Then*

$$N(d) \geq \min \{N(p_1^{a_1}), N(p_2^{a_2}), \dots, N(p_r^{a_r})\}.$$

Proof. We denote the minimum by $m = \min_i N(p_i^{a_i})$. Choose m mutually unbiased bases $B_1^{(i)}, \dots, B_m^{(i)}$ of $\mathbb{C}^{p_i^{a_i}}$, for all i in the range $1 \leq i \leq r$. Then

$$\{B_k^{(1)} \otimes \dots \otimes B_k^{(r)} : k = 1, \dots, m\}$$

is a set of m mutually unbiased bases of $\mathbb{C}^d$. □

An easily memorable form of the above lemma is $N(nm) \geq \min\{N(n), N(m)\}$ for all $m, n \geq 2$. A simple consequence is that $N(d) \geq 3$ for all dimensions $d \geq 2$, that is, in each dimension there are at least three mutually unbiased bases.

Many researchers in the quantum physics community seem to be under the impression that $N(d) = d + 1$ for all integers $d \geq 2$. However, there is some numerical evidence that considerably fewer mutually unbiased bases might be possible if the dimension is not a prime power. In fact, a conjecture by Zauner on the existence of affine quantum designs implies that $N(6) = 3$ rather than $N(6) = 7$, see [24].

Conjecture 1 (Zauner). The number of mutually unbiased bases in dimension 6 is given by $N(6) = 3$.

Apparently, Zauner did considerable numerical computations to bolster his conjecture. Our computational experiments indicate that $N(d)$ is in general smaller than $d + 1$ when d is not a prime power.

Problem 1. Does $N(d) = d+1$ hold for any dimension $d \geq 2$ that is not a prime power?

Another interesting problem concerns lower bounds on $N(d)$. Recall that for mutually orthogonal Latin squares, $M(d) \to \infty$ for $d \to \infty$, as shown by Chowla, Erdős, and Strauss [10]. It is natural to ask whether a similar property holds for the number of mutually unbiased bases:

Problem 2. Does $N(d) \to \infty$ for $d \to \infty$ hold?

More constructions of mutually unbiased bases are needed to prove such a result. A result similar to Wilson's theorem on the number of mutually orthogonal Latin squares [21] would be particularly interesting.

5   Conclusions

Mutually unbiased bases are basic primitives in quantum information theory. They have applications in quantum cryptography and the design of optimal measurements. It is known that in dimension d at most $d+1$ mutually unbiased bases can exist. In this paper, we gave a simplified proof of the fact that $d+1$ mutually unbiased bases exist in $\mathbb{C}^d$ when d is a prime power.

Specifically, we were able to generalize the construction by Alltop to powers of a prime $p \geq 5$. Elementary estimates of Weil sums allowed us to derive a particularly short proof of a theorem by Wootters and Fields. For dimensions $d = 2^n$, we took advantage of known properties of exponential sums over $GR(4, n)$ to obtain extremal sets of mutually unbiased bases.

An open problem is to determine the maximal number of mutually unbiased bases when the dimension is not a prime power. We derived an elementary lower bound for abritrary dimensions and discussed some conjectures and open problems. Finally, we recommend the mean king's problem [2, 12, 19] as an enjoyable application of mutually unbiased bases.

Acknowledgments. This research was supported by NSF grant EIA 0218582, Texas A&M TITF, as well as ARDA, CFI, ORDFC, MITACS, and NSA. We thank Hilary Carteret, Chris Godsil, Bruce Richmond, and Igor Shparlinski for useful discussions.

References

1. W.O. Alltop. Complex sequences with low periodic correlations. *IEEE Transactions on Information Theory*, 26(3):350–354, 1980.
2. P.K. Aravind. Solution to the king's problem in prime power dimensions. *Z. Naturforschung*, 58a:2212, 2003.
3. S. Bandyopadhyay, P.O. Boykin, V. Roychowdhury, and F. Vatan. A new proof of the existence of mutually unbiased bases. *Algorithmica*, 34:512–528, 2002.
4. H. Bechmann-Pasquinucci and W. Tittel. Quantum cryptography using larger alphabets. *Phys. Rev. A*, 61(6):062308, 2000.
5. C.H. Bennett and G. Brassard. Quantum cryptography: Public key distribution and coin tossing. In *Proceedings of the IEEE Intl. Conf. Computers, Systems, and Signal Processing*, pages 175–179. IEEE, 1984.
6. T. Beth, D. Jungnickel, and H. Lenz. *Design Theory*, volume 2. Cambridge University Press, Cambridge, 2nd edition, 1999.
7. B. Bollobás. *Random Graphs*. Academic Press, London, 1985.
8. C. Carlet. One-weight Z_4-linear codes. In J. Buchmann, T. Høholdt, H. Stichtenoth, and H. Tapia-Recillas, editors, *Coding Theory, Cryprography and Related Areas*, pages 57–72. Springer, 2000.
9. S. Chaturvedi. Aspects of mutually unbiased bases in odd-prime-power dimensions. *Phys. Rev. A*, 65:044301, 2002.
10. S. Chowla, P. Erdős, and E.G. Strauss. On the maximal number of pairwise orthogonal latin squares of given order. *Canadian J. Math.*, 12:204–208, 1960.
11. P. Delsarte, J.M. Goethals, and J.J. Seidel. Bounds for systems of lines, and Jacobi polynomials. *Philips Res. Repts.*, pages 91*–105*, 1975.

12. B.-G. Englert and Y. Aharonov. The mean king's problem: Prime degrees of freedom. *Phys. Letters*, 284:1–5, 2001.
13. S.G. Hoggar. *t*-designs in projective spaces. *Europ. J. Combin.*, 3:233–254, 1982.
14. I.D. Ivanović. Geometrical description of quantal state determination. *J. Phys. A*, 14:3241–3245, 1981.
15. G.A. Kabatiansky and V.I. Levenshtein. Bounds for packings on a sphere and in space. *Problems of Information Transmission*, 14(1):1–17, 1978.
16. C.F. Laywine and G.L. Mullen. *Discrete Mathematics Using Latin Squares*. John Wiley, New York, 1998.
17. R. Lidl and H. Niederreiter. *Introduction to finite fields and their applications*. Cambridge University Press, 2nd edition, 1994.
18. J. Schwinger. Unitary operator bases. *Proc. Nat. Acad. Sci. U.S.A.*, 46:570–579, 1960.
19. L. Vaidman, Y. Aharonov, and D.Z. Albert. How to ascertain the values of σ_x, σ_y, and σ_z. *Phys. Rev. Lett.*, 58:1385–1387, 1987.
20. Z.-X. Wan. *Quaternary Codes*. World-Scientific, Singapore, 1997.
21. R.M. Wilson. Concerning the number of mutually orthogonal Latin squares. *Discr. Math.*, 9:181–198, 1974.
22. W.K. Wootters and B.D. Fields. Optimal state-determination by mutually unbiased measurements. *Ann. Physics*, 191:363–381, 1989.
23. K. Yang, T. Helleseth, P. V. Kumar, and A. G. Shanbhag. On the weight hierarchy of Kerdock codes over Z_4. *IEEE Transactions on Information Theory*, 42(5):1587–1593, 1996.
24. G. Zauner. *Quantendesigns – Grundzüge einer nichtkommutativen Designtheorie*. PhD thesis, Universität Wien, 1999.

A Construction of Matrices With No Singular Square Submatrices

Jérôme Lacan and Jérôme Fimes

ENSICA / TéSA
Département de Mathématiques Appliquées et d'Informatique,
1, Place E. Blouin, 31053, Toulouse Cedex 5, France
{jerome.lacan,jerome.fimes}@ensica.fr

Abstract. This paper presents a new construction of matrices with no singular square submatrix. This construction allows designing erasure codes over finite fied with fast encoding and decoding algorithms.

1   Systematic MDS Erasure Codes

It is well known that a $[n, k, d]$-error correcting code is Maximum Distance Separable (MDS) if and only if its $k \times n$-generator matrix does not contain any singular $k \times k$-square submatrix [1, p. 319, Cor. 3]. Similarly, a MDS code is systematic if and only if its generator matrix has the form $(I_k|R)$, where I_k is the $k \times k$-identity matrix and R is a $(k \times (n - k))$-matrix such that any $r \times r$-square submatrix is nonsingular [1, p. 321, Th. 8], for $r \le k$.

It should be noted that systematic MDS codes built from this property (i.e. from the matrix R) are used in practical computer communications to cope with losses of data packets [2].

Cauchy matrices are generally used to build matrices over finite fields whose any square submatrix is non-singular [3]. A $r \times r$-Cauchy matrix is defined as $(\frac{1}{a_i - b_j})_{i,j=0}^{r-1}$, where $(a_i)_{i=0}^{r-1}$ and $(b_j)_{j=0}^{r-1}$ are given vectors of $(\mathbb{F}_q)^r$. Such a matrix is nonsingular if and only if the elements a_i, $i = 1, \ldots, r$ are distinct, the elements b_j, $j = 1, \ldots, r$ are distinct and $a_i + b_j \ne 0$, $1 \le i \le j \le j$. It can be easily verified that any submatrix of a Cauchy matrix is a Cauchy matrix, and then any square submatrix of a nonsingular Cauchy matrix is nonsingular. It should be noted that the Vandermonde matrices defined over finite field can contain singular square submatrices [1, p. 323, Problem 7].

2   A New Class of Matrices with No Singular Square Submatrices

Theorem 1. *Let us denote by A and B two $r \times r$-matrices of rank r over a given field such that any $r \times r$-submatrix of the $r \times 2r$-matrix $(A|B)$ has a rank r. Then, the matrix $A^{-1}B$ is such that any of its square submatrices is nonsingular.*

G. Mullen, A. Poli and H. Stichtenoth (Eds.): Fq7 2003, LNCS 2948, pp. 145–147, 2003.

Proof. Let us denote by W the $r \times 2r$-matrix $(A|B)$. The product $A^{-1}W$ is of the form $(I_r|C)$, where $C = A^{-1}B$. Since, by construction, any $r \times r$-submatrix of W is nonsingular then any $r \times r$-submatrix of the product $A^{-1}W$ is nonsingular.

By combining [1, p. 319, Cor. 3] and [1, p. 321, Th. 8], it can be stated that a $r \times 2r$-matrix on the form $[I_r|C]$ whose any $r \times r$-matrix is nonsingular is necessarily such that C does not contain any $r' \times r'$-singular submatrix for $r' \le r$. This concludes the proof.

Note that it can be verified that these matrices are not Cauchy matrices. Let us give a counterexample. Let us work in the field $\mathbb{F}_5$ and let us consider the matrices A and B respectively equal to $\begin{pmatrix} 1 & 1 \\ 1 & 2 \end{pmatrix}$ and $\begin{pmatrix} 1 & 1 \\ 3 & 4 \end{pmatrix}$. Then, the product $A^{-1} \times B$ is equal to $\begin{pmatrix} 2 & 4 \\ 4 & 1 \end{pmatrix} \times \begin{pmatrix} 1 & 1 \\ 3 & 4 \end{pmatrix} = \begin{pmatrix} 4 & 3 \\ 2 & 3 \end{pmatrix}$. From the definition of Cauchy matrices, it can be easily verified that this product cannot be a Cauchy matrix.

The construction was presented with square matrices, but it can be generalized when A is a $k \times k$-matrix and B is a $k \times (n-k)$ matrix. One can then build a $k \times (n-k)$ matrix, for any "suitable" n ($n \ge k$), whose any square submatrix is nonsingular. Such matrix can be directly used to build the generator matrix of a systematic MDS codes (see Section 1).

3   Application of this Construction to Build Fast Erasure Codes

In order to apply this construction to design efficient erasure codes for computer communications, one must consider matrices for which there exist fast matrix-vector multiplication and fast inversion algorithms. The Vandermonde matrices have these properties. These matrices are defined from a vector of r distinct elements $(a_1, \ldots, a_r)$ of $(\mathbb{F}_q)^r$ as $\left(a_i^{j-1} \right)_{i,j=1}^r$.

The determinant of this matrix is equal to $\prod_{1 \le i < j \le r}(a_i - a_j)$.

As recalled in Section 1, the Vandermonde matrices defined over finite field contains singular square submatrices [1, p. 323, Problem 7]. An upper bound of the number of singular submatrices of Vandermonde matrices is given in [4]. Then, these matrices cannot be directly used to design systematic erasure codes.

However, they can be used in our construction. Indeed, let us define the $k \times k$-matrix A as the Vandermonde matrix $V(\alpha_1, \ldots, \alpha_k)$, where $(\alpha_1, \ldots, \alpha_k)$ is a vector of k distinct elements of $\mathbb{F}_q$. Let us now define the $k \times (n-k)$-matrix $B = \left(b_{i,j} \right)_{i=1,\ldots,k;j=1,\ldots,n-k}$ such that $b_{i,j} = \beta_j^i$, where $(\beta_1, \ldots, \beta_{n-k})$ is a vector of $n-k$ distinct elements of $\mathbb{F}_q$ and such that $\alpha_i \ne \beta_j$, for any $i = 1, \ldots, k$ and any $j = 1, \ldots, n-k$. It can be easily verified that A and B verify the conditions given in Theorem 1. These matrices can therefore be used to design a systematic MDS erasure code.

To optimize this construction, we make the assumption that k is a divisor of $q-1$ and we suggest to take the vector $(\alpha_1, \ldots, \alpha_k)$ equal to $(1, \alpha, \alpha^2, \ldots, \alpha^{k-1})$, where α is an element of $\mathbb{F}_q$ of order k. The corresponding Vandermonde matrix is then denoted by $V(\alpha)$. This choice has two objectives. First, the inversion of A is direct since $(V(\alpha))^{-1} = \frac{1}{k} \times V(\alpha^{-1})$. Then, the matrix-vector multiplication with $V(\alpha)$ (and $V(\alpha^{-1})$) can be performed in $O(k \log k)$ operations by using Fast Fourier Transform (FFT). The general term of the product $\Pi = V(\alpha)^{-1} \times B$ is then $\pi_{i,j} = k^{-1} \times \frac{1-b_j^k}{1-b_j/\alpha^i}$

Let us now describe the coding and the decoding algorithms and evaluate their complexities. As explained in Section 1, the generator matrix of the code is $G = (I_k|R)$, where I_k is the $k \times k$-identity matrix and R is the constructed matrix.

The coding operation consists simply in multiplying the information vector u by the matrix G. It should be first noted that the generator matrix can be factorized into the form $G = \frac{1}{k}V(\alpha) \times [V(\alpha^{-1})|\ B]$. The coding is done by multiplying u and $V(\alpha)$, then by multiplying the resulting vector and $[V(\alpha^{-1})|B]$. The first vector-matrix multiplication can be processed in $O(k \log k)$ operations by using FFT. Since the matrix $[V(\alpha^{-1})|B]$ can be considered as the k first rows of the $n \times n$-matrix $V(1, \alpha^{-1}, \ldots, \alpha^{-(k-1)}, \beta_1, \ldots, \beta_{n-k})$, the second matrix-vector multiplication can be performed in $O(n \log n)$ if FFT can be used (i.e. if the set $\{1, \alpha^{-1}, \ldots, \alpha^{-(k-1)}, \beta_1, \ldots, \beta_{n-k}\}$ can be expressed as the first n powers of an element of $\mathbb{F}_q$ of order n) and in $O(n \log^2 n)$ otherwise [5].

The first step of the decoding consists in considering the $k \times k$-submatrix of G corresponding to the k received symbols. This matrix is then inverted and multiplied by the received vector to obtain the information vector. Since any $k \times k$-submatrix of G can be considered as the product of two Vandermonde matrices, the decoding complexity is $O(k \log^2 k)$ operations for the inversion and $O(k \log^2 k)$ for the matrix-vector multiplication [5]. It should be noted that, in all the cases, the presented complexities are better than the coding and decoding complexities of erasure codes constructed from Cauchy matrices, even when FFT can not be used (i.e. when k and n cannot be chosen accordingly).

Acknowledgments

The authors wish to thank the referee for his/her helpful comments.

References

1. F. J. McWilliams, N.J.A. Sloane, "The Theory of Error Correcting Codes", *Amsterdam, The Netherlands: North Holland, 1977*.
2. L. Rizzo, Effective erasure codes for reliable computer communication protocols, ACM Computer Communication Review, April 1997.
3. J. Bloemer, M. Kalfane, M. Karpinski, R. Karp, M. Luby, D. Zuckerman,"An XOR-Based Erasure-Resilient Coding Scheme", *ICSI TR-95-048, August 1995*.
4. I. Shparlinski, "On singularity of Generalized Vandermonde Matrices Over Finite Fields", *submitted for publication*.
5. I. Gohberg, V. Olshevsky, "Fast Algorithms with preprocessing for matrix-vector multipication problems", *Journal of Complexity, 10, 1994*

Everywhere Ramified Towers of Global Function Fields

Iwan Duursma[1], Bjorn Poonen[*,2], and Michael Zieve[3]

[1] Department of Mathematics, University of Illinois, Urbana, IL 61801–2975, USA
duursma@math.uiuc.edu
[2] Department of Mathematics, University of California, Berkeley, CA 94720–3840, USA
poonen@math.berkeley.edu
[3] Center for Communications Research, 805 Bunn Drive, Princeton, NJ 08540–1966, USA
zieve@idaccr.org

Abstract. We construct a tower of function fields $F_0 \subset F_1 \subset \ldots$ over a finite field such that every place of every F_i ramifies in the tower and $\lim \operatorname{genus}(F_i)/[F_i : F_0] < \infty$. We also construct a tower in which every place ramifies and $\lim N_{F_i}/[F_i : F_0] > 0$, where N_{F_i} is the number of degree-1 places of F_i. These towers answer questions posed by Stichtenoth at Fq7.

1   Introduction

Let q be a prime power, and let $\mathbb{F}_q$ be a finite field of size q. By a *function field over* $\mathbb{F}_q$, we mean a finitely generated extension $K/\mathbb{F}_q$ of transcendence degree 1 in which $\mathbb{F}_q$ is algebraically closed. By an *extension of function fields* K'/K, we mean a finite separable extension such that K and K' are function fields over the same $\mathbb{F}_q$. Let g_K be the genus of K. Let N_K be the number of degree-1 places of K (the number of $\mathbb{F}_q$-rational points on the corresponding curve). A *tower* of function fields over $\mathbb{F}_q$ is a sequence of extensions of such function fields

$$K_0 \subset K_1 \subset K_2 \subset \ldots$$

such that $g_i := g_{K_i} \to \infty$ as $i \to \infty$. Define $N_i = N_{K_i}$, and $d_i = [K_i : K_0]$. Since N_i/d_i is decreasing while $(g_i-1)/d_i$ is increasing (Hurwitz), $\lim N_i/d_i$ and $\lim g_i/d_i$ exist. (The latter can be ∞.)

The Weil bound $N_K \le q + 1 + 2g_K\sqrt{q}$ implies

$$\lim N_i/g_i \le 2\sqrt{q}.$$

This was improved by Drinfeld and Vladut [4] (following Ihara [19]) to

$$\lim N_i/g_i \le \sqrt{q} - 1.$$

---

[*] Supported in part by NSF grant DMS-0301280 and a Packard Fellowship.

G. Mullen, A. Poli and H. Stichtenoth (Eds.): Fq7 2003, LNCS 2948, pp. 148–153, 2003.

Ihara also showed that, for any square q, there are towers of Shimura curves with $\lim N_i/g_i = \sqrt{q}-1$ [15–19]. Subsequent authors have given further constructions of 'asymptotically good' towers, i.e., towers with $\lim N_i/g_i > 0$ [1–3, 5–14, 20–31, 33–36].

Every known asymptotically good tower has two special properties: there is some place of some K_i which splits completely in the tower, and there are only finitely many places of K_0 which ramify in the tower. (We say that a place of K_i *splits completely in the tower* if it splits completely in K_j/K_i for every $j \geq i$. We say that a place of K_0 *ramifies in the tower* if there exists i such that it ramifies in K_i/K_0.) But it is difficult to study asymptotically good towers directly since one must control both the genus and the number of rational places. With this as motivation, Stichtenoth posed the following two questions in his talk at Fq7 (the Seventh International Conference on Finite Fields and Their Applications):

Question 1. If $\lim N_i/d_i > 0$, must some K_i have a place that splits completely in the tower?

Question 2. If $\lim g_i/d_i < \infty$, must only finitely many places of K_0 ramify in the tower?

Our Theorems 1 and 2 imply negative answers to these two questions. Call a tower $K_0 \subset K_1 \subset \ldots$ of function fields over $\mathbb{F}_q$ *everywhere ramified* if for each place P of each K_i, there exists $j > i$ such that P ramifies in K_j/K_i.

Theorem 1. *Given a function field K_0 over $\mathbb{F}_q$ with a rational place, there exists an everywhere ramified tower $K_0 \subset K_1 \subset \ldots$ such that $\lim N_i/d_i > 0$.*

Theorem 2. *Given a function field K_0 over $\mathbb{F}_q$, there exists an everywhere ramified tower $K_0 \subset K_1 \subset \ldots$ such that $\lim g_i/d_i < \infty$.*

2   Proof of Theorem 1

Lemma 1. *Let K be a function field over $\mathbb{F}_q$. Then there is a nontrivial extension K'/K in which all rational places of K split completely.*

Proof. Weak approximation (or Riemann-Roch) gives $f \in K^*$ having a zero at each rational place of K and a simple pole at some other place of K. Adjoin a root of $y^q - y = f$ to obtain K'. Then K'/K is totally ramified above the simple pole of f, so K' is another function field over $\mathbb{F}_q$ and $[K' : K] = q > 1$.

Lemma 2. *Let K be a function field over $\mathbb{F}_q$ with $N_K > 0$, and let P be a place of K. For any $\varepsilon > 0$, there is an extension L/K such that $N_L/N_K > (1 - \varepsilon)[L : K]$ and P ramifies in L/K.*

Proof. We first reduce to the case where $1/N_K < \varepsilon$. Repeated application of Lemma 1 yields K'/K such that $1/([K' : K]N_K) < \varepsilon$ and all rational places of

K split completely. Then $N_{K'} = [K' : K]N_K$. Pick a place P' of K' above P. If we could find L/K' satisfying the conditions of the lemma for (K', P'), then

$$\frac{N_L}{N_K} = \frac{N_L}{N_{K'}} \frac{N_{K'}}{N_K} > (1 - \varepsilon)[L : K'][K' : K] = (1 - \varepsilon)[L : K],$$

so L/K would work for (K, P). Thus, renaming K' as K, we may assume $1/N_K < \varepsilon$.

Weak approximation gives $f \in K^*$ having a simple pole at P and zeros at all rational places not equal to P. Adjoin a root of $y^q - y = f$ to obtain L. Then P ramifies in L/K, but all other rational places of K split completely, so $N_L \geq (N_K - 1)q$. Thus $N_L/N_K \geq q(1 - 1/N_K) > [L : K](1 - \varepsilon)$.

Proof (Proof of Theorem 1). Fix a sequence of positive numbers $\varepsilon_m \to 0$ such that $\prod_{m=1}^{\infty}(1 - \varepsilon_m)$ converges to a positive number. In our proof we will apply Lemma 2 infinitely often, using ε_1 in the first application, ε_2 in the second application, and so on.

Let $P_0, P_1, \ldots$ be an enumeration of the places of K_0 (of all degrees). Given K_i, we construct K_{i+1} in stages so that all places of K_i lying above $P_0, \ldots, P_i$ ramify in K_{i+1}/K_i. Namely, if $Q_1, \ldots, Q_I$ are all the places of K_i lying above $P_0, \ldots, P_i$, we set $K_{i,0} = K_i$ and then for $j = 1, \ldots, I$ in turn, apply Lemma 2 with the first unused ε_m to find $K_{i,j}/K_{i,j-1}$ in which some place of $K_{i,j-1}$ above Q_j ramifies and $N_{K_{i,j}}/N_{K_{i,j-1}} > (1-\varepsilon_m)[K_{i,j} : K_{i,j-1}]$. Finally, set $K_{i+1} = K_{i,I}$.

If R is a place of some K_r, then R lies over some P_j of K_0. By construction, for all $i \geq \max\{j, r\}$, all places of K_i above R ramify in K_{i+1}/K_i. Thus R is ramified in K_{i+1}/K_r.

The inequality in Lemma 2 guarantees that the value of N/d for $K_{i,j}$ is at least $1 - \varepsilon_m$ times the value of N/d for $K_{i,j-1}$. Thus N_i/d_i is at least $\left(\prod_{m \leq M}(1 - \varepsilon_m)\right) N_0/d_0$, if M is the number of applications of Lemma 2 used in the construction up to K_i. Since $N_0/d_0 > 0$ and $\prod_{m=1}^{\infty}(1 - \varepsilon_m)$ converges, the decreasing sequence N_i/d_i is bounded below by

$$\left(\prod_{m=1}^{\infty}(1 - \varepsilon_m)\right) N_0/d_0,$$

which is positive. So N_i/d_i has a positive limit. Finally, $N_i \to \infty$ implies $g_i \to \infty$.

Remark 1. A slight modification of the argument shows that, given K_0, we can construct an everywhere ramified tower in which N_i/d_i converges to any prescribed value less than N_0. This is because weak approximation lets us prescribe the ramification and splitting of any finite number of places at each step.

3   Proof of Theorem 2

Let p be the characteristic of $\mathbb{F}_q$.

Lemma 3. *Let K be a function field over $\mathbb{F}_q$ of genus > 1, and let P be a place of K. Then there exist unramified extensions K'/K of arbitrarily high genus such that for some place Q of K' lying over P, the residue field extension for Q/P is trivial.*

Proof. Let C be the smooth, projective, geometrically integral curve with function field K. Let J be the Jacobian of C. There exists a degree-1 divisor D on C [32, V.1.11]. Use D to identify C with a closed subvariety of J.

The place P corresponds to a Galois conjugacy class of points in $C(\mathbb{F}_{q^f})$, where $\mathbb{F}_{q^f}$ is the residue field. Choose P_0 in this conjugacy class. Choose $n \in \mathbb{Z}_{>0}$ such that $n \equiv 1 \pmod{p \cdot \#J(\mathbb{F}_{q^f})}$. Then the multiplication-by-n map $[n]\colon J \to J$ is étale, and maps P_0 to itself. Let $C' = [n]^{-1}C$, so C' is an étale cover of C. Then C' corresponds to a function field K' that is unramified over K. Also $P_0 \in C'(\mathbb{F}_{q^f})$ represents a place Q of K' lying over P, having the same residue field as P. By choosing n large, we can make $g_{K'}$ as large as desired, by the Hurwitz formula.

Lemma 4. *Let K be a function field over $\mathbb{F}_q$ of genus > 1, let P be a place of K, and let $\varepsilon > 0$. Then there exists an extension L/K with $(g_L - 1)/(g_K - 1) < (1 + \varepsilon)[L : K]$ such that P ramifies in L/K.*

Proof. Let f be the degree of P over $\mathbb{F}_q$. For an unramified extension K'/K, we have $(g_{K'} - 1)/(g_K - 1) = [K' : K]$ by Hurwitz. By applying Lemma 3, we may replace (K, P) by some (K', Q) in order to assume that g_K is arbitrarily large, without changing f.

When g_K is sufficiently large, an easy estimate (e.g. cf. [32, V.2.10]) based on the Weil bounds implies there exist places Q, Q' of K of degrees $d, d + f$ respectively, where d is the smallest integer $> \sqrt{g_K}$ and not equal to f. Choose a prime $\ell \nmid p \cdot \#G$, where G is the group of degree-zero divisor classes of K. Then every element of G, and in particular $[Q' - Q - P]$, is divisible by ℓ. Thus, there exists a divisor D of degree 0 and an element h of K such that $(h) = Q' - Q - P - \ell D$. Let $L = K(h^{1/\ell})$, so $[L : K] = \ell$. Hurwitz gives

$$2g_L - 2 = \ell(2g_K - 2) + (\ell - 1)((d + f) + d + f),$$

so

$$\frac{g_L - 1}{[L : K](g_K - 1)} = 1 + \frac{\ell - 1}{\ell}\left(\frac{d + f}{g_K - 1}\right) = 1 + O(g_K^{-1/2}).$$

The $O(g_K^{-1/2})$ term will be $< \varepsilon$ if g_K is sufficiently large.

Proof (Proof of Theorem 2). Given K_0, let K_1/K_0 be an extension with $g_1 > 1$. Just as Lemma 2 let us prove Theorem 1, Lemma 4 now lets us construct an everywhere ramified tower $K_1 \subset K_2 \subset \ldots$ such that at the ith step the value of $(g_i - 1)/d_i$ increases by a factor at most $1 + \varepsilon_i$ for a prescribed $\varepsilon_i > 0$. By choosing ε_i so that $\prod(1 + \varepsilon_i)$ converges, we obtain such a tower with $\lim(g_i - 1)/d_i < \infty$. Since $d_i \to \infty$, this limit equals $\lim g_i/d_i$.

4   Question

Can one combine Theorems 1 and 2? In particular, does there exist an everywhere ramified tower in which both $\lim N_i/d_i > 0$ and $\lim g_i/d_i < \infty$?

References

1. Angles, B., Maire, C.: A note on tamely ramified towers of global function fields. Finite Fields Appl. **8** (2002) 207–215
2. Beelen, P.: Graphs and recursively defined towers of function fields. Preprint
3. Bezerra, J., Garcia, A.: A tower with non-Galois steps which attains the Drinfeld-Vladut bound. Preprint
4. Drinfeld, V.G., Vladut, S.G.: The number of points of an algebraic curve. Funktsional. Anal. i Prilozhen. **17** (1983) 68–69 [Funct. Anal. Appl. **17** (1983) 53–54]
5. Elkies, N.D.: Explicit modular towers. In: Başar, T., Vardy, A. (eds.): Proceedings of the Thirty-Fifth Annual Allerton Conference on Communication, Control and Computing. Univ. of Illinois at Urbana-Champaign (1998) 23–32
6. Elkies, N.D.: Explicit towers of Drinfeld modular curves. In: Casacuberta, C., et al. (eds.): European Congress of Mathematics, Vol. II. Birkhauser, Basel (2001) 189–198. arXiv:math.NT/0005140
7. Frey, G., Kani, E., Völklein, H.: Curves with infinite K-rational geometric fundamental group. In: Völklein, H. et al. (eds.): Aspects of Galois Theory. London Mathematical Society Lecture Note Series, Vol. 256. Cambridge University Press, Cambridge (1999) 85–118
8. Frey, G., Perret, M., Stichtenoth, H.: On the different of abelian extensions of global fields. In: Stichtenoth, H., Tsfasman, M.A. (eds.): Coding Theory and Algebraic Geometry. Lecture Notes in Mathematics, Vol. 1518. Springer-Verlag, New York (1992) 26–32
9. Garcia, A., Stichtenoth, H.: A tower of Artin-Schreier extensions of function fields attaining the Drinfeld-Vladut bound. Invent. Math. **121** (1995) 211–222
10. Garcia, A., Stichtenoth, H.: On the asymptotic behavior of some towers of function fields over finite fields. J. Number Theory **61** (1996) 248–273
11. Garcia, A., Stichtenoth, H.: On tame towers over finite fields. J. Reine Angew. Math. **557** (2003) 53–80
12. Garcia, A., Stichtenoth, H., Thomas, M.: On towers and composita of towers of function fields over finite fields. Finite Fields Appl. **3** (1997) 257–274
13. van der Geer, G., van der Vlugt, M.: An asymptotically good tower of curves over the field with eight elements. Bull. London Math. Soc. **34** (2002) 291–300
14. Hajir, F., Maire, C.: Asymptotically good towers of global fields. In: Casacuberta, C., et al. (eds.): European Congress of Mathematics, Vol. II. Birkhauser, Basel (2001) 207–218
15. Ihara, Y.: Algebraic curves mod p and arithmetic groups. In: Borel, A., Mostow, G.D. (eds.): Algebraic Groups and Discontinuous Subgroups. Proceedings of Symposia in Pure Mathematics, Vol. IX. American Mathematical Society, Providence (1966) 265–271
16. Ihara, Y.: On Congruence Monodromy Problems. Vol. 2. Department of Mathematics, University of Tokyo, 1969.
17. Ihara, Y.: On modular curves over finite fields. In: Discrete Subgroups of Lie Groups and Applications to Moduli (Internat. Colloq., Bombay, 1973). Oxford Univ. Press, Bombay (1975) 161–202

18. Ihara, Y.: Congruence relations and Shimūra curves. In: Borel, A., Casselman, W. (eds.): Automorphic Forms, Representations, and L-functions. Part 2. Proceedings of Symposia in Pure Mathematics, Vol. XXXIII. American Mathematical Society, Providence (1979) 291–311

19. Ihara, Y.: Some remarks on the number of rational points of algebraic curves over finite fields. J. Fac. Sci. Univ. Tokyo **28** (1981) 721–724

20. Li, W.-C. W., Maharaj, H.: Coverings of curves with asymptotically many rational points. J. Number Theory **96** (2002) 232–256. arXiv:math.NT/9908152

21. Li, W.-C. W., Maharaj, H., Stichtenoth, H.: New optimal towers over finite fields. In: Fieker, C., Kohel, D. (eds.): Algorithmic Number Theory. Lecture Notes in Computer Science, Vol. 2369. Springer-Verlag, New York (2002) 372–389

22. Manin, Y.I., Vladut, S.G.: Linear codes and modular curves. Itogi Nauki i Tekhniki **25** (1984) 209–257 [J. Soviet Math. **30** (1985) 2611–2643]

23. Niederreiter, H., Xing, C.: Towers of global function fields with asymptotically many rational places and an improvement on the Gilbert-Varshamov bound. Math. Nachr. **195** (1998) 171–186

24. Niederreiter, H., Xing, C.: Curve sequences with asymptotically many rational points. In: Fried, M. (ed.): Applications of Curves over Finite Fields. Contemporary Mathematics, Vol. 245. American Mathematical Society, Providence (2000) 3–14

25. Niederreiter, H., Xing, C.: Global function fields with many rational places and their applications. In: Mullin, R., Mullen, G. (eds.): Finite Fields: Theory, Applications, and Algorithms. Contemporary Mathematics, Vol. 225. American Mathematical Society, Providence (1999) 87–111

26. Niederreiter, H., Xing, C.: Rational Points on Curves over Finite Fields: Theory and Applications. London Mathematical Society Lecture Note Series, Vol. 285. Cambridge University Press, Cambridge (2001)

27. Perret, M.: Tours ramifiées infinies de corps de classes. J. Number Theory **38** (1991) 300–322

28. Schoof, R.: Algebraic curves over $\mathbb{F}_2$ with many rational points. J. Number Theory **41** (1992) 6–14

29. Serre, J.-P.: Sur le nombre des points rationnels d'une courbe algébrique sur un corps fini. C. R. Acad. Sci. Paris **296** (1983) 397–402; = Œuvres [128]

30. Serre, J.-P.: Rational points on curves over finite fields. unpublished lecture notes by Gouvêa, F. Q., Harvard University, 1985

31. Stichtenoth, H.: Explicit constructions of towers of function fields with many rational places. In: Casacuberta, C., et al. (eds.): European Congress of Mathematics, Vol. II. Birkhauser, Basel (2001) 219–224

32. Stichtenoth, H.: Algebraic Function Fields and Codes. Springer-Verlag, Berlin (1993)

33. Temkine, A.: Hilbert class field towers of function fields over finite fields and lower bounds for $A(q)$. J. Number Theory **87** (2001) 189–210

34. Wulftange, J.: Zahme Türme algebraischer Funktionenkörper. Ph.D. thesis (Essen, 2003)

35. Xing, C.: The number of rational points of algebraic curves over finite fields. Acta Math. Sinica **37** (1994) 584–589

36. Zink, Th.: Degeneration of Shimura surfaces and a problem in coding theory. In: Budach, L., (ed.): Fundamentals of Computation Theory. Lecture Notes in Computer Science, Vol. 199. Springer-Verlag, New York (1985) 503–511

On the Construction of Some Towers over Finite Fields

Universität Essen
Germany

Abstract. The explicit construction of towers with many rational places plays a key role for the construction of asymptotically good algebraic geometric codes. One way of explicitly constructing towers is given by defining them recursively via a single equation. In this paper we discuss conditions on the defining equation to give towers with many rational places and introduce a new family of such towers.

1 Introduction

Throughout the whole paper, let $f(x, y)$ be an absolutely irreducible polynomial over $\mathbb{F}_q$ with $\deg_y f =: m > 1$. We call a field $\mathcal{T} := \cup_{k \geq 0} F_k$ recursively defined by $f(x, y)$ if $F_0 := \mathbb{F}_q(x_0)$ is the rational function field and $F_k := F_{k-1}(x_k)$ where $f(x_{k-1}, x_k) = 0$. If $F_{k-1} \neq F_k$ and $\mathbb{F}_q$ is the exact constant field for all $k \geq 0$, and $g(F_k) > 1$ for some $k \geq 0$, we call $\mathcal{T}$ a tower. In the first part of the paper, we are interested in conditions on $f(x, y)$ to define recursively a tower.

A tower $\mathcal{T}$ is called asymptotically good if

$$\lambda(\mathcal{T}) := \lim_{k \to \infty} \frac{N(F_k)}{g(F_k)} > 0,$$

where $N(F_k)$ denotes the number of $\mathbb{F}_q$-rational places of F_k and $g(F_k)$ denotes the genus of F_k. The value $\lambda(\mathcal{T})$ is called the limit of $\mathcal{T}$. The construction of asymptotically good towers is of general interest, and in particular it is motivated from coding theory (cf. e.g. [7], [2]). In this paper we are only considering towers with F_{k+1}/F_k tame. Thus all extensions considered are tame. In the second section we prove some facts about a family of recursively defined towers given by an equation of the type

$$y^m = a(x + b)^m + c \text{ with } a, b, c \in \mathbb{F}_q^* \text{ and } (m, q) = 1.$$

Among these so-called Fermat towers, there are some asymptotically good ones (cf. e.g. [3]).

In the third section we discuss a (finite) family of new asymptotically good towers with the property that all F_k are unramified over F_2 for $k \geq 2$. In general, we call relatively unramified the towers that are unramified above some field F_i.

G. Mullen, A. Poli and H. Stichtenoth (Eds.): Fq7 2003, LNCS 2948, pp. 154–165, 2003.
© Springer-Verlag Berlin Heidelberg 2003

2   On the Construction of Towers

Let $F := \bar{\mathbb{F}}_q(x, y)$ be an algebraic function field with $f(x, y) = 0$. For elements $\mu, \nu \in \mathbb{P}^1 := \bar{\mathbb{F}}_q \cup \{\infty\}$ we write $\nu \leftarrow \mu$ if there is a place $Q \subset F$ with $x(Q) = \nu$ and $y(Q) = \mu$. We define the sets

$$M := \Big\{ \alpha \in \mathbb{P}^1 | P \in \mathbb{P}_{\bar{\mathbb{F}}_q(x)} \text{ with } x(P) = \alpha \text{ is ramified in} \\ F/\bar{\mathbb{F}}_q(x) \text{ of index } m \Big\} \tag{1}$$

and

$$N := \Big\{ \alpha \in \mathbb{P}^1 | Q \in \mathbb{P}_{\bar{\mathbb{F}}_q(y)} \text{ with } y(Q) = \alpha \text{ has at least one} \\ \text{place above in } F, \text{ which is ramified of index } e \tag{2} \\ \text{with } \gcd(e, m) > 1 \Big\}.$$

For $\mu \in K \cup \{\infty\}$ ($K = \mathbb{F}_q$ or $K = \bar{\mathbb{F}}_q$), we define P_μ to be the place P of $K(x_0)$ with $x_0(P_\mu) = \mu$. By $\mathcal{O}_{P_\mu}$ we denote the valuation ring of a place P_μ.

Theorem 1. *Suppose there exists a sequence $(\mu_i)_{i \geq 0}$ in $\mathbb{P}^1 \setminus N$ with $\mu_0 \in M \setminus N$, such that $f(x, y)$ is integral over $\mathcal{O}_{P_{\mu_i}}$ and $\mu_{i+1} \leftarrow \mu_i$ for all $i \geq 0$. Then, $f(x, y)$ defines recursively a tower $\mathcal{T}$.*

Proof. Let $\bar{\mathbb{F}}_q$ be an algebraic closure of $\mathbb{F}_q$. First we consider the field $\mathcal{T}$ recursively defined over $\bar{\mathbb{F}}_q$ (i.e. $F_0 = \bar{\mathbb{F}}_q(x_0)$). Inductively we prove the claim that for each field F_k, there is at least one place which is ramified in F_{k+1} of index m. Then, it follows from the degree formula that $[F_{k+1} : F_k] = m$.

The claim is true for F_0, because the place $P \in \mathbb{P}_{F_0}$ with $x_0(P) = \mu_0$ is totally ramified in F_1. Now, suppose the claim is true for $0 \leq i < k$. Since $\mu_{i+1} \leftarrow \mu_i$ for all $0 \leq i \leq k$ and the polynomial $f(x_0, x_1)$ is integral over $\mathcal{O}_{P_{\mu_i}}$, it follows from the theorem of Kummer (cf. [6, III.3.7]), that there exists a place $P \in \mathbb{P}_{F_k}$ with $x_i(P) = \mu_{k-i}$ for $0 \leq i \leq k$. As $x_i(P) := \mu_{k-i} \notin N$ and the tower is recursively defined, the place $P \cap \bar{\mathbb{F}}_q(x_i)$ is unramified or ramified with index e satisfying $\gcd(m, e) = 1$ in $\bar{\mathbb{F}}_q(x_{i-1}, x_i)$. Thus, repeated application of Abhyankar's lemma (cf. [6, III.8.9]) yields that P is unramified or ramified with index e satisfying $\gcd(e, m) = 1$ over $P \cap \bar{\mathbb{F}}_q(x_k)$. Since $x_k(P) = \mu_0 \in M$, the place $P \cap \bar{\mathbb{F}}_q(x_k)$ is ramified in $\bar{\mathbb{F}}_q(x_k, x_{k+1})$ of index m. Again by Abhyankar's lemma, there exists a place P' of $F_k(x_{k+1})$, which is ramified of index m over P (cf. Figure 1).

Now, let $F_0 = \mathbb{F}_q(x_0)$. Then, $\mathbb{F}_q$ is the exact constant field for all F_k, as there is a place of F_k which is totally ramified in F_{k+1} but constant field extensions are unramified.

The theorem above relies essentially on the existence of a totally ramified place in each step of the recursive construction. Obviously, its proof does not work for recursively defined towers which are unramified after a few steps. In the situation of such relatively unramified towers, we can prove a result for the construction of towers by guaranteeing the existence of an inert place in each step.

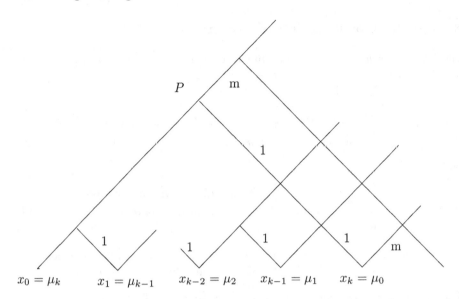

Fig. 1. Ramification index of a place $P \in \mathbb{P}_{F_k}$

Theorem 2. *Let $F = \mathbb{F}_q(x_0, x_1)$ be a function field with defining equation $f(x_0, x_1) = 0$. Suppose there is an $\mathbb{F}_q$-rational place $P \in \mathbb{P}_{\mathbb{F}_q(x_0)}$ with the following properties:*

(*) *P is completely splitting in F, and there is a place $P'|P, P' \in \mathbb{P}_F$, with $x_1(P') = x_0(P)$.*

(**) *For some $n \geq 1$, there is a finite sequence $(\mu_i)_{0 \leq i \leq n}$ with $\mu_n = x_0(P)$, such that $\mu_{i+1} \leftarrow \mu_i$, the place $P_{\mu_0} := Q$ of F_0 has only one totally inert extension in F, and the places P_{μ_i} of F_0 are completely splitting for $1 \leq i \leq n$ in F.*

Then, $f(x, y)$ defines recursively a tower $\mathcal{T} = \cup_{k \geq 0} F_k$, where the field extensions F_{k+1}/F_k satisfy:

1. *$[F_{k+1} : F_k] = m$.*
2. *$\mathbb{F}_q$ is the exact constant field of F_{k+1}.*
3. *F_{k+1} has an $\mathbb{F}_q$-rational place Q_{k+1} with $x_{k+1}(Q_{k+1}) = x_0(Q) = \mu_0$.*

Proof. We prove the theorem by induction on k. For $k = 0$ all the claims above follow from the assumptions.

Let the claims be true for $k - 1$. By induction it follows from (3), that there exists an $\mathbb{F}_q$-rational place $Q_k \in \mathbb{P}_{F_k}$ with $x_k(Q_k) = x_0(Q)$. We define $R := Q_k \cap \mathbb{F}_q(x_k)$. Then it follows from assumption (**), that there is only one place R' over R in $\mathbb{F}_q(x_k, x_{k+1})$, which is inert of index $f(R'|R) = m$. Now, choose a place $\tilde{Q}$ above Q_k in $F_k(x_{k+1})$. The place $\tilde{Q}$ lies above R', because R' is the only extension of R in $\mathbb{F}_q(x_k, x_{k+1})$. Thus, we have $\deg \tilde{Q} \geq m$ and $[F_{k+1} : F_k] = m$ by the degree formula.

Since $\mu_n = x_0(P)$ and $x_0(P) \leftarrow x_0(P)$, by assumption $(*)$, we can extend the sequence $(\mu_i)_{0 \leq i \leq n}$ by elements $\mu_{n+1} = x_0(P), \mu_{n+2} = x_0(P), \ldots$ and the sequence keeps the property $\mu_{k+1} \leftarrow \mu_k$. We choose the first $k+2$ elements of the possibly extended sequence. Since $[F_{j+1} : F_j] = m$ for all $0 \leq j \leq k$ and the places P_{μ_i} are completely splitting by assumption $(**)$, we obtain by repeated application of [6, III.8.4] a place Q_{k+1} with $x_i(Q_{k+1}) = \mu_{k+1-i}$ for $i = 0, \ldots, k+1$. In particular, we have $x_{k+1}(Q_{k+1}) = x_0(Q)$.

The second claim is a consequence of the fact, that the $\mathbb{F}_q$-rational places P_{μ_i} for $i > 0$ are totally splitting and $[F_{k+1} : F_k] = m$.

3 On Fermat Towers

In this section we consider tame towers of the following type.

Definition 1. *A tower $\mathcal{F}$ recursively defined by equation*

$$y^m = a(x + b)^m + c \text{ with } a, b, c \in \mathbb{F}_q \text{ and } \gcd(m, q) = 1 \tag{3}$$

is said to be a tower of Fermat type or a Fermat tower.

The discussion of these towers is motivated from the fact, that some among them are asymptotically good (cf. e.g. [3]). For some parameters it is only known that the corresponding equation yields an asymptotically good tower, provided the equation defines recursively a tower in the first place.

Theorem 3. *Equation (3) defines recursively a tower $\mathcal{F}$, if and only if $a, b, c \in \mathbb{F}_q^*$.*

Proof. First, we consider the case $a, b, c \in \mathbb{F}_q^*$. We want to apply Theorem 1. Let $F := \bar{\mathbb{F}}_q(x, y)$ be defined by Equation (3). Then, $F/\bar{\mathbb{F}}_q(x)$ is a Kummer extension. Exactly the zeroes P_α of $x - \alpha$ with $a(\alpha + b)^m + c = 0$ are ramified, each of index m. Thus, with the notation from section 1, we have $M := \{\alpha \in \bar{\mathbb{F}}_q | a(\alpha + b)^m + c = 0\}$. The extension $F/\bar{\mathbb{F}}_q(y)$ is a Kummer extension, too. Exactly the zeroes Q_β of $y - \beta$, where $\beta^m - c = 0$, are ramified, each of index m. Thus, we obtain $N := \{\beta \in \bar{\mathbb{F}}_q | \beta^m = c\}$.

Next, we want to prove the following claim.

Claim. For each $d \in \bar{\mathbb{F}}_q$ with $d \neq c$, there is a $\mu \in \bar{\mathbb{F}}_q$ with

$$a(\mu + b)^m + c = d \text{ and } \mu^m \neq c.$$

Once we have proved the claim, we can choose elements $(\mu_i)_{i \geq 0}$ with $\mu_0 \in M \setminus N$, $a(\mu_{i+1} + b)^m + c = \mu_i^m$ and $\mu_{i+1}^m \neq c$. Equation (3) is integral over all places P which are different from the pole of the function x. Thus, Equation (3) defines recursively a tower by Theorem 1.

Proof of the claim. The two polynomials

$$\varphi(T) := (T + b)^m + a^{-1}(c - d) \in \bar{\mathbb{F}}_q[T]$$

and

$$\Psi(T) := T^m - c \in \bar{\mathbb{F}}_q[T]$$

are monic and separable (as $c \neq 0$ and $c \neq d$). Since

$$\varphi(T) = T^m + mbT^{m-1} + \dots$$

and $b \neq 0$, we have $\varphi(T) \nmid \Psi(T)$. As $\deg \varphi(T) = \deg \Psi(T)$, there exists a $\mu \in \bar{\mathbb{F}}_q$ with $\varphi(\mu) = 0$ and $\Psi(\mu) \neq 0$. This proves the claim.

Now, let $abc = 0$. If $a = 0$ or $c = 0$, it is obvious that Equation (3) does not define recursively a tower. Thus, assume the equation

$$y^m = ax^m + c \text{ with } a, c \in \mathbb{F}_q \text{ and } (m, q) = 1$$

defines recursively a tower $\mathcal{F} := \cup_{k \geq 0} F_k$ over $\bar{\mathbb{F}}_q$. Then, $[F_k : F_{k-1}] > 1$ for all $k \geq 0$. We have $F_k = F_{k-1}(x_k)$, where

$$x_k^m := a^k x_0^m + (1 + a + a^2 + \dots + a^{k-1})c.$$

Thus, $[F_k : F_{k-1}] = 1$ for $k = char\ \mathbb{F}_q$, if $a = 1$, and for $k = ord(a)$, if $a \neq 1$.

Now, we have the following corollary.

Corollary 1. *Let l be a power of a prime and let $q = l^r$ with $r \geq 2$. Then the equation*

$$y^{(q-1)/(l-1)} = a(x + b)^{(q-1)/(l-1)} + c, \text{ with } a, c \in \mathbb{F}_l^* \text{ and } b \in \mathbb{F}_q^*, \quad (4)$$

defines an asymptotically good Fermat tower $\mathcal{F}$ over $\mathbb{F}_q$, and its limit satisfies

$$\lambda(\mathcal{F}) \geq \frac{2}{q-2}.$$

Proof. By Theorem 3, Equation (4) defines recursively a tower. Now the corollary follows from [3, Proposition 3.6]. $\quad\square$

Remark:

1. If $ab^m + c = 0, m = (q-1)/(l-1)$, it has already been known, that Equation (4) defines a tower. If $q = l^2$ it has already been proved by Thomas and Özbudak (cf. [5]).
2. In the case $ab^m + c = 0, m = (q-1)/(l-1)$, it can be proved that $\lambda(\mathcal{F}) = 2/(q-2)$ (cf. [8]).

Only for few choices of the parameters a, b, c and m it has been proved that Fermat equations define recursively asymptotically good towers. For most choices of the parameters it is not known, whether the resulting towers are asymptotically good or not (over appropriately chosen constant fields). For towers with $ab^m + c = 0$ the ramification locus is infinite in many cases. Here, the ramification locus $V_{F_0}(\mathcal{T})$ of a tower $\mathcal{T}$ over F_0 is defined as usual:

$$V_{F_0}(\mathcal{T}) := \{P \in \mathbb{P}_{F_0} | \text{ There is an } i \in \mathbb{N} \text{ such that } P \text{ ramifies in } F_i/F_0\}.$$

In order to prove this result we need the following polynomial identity, which was first found by Lenstra (cf. [4]). Since we need a slightly different form, we give the proof as well.

Lemma 1. *Let T be a recursively defined tower with defining equation*

$$y^m = g(x) \in \mathbb{F}_q[x] \text{ where } \deg g(x) = m \text{ and } \gcd(m, q) = 1.$$

Moreover, let $g(x) = x^d g_1(x)$ with $g_1(0) \neq 0$ and $\gcd(d, m) = 1$. Suppose the tower T has finite ramification locus $V_{F_0}(T)$. We define $T := \{\alpha^m | P_\alpha \in V_{F_0}(T)\}$ and $t := |T|$.
Then we have

$$mx^{m-1} \prod_{\alpha \in T} (g(x) - \alpha) = a^{t-1} g'(x) \prod_{\alpha \in T} (x^m - \alpha), \tag{5}$$

where a is the leading coefficient of $g(x)$ and $g'(x)$ is the formal derivative of $g(x)$.

Proof. We define $V_0 := \{P_0\}$ and $V_{k+1} := \{P_\alpha \in \mathbb{P}_{\mathbb{F}_q(x_0)} |$ There is a $P_\beta \in V_k$ with $f(\alpha) = \beta^m\}$. First, we prove $V_{F_0}(T) = \cup_{k \geq 0} V_k$. The inclusion $V_{F_0}(T) \subseteq \cup_{k \geq 0} V_k$ follows from the recursive definition of T. In order to prove the reverse inclusion, we use induction on k. $V_0 \subseteq V_{F_0}(T)$ follows from Kummer theory, as $\gcd(d, m) = 1$. Now, suppose $\cup_{0 \leq i \leq k} V_k \subseteq V_{F_0}(T)$. Let $P_\alpha \in V_{k+1}$. By definition of V_{k+1}, there is a $\beta \in \bar{\mathbb{F}}_q$, with $\beta^m = g(\alpha)$ and $P_\beta \in V_k$. It follows from the theorem of Kummer (cf. [6, III.3.7]), that there is a place R of F_1, which lies above P_α and the zero $\tilde{P}$ of $x_1 - \beta$ in $\bar{\mathbb{F}}_q(x_1)$. Due to the induction hypothesis, there is a place $Q \in \mathbb{P}_{F_j}$ with $e(Q|P_\beta) > 1$ for some $j \in \mathbb{N}$. Since T is recursively defined, there is a place $\tilde{Q}$ in $\bar{\mathbb{F}}_q(x_1, x_2, \ldots, x_{j+1})$ over the zero $\tilde{P}$ of $x_1 - \beta$ in $\bar{\mathbb{F}}_q(x_1)$ with $e(\tilde{Q}|\tilde{P}) > 1$. Since $\gcd(d, m) = 1$, the place $\tilde{Q}$ is totally ramified in $\bar{\mathbb{F}}_q(x_1, \ldots, x_l)/\bar{\mathbb{F}}_q(x_1, \ldots, x_j)$ for all $l > j$. In particular, there exists an $l \in \mathbb{N}$, such that $e(\tilde{Q}_l|\tilde{P}) \nmid e(R|\tilde{P})$. Again from the theorem of Kummer, there exists a place $S \in \mathbb{P}_{F_s}$ with $S|R$ and $S|\tilde{Q}_k$. From Abhyankar's lemma, we obtain $e(S|R) > 1$. Thus, $e(S|P_\beta) > 1$.

Next, we show that the left hand side of Equation (5) divides the right hand side. Let $\beta \in \bar{\mathbb{F}}_q$ be a zero of the left hand side. If $\beta = 0$, β is a zero of the right hand side of multiplicity at least m, because $P_0 \in V_{F_0}(T)$. If $\beta \neq 0$, there is a $P_\gamma \in V_{F_0}(T)$ with $\gamma^m = g(\beta)$. Thus, $P_\beta \in V_{F_0}(T)$ due to the first part of the proof, and β is a zero of the right hand side. The multiplicity of β in $g(x) - \alpha$ is exceeding the multiplicity of the zero β in $g'(x)$ at most by one.

Comparison of the degrees and the leading coefficients in Equation (5) proves the lemma. $\qed$

Theorem 4. *Let $\mathcal{F}$ be a Fermat tower in characteristic p recursively defined by*

$$y^m = a(x + b)^m + c \text{ with } m > 2 \text{ and } ab^m + c = 0.$$

If the ramification locus $V_{F_0}(\mathcal{F})$ is finite, we have $(t - 1)m \equiv -1 \pmod{p}$, where $t := |\{\alpha^m | P_\alpha \in V_{F_0}(\mathcal{F})\}|$.

Proof. Due to lemma 1 we have

$$x^{m-1} \prod_{\alpha \in T} (a(x+b)^m + c - \alpha) = a^t(x+b)^{m-1} \prod_{\alpha \in T} (x^m - \alpha).$$

W.l.o.g. we can assume $b = 1$ carrying out the transformation $\tilde{x}_i = b^{-1}x_i$. Therefore, we obtain

$$(x+1) \prod_{c \neq \alpha \in T}((x+1)^m + \tfrac{c-\alpha}{a}) = x \prod_{0 \neq \alpha \in T}(x^m - \alpha)$$

$$\Leftrightarrow \ x \prod_{c \neq \alpha \in T}(x^m + mx^{m-1} + \binom{m}{2} x^{m-2} + \ldots + mx + 1 + \tfrac{c-\alpha}{a}) +$$

$$\prod_{c \neq \alpha \in T}(x^m + mx^{m-1} + \binom{m}{2} x^{m-2} + \ldots + mx + 1 + \tfrac{c-\alpha}{a})$$

$$= x^{(t-1)m+1} + \left(\sum_{0 \neq \alpha \in T} -\alpha\right) x^{(t-2)m+1} + \ldots + (-1)^{t-1} \prod_{0 \neq \alpha \in T} \alpha x.$$

Comparing the coefficients (at the exponent $(t-1)m$) yields

$$(t-1)m \equiv -1 \pmod{p}$$

(as $(t-2)m+1 < (t-1)m$).

4   On Relatively Unramified Towers

Let $q \neq 2,3$ be a prime power and $m = q - 1$. Moreover, let

$$K = \begin{cases} \mathbb{F}_{q^m} & \text{if } q \equiv 0 \pmod{4} \text{ or } q \equiv 3 \pmod{4} \\ \mathbb{F}_{q^{2m}} & \text{if } q \equiv 1 \pmod{4}. \end{cases}$$

We discuss recursively defined towers $\mathcal{T} = \cup_{i \geq 0} F_i$ over K with defining equation

$$y^m = 1 - \frac{x^m}{(x-1)^m}. \tag{6}$$

Remark: In case $q = 2,3$ one easily sees that Equation (6) does not define recursively a tower.

First, we want to use Theorem 2 in order to prove that Equation (6) defines recursively towers over specific constant fields.

Proposition 1. *The equation*

$$y^3 = 1 + \frac{x^3}{(x+1)^3} \tag{7}$$

defines recursively a tower $\mathcal{T} := \cup_{i \geq 0} F_i$ *over* $\mathbb{F}_{4^3}$.

Proof. Let α be a fixed primitive element of $\mathbb{F}_4$. Let P be a zero of $x_0^3 + x_0 + 1$ in F_0. Then P splits completely in the extension F_1/F_0 and we denote by P', Q_1 and Q_2 the places of F_1 above P. They satisfy: $x_1(P') = x_0(P)$, $x_1(Q_1) = \alpha x_0(P)$ and $x_1(Q_2) = \alpha^2 x_0(P)$. Setting $\mu_1 = x_0(P)$ and $\mu_0 = \alpha^2 x_0(P)$ we have $\mu_1 \leftarrow \mu_0$ and, moreover, that μ_0 is a root of $x_0^3 + \alpha x_0 + 1 = 0$. One checks that the place P_{μ_0} of F_0 is totally inert in F_1 and now the proposition follows from Theorem 2.

Remark: Equation (7) coincides with an equation for a Shimura tower given by Elkies (cf. [1, eq. 49,50]).

Using Theorem 2 we can prove for small prime powers, that Equation (6) defines towers. The results are listed in the tables below. The left column gives the size q of the constant field $\mathbb{F}_q$. The zeroes of the polynomial in the second column are totally inert and can be used as places P_{μ_0} from condition (∗∗) in Theorem 2. The third column gives the length $(n+1)$ of a shortest sequence $(\mu_k)_{0 \leq k \leq n}$ between a place from condition (∗) and a zero from the second column. In all the discussed examples, Theorem 2 guarantees that Equation (6) defines recursively a tower.

q	inert zeroes	length of the seq. (μ_k)
5	$x^2 + 4x + 2$	2
7	$x^{12} + 2x^{11} + 6x^{10} + 5x^8 + x^7 + 2x^6$ $+2x^5 + x^4 + 3x^3 + 2x^2 + 6$	3
11	$x^2 + 5x + 1$	2
13	$x^{20} + 2x^{19} + 2x^{18} + 4x^{17} + 2x^{16}$ $+4x^{15} + 4x^{14} + 5x^{13} + 10x^{12} + 10x^{11}$ $+6x^{10} + 7x^9 + 8x^8 + 9x^7 + 5x^6 + 2x^5$ $+7x^4 + 10x^3 + 8x^2 + 7x + 11$	3

q	inert zeroes	length of the seq. (μ_k)
4	$x^6 + x^4 + x^2 + x + 1$	2
8	$x^6 + x^4 + x^2 + x + 1$	2
16	$x^{20} + x^{17} + x^{14} + x^{12} + x^{11} + x^{10}$ $+x^9 + x^8 + x^7 + x^6 + 1$	2
32	$x^{10} + x^5 + x^4 + x^2 + 1$	2
64	$x^{42} + x^{36} + x^{30} + x^{29} + x^{21} + x^{18}$ $+x^{17} + x^{15} + x^{11} + x^{10} + x^9 + x^8$ $+x^6 + x^4 + x^3 + x + 1$	2
128	$x^{14} + x^7 + x^2 + x + 1$	2
256	$x^{24} + x^{19} + x^{18} + x^{16} + x^{14} + x^{13}$ $+x^{11} + x^6 + x^4 + x^2 + 1$	2

In the following, we suppose that Equation (6) defines recursively a tower $\mathcal{T} = \cup_{k \geq 0} F_k$.

Next, we want to show, that the towers defined by Equation (6) are unramified after a few steps.

Lemma 2. *Let $F := K(x, y)$ be defined by Equation (6).*

1. *Over $K(x)$ exactly the zeroes of $x - \alpha, \alpha \in \mathbb{F}_q^* \setminus \{1\}$, and the pole of x are ramified in F, each of index m.*
2. *Over $K(y)$ exactly the zeroes of $y - \alpha, \alpha \in \mathbb{F}_q^*$, are ramified in F, each of index m.*

Proof. The extensions $F/K(x)$ and $F/K(y)$ are Kummer extensions. Thus, the claims follow from (note that $m = q - 1$)

$$y^m = \frac{x^{m-1} + x^{m-2} + x^{m-3} + \ldots + x + 1}{(x-1)^m} = 1 - \left(\frac{x}{x-1}\right)^m.$$

Proposition 2. *Let T be defined recursively by Equation (6).*

1. *T is unramified over F_2, i.e. F_n/F_2 is unramified for all $n \geq 2$.*
2. *In F_2/F_1 exactly the places of F_1 over the place P_1 satisfying $x_0(P_1) = 1$ are ramified, each of index m.*

Proof. First, we prove (2). Let $Q \in \mathbb{P}_{F_1}$ be a place which is ramified in F_2. Then, $P := Q \cap K(x_1)$ ramifies in $K(x_1, x_2)$. Assume $x_1(P) = \alpha \in \mathbb{F}_q^* \setminus \{1\}$. Then, P ramifies in F_1 of index m, and due to Abhyankar's lemma the place Q is unramified in F_2. Thus, P is a pole of x_1. Then, again by Abhyankar's lemma, the place Q is ramified in F_2/F_1 of index m. The poles of x_1 in F_1 are exactly the places over the zero of $x_0 - 1$.

Assume there is a place $Q \in \mathbb{P}_{F_2}$ which is ramified in T. Then, Q has an extension $R \in \mathbb{P}_{F_j}$ for some $j \geq 2$, such that R ramifies in F_{j+1}. Analogous to the first part of the proof, we see that R is a pole of x_j. Then, $x_{j-1}(R) = 1$ and by Abhyankar's lemma the place R is unramified in F_{j+1} (since $j \geq 2$), a contradiction (cf. Figure 2).

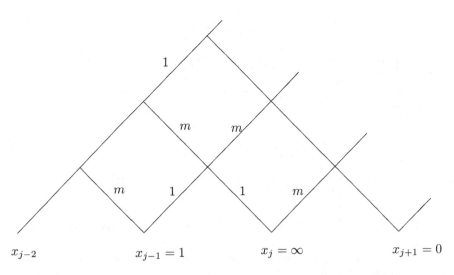

Fig. 2. Ramification of a place $R \in \mathbb{P}_{F_j}$

Corollary 2. *The genus of F_2 defined by Equation (6) is equal to*

$$g(F_2) = \frac{(m-2)(m-1)(m+1)}{2}.$$

Proof. The claim follows from the Hurwitz genus formula.

A tower $\mathcal{T} = \cup_{k \geq 0} F_k$ is called completely splitting, if there is an algebraic function field $F \subset \mathcal{T}$, such that at least one place of F is completely splitting in all $F_k \supset F$ (cf. [3]).

Proposition 3. *The tower $\mathcal{T}$ defined by Equation (6) is completely splitting over K. More precisely: Let Q be a place in F_2 over an $\mathbb{F}_q$-rational place of F_0. Then, the place Q is completely splitting in $\mathcal{T}/F_2$.*

Proof. First, we will investigate the places over $P_\alpha \in \mathbb{P}_{F_0}$ in F_3 ($\alpha \in \mathbb{F}_q \cup \{\infty\}$). In order to do so, we distinguish the possible cases and suppose that $2 \nmid q$:

1. Let $\alpha \in \mathbb{F}_q^* \setminus \{1\}$ or $\alpha = \infty$. There is exactly one totally ramified $\mathbb{F}_q$-rational place P' in F_1 over P_α with $x_1(P') = 0$. The place P' splits completely in F_2 into different $\mathbb{F}_q$-rational places Q in F_2 with $x_2(Q) \in \mathbb{F}_q^*$ (due to the theorem of Kummer, cf. [6, III.3.7]). Since F_3/F_2 is Galois, each of the places Q splits in F_3 into $\mathbb{F}_{q^m}$-rational places (cf. Proposition 2 and the degree formula).

2. Let $\alpha = 1$. We are substituting $z := x_1(x_0 - 1)$. Then we have $F_1 = F_0(z)$ and

$$z^m \equiv x_0^{m-1} + x_0^{m-2} + x_0^{m-3} + \ldots + x_0 + 1 \equiv -1 \pmod{P'}.$$

It follows, that P_α splits into $\mathbb{F}_{q^2}$-rational places P' in F_1, all of them having x_1 as a simple pole. Over each place P' there is exactly one totally ramified place Q in F_2 (cf. Proposition 2). Each of these places Q splits into $\mathbb{F}_{q^2}$-rational places in F_3 (due to the theorem of Kummer).

3. Now, let $\alpha = 0$. The place P_α splits into $\mathbb{F}_q$-rational places P' in F_1 with $x_1(P') = \beta \in \mathbb{F}_q^*$ and $v_{P'}(x_1 - \beta) = m$. First, we consider the case $\beta \neq 1$. Then, the place P' splits in F_2 into $\mathbb{F}_{q^m}$-rational places (with $x_2(Q) = 0$), which splits into $\mathbb{F}_{q^m}$-rational places in F_3. Now, we consider the case $\beta = 1$. The place P' splits into $\mathbb{F}_{q^2}$-rational places in F_2. Since F_3/F_2 is Galois, these places split into $\mathbb{F}_{q^{2m}}$-rational places Q' in F_3. Because of the definition of K we have to prove, that the places Q' are $\mathbb{F}_{q^m}$-rational, if $q \equiv 3 \pmod 4$. To this end, let $Q' \in \mathbb{P}_{F_3}$ with $x_0(Q') = 0, x_1(Q') = 1, x_2(Q') = \infty$ and $x_3(Q') = 0$. We use the notation $z = x + \mathcal{O}(nQ')$, if $v_{Q'}(z - x) \geq n$ holds. Let $\alpha \in \mathbb{F}_{q^2} \leq \mathbb{F}_{q^m}$ with $\alpha^m = -1$. Then, we have

$$x_1^m = 1 + t^m, \text{ where } t = \frac{\alpha x_0}{x_0 - 1},$$

thus $x_1 = 1 - t^m + t^{2m} + \mathcal{O}(3mQ')$ and $(x_1 - 1)^{-1} = -t^{-m}(1 + t^m + \mathcal{O}(2mQ'))$. As

$$((x_1 - 1)x_2)^m = (-t^m + t^{2m} + \mathcal{O}(3mQ'))^m - 1 - t^m,$$

we have moreover

$$\frac{(x_1 - 1)x_2}{\alpha} = 1 - t^m + t^{2m} + \mathcal{O}(3mQ')$$

(as $q \neq 3$). We obtain

$$x_2 = \frac{\alpha - \alpha t^m + \alpha t^{2m} + \mathcal{O}(3mQ')}{(x_1 - 1)} = -\alpha t^{-m} + \mathcal{O}((-m+1)Q'),$$

and $(x_2 - 1)^{-1} = -\alpha^{-1} t^m + \mathcal{O}((m+1)Q')$. Now,

$$x_3^m = \frac{(x_2 - 1 + 1)^{m-1} + (x_2 - 1 + 1)^{m-2} + \mathcal{O}(-m(m+3)Q')}{(x_2 - 1)^m}$$

$$= \frac{1}{(x_2 - 1)} - \frac{1}{(x_2 - 1)^2} + \mathcal{O}(3mQ').$$

This gives

$$\left(\frac{x_3}{t}\right)^m = -\alpha^{-1} + \mathcal{O}(Q').$$

$-\alpha^{-1}$ is an m-th power in $\mathbb{F}_{q^m}$, if and only if α is an m-th power in $\mathbb{F}_{q^m}$. We have $\alpha^q = -\alpha$ and $\alpha^{q^2} = \alpha$, which means $\alpha^{q^i} = (-1)^i \alpha$. Now, α is an m-th power in $\mathbb{F}_{q^m}$, if there is a $\beta \in \mathbb{F}_{q^m}$ with $\beta^m = \alpha$. As $(-1)(-1)^{\frac{m}{2}} = \alpha^{\frac{q^m-1}{q-1}} = \beta^{q^m-1}$ we obtain that β is in $\mathbb{F}_{q^m}$ if and only if $q \equiv 3 \pmod 4$.

The case $2|q$ is analogous; since $1 \equiv -1 \pmod 2$ all the places are completely splitting over $\mathbb{F}_{q^m}$.

Now, let $k \geq 3$ and $Q'|P_\alpha$ be a K-rational place in F_k. Then, we have $x_{k-2}(Q') \in \mathbb{F}_q \cup \{\infty\}$, and the place $Q' \cap K(x_{k-2}, x_{k-1}, x_k)$ is completely splitting (over K) in $K(x_{k-2}, x_{k-1}, x_k, x_{k+1})$, which follows from the discussion above. Due to [6, III.8.4] the place Q' is completely splitting in F_{k+1}.

Corollary 3. *In the tower T/K there are at least $2m^2 + m$ places of F_2 completely splitting in F_k/F_2 for all $k \geq 2$.*

Theorem 5. *Let T be a relatively unramified recursively defined tower over $\mathbb{F}_q$. Let $F < T$ be a function field, such that T/F is unramified. Then we have:*

$$\lambda(T) \geq \frac{t}{g(F) - 1},$$

where t is the number of places of F which are completely splitting.

Proof. The claim follows directly from the Hurwitz genus formula and the definition of $\lambda(T)$ (cf. [3, Theorem 2.24]).

Theorem 6. *Suppose that Equation (6) defines recursively a tower T over the finite field K. Then its limit satisfies:*

$$\lambda(T) \geq \frac{4m + 2}{m^2 - 2m - 1}.$$

Proof. The claim follows from Theorem 5, from Corollary 3 and Corollary 2.

References

1. N. D. Elkies: Explicit modular towers. In: Proceedings of the Thirty-Fifth Annual Allerton Conference on Communication, Control and Computing. T. Basar and A. Vardy (Eds.), 1997, 23-32.
2. A. Garcia, H. Stichtenoth: A tower of Artin-Schreier extensions of function fields attaining the Drinfeld-Vladut bound. Invent. Math. 121 (1995), 211-222.
3. A. Garcia, H. Stichtenoth: On Tame Towers over finite fields. In: J. Reine Angew. Math. 557 (2003), 53 - 80.
4. H.W. Lenstra: On a problem of Garcia, Stichtenoth, and Thomas. In: Finite Fields Appl. 8 (2001), 166 - 170.
5. F. Özbudak, M. Thomas: A Note on Towers Of Function Fields Over Finite Fields. In: Communications in Algebra, 26, 11 (1998).
6. H. Stichtenoth: Algebraic Function Fields and Codes. Springer Verlag. Berlin, New York and Heidelberg, 1993.
7. M. A. Tsfasman, S. G. Vladut and T. Zink: Modular Curves, Shimura curves and Goppa codes better than the Varshamov-Gilbert bound. Math. Nachr. 109 (1982), 21 - 28.
8. J. Wulftange: Zahme Türme algebraischer Funktionenkörper. Dissertation. Essen, 2002.

The Covering Radius of Some Primitive Ternary BCH Codes

Ralf Franken and Stephen D. Cohen

Department of Mathematics, University of Glasgow, University Gardens, Glasgow
G12 8QW, Scotland rf@maths.gla.ac.uk, sdc@maths.gla.ac.uk

Abstract. Let $\mathcal{C}$ be the primitive ternary BCH code of length $3^m - 1$ with designed distance δ. It is shown that, when $\delta = 8$, then the covering radius of $\mathcal{C}$ is 7 whenever $m \geq 20$ and m is even, and when $\delta = 14$, then the covering radius of $\mathcal{C}$ is 13 whenever $m \geq 46$. The technique involves Galois-theoretic criteria on the splitting of polynomials over finite fields.

1 The Problem

Let $q = 3^m$ and $\mathcal{C} = \mathcal{C}(m, \delta)$ be the primitive ternary (narrow sense) BCH code of length $q-1$ with designed distance δ. We consider here only the case $\delta = 3t-1$. For this it has been shown by Kaipainen ([9], Theorem 3.0.1) that the covering radius ρ of $\mathcal{C}$ is at most δ whenever $q > q_0$, where q_0 depends only on δ. On the other hand it follows from the "Supercode Lemma" (cf. [2]) that $\rho \geq \delta - 1$ for sufficiently large q.

In this paper we identify two situations where ρ attains this lower bound for an infinite range of q exceeding an explicit value.

Theorem 1.

  (i) *Suppose $\delta = 8$. Then the covering radius of $\mathcal{C}$ is 7 whenever $m \geq 20$ and m is even.*
 (ii) *Suppose $\delta = 14$. Then the covering radius of $\mathcal{C}$ is 13 whenever $m \geq 46$.*

To prove this it is necessary to modify significantly a method that was originally used by the second author in [4] to establish the corresponding results about the covering radius of binary primitive BCH codes.

We remark that with the given small values of t it is convenient to settle certain problems by explicit calculations, e.g. with Maple. We have in hand, however, a study of the whole problem for designed distance $\delta = 3t - 1$ and $\delta = 3t$ for general t that avoids the use of a computer.

2 Algebraic Formulation

The first step of the method is the transformation of the coding theoretical problem into one of the splitting of a certain polynomial over $\mathbb{F}_q$. This part is analogous to the procedure in the binary case [4]. We describe it rather briefly.

G. Mullen, A. Poli and H. Stichtenoth (Eds.): Fq7 2003, LNCS 2948, pp. 166–180, 2003.

We start from an idea that goes back to Helleseth [8]: to prove $\rho \leq \delta - 1 = 3t - 2$ it suffices to show that for any choice of $a_k \in \mathbb{F}_q$ $(1 \leq k \leq 3t - 2, 3 \nmid k)$ the system of equations

$$
\begin{array}{lllll}
\varepsilon_1 x_1 & + \varepsilon_2 x_2 & + \ldots & + \varepsilon_{3t-2} x_{3t-2} = a_1 \\
\varepsilon_1 x_1^2 & + \varepsilon_2 x_2^2 & + \ldots & + \varepsilon_{3t-2} x_{3t-2}^2 = a_2 \\
\vdots & \vdots & & \vdots & \vdots \\
\varepsilon_1 x_1^{3t-2} & + \varepsilon_2 x_2^{3t-2} & + \ldots & + \varepsilon_{3t-2} x_{3t-2}^{3t-2} = a_{3t-2}
\end{array}
\tag{1}
$$

has a solution $(\varepsilon_1, \ldots, \varepsilon_{3t-2}), (x_1, \ldots, x_{3t-2}) \in \mathbb{F}_q^{3t-2}$.

To fill the "gaps" for $3 | k$ in the system we add redundant equations of the same form with $a_{3j} := a_j^3$ $(j = 1, \ldots, t - 1)$. Although there is freedom in the choice of the ε_i, in view of subsequent manipulations we fix $\varepsilon_i := 1$ for all $i = 1, \ldots, 3t - 2$. Then, by replacing all x_i with $x_i - a_1$, it can be assumed that $a_1 = 0$. Writing $\sigma_k := x_1^k + \cdots + x_{3t-2}^k$, we may abbreviate the system (1) to

$$
\boxed{\sigma_k = a_k \quad (k = 1, \ldots, 3t - 2), \quad a_1 = 0.}
\tag{2}
$$

Now let a fixed system (2) be given. Let s_k denote the k-th elementary symmetric polynomial in $x_1, \ldots, x_{3t-2}$, $s_0 := 1$, $\sigma_0 := 1$. Newton's identities imply that a solution to (2) satisfies

$$
s_k = \frac{1}{k} \sum_{l=0}^{k-1} (-1)^{k-l-1} a_{k-l} s_l, \qquad 1 \leq k \leq 3t - 2, \ 3 \nmid k,
\tag{3}
$$

with $s_k = s_k(x_1, \ldots, x_{3t-2})$. Conversely it is evident that if we start from arbitrary elements $s_3, \ldots, s_{3t-3} \in \mathbb{F}_q$ and use (3) and $s_1 := a_1 = 0$ to define $s_1, s_2, s_4, \ldots, s_{3t-2} \in \mathbb{F}_q$ recursively, then the roots of the polynomial

$$
f(x) := \sum_{i=0}^{3t-2} (-1)^i s_i x^{(3t-2)-i} \in \mathbb{F}_q[x]
$$

in the algebraic closure $\overline{\mathbb{F}}_q$ of $\mathbb{F}_q$ form a solution to (2). Hence it suffices if for any choice of $a_k \in \mathbb{F}_q$ $(3 \nmid k)$, $a_1 = 0$, we manage to find $s_3, \ldots, s_{3t-3} \in \mathbb{F}_q$ such that $f(x)$ splits completely over $\mathbb{F}_q$.

We re-write $f(x)$ expressing the s_k for which $3 \nmid k$ in terms of the s_k for which $3 | k$. Set $A_0 := 1$ and define recursively for $k = 1, \ldots, 3t - 2$,

$$
A_k := \begin{cases} 0, & \text{if } 3 | k, \\ \dfrac{1}{k} \sum_{l=0}^{k-1} (-1)^{k-l-1} a_{k-l} A_l, & \text{if } 3 \nmid k. \end{cases}
$$

(This transition from $a_1, \ldots, a_{3t-2}$ to $A_0, \ldots, A_{3t-2}$ is reversible; we can therefore assume that (2) is given in the latter form. Also note that $A_1 = 0$.) Then, from (3),

$$s_k = \sum_{\mu=0}^{\lfloor k/3 \rfloor} A_{k-3\mu} s_{3\mu} \qquad \text{for all } 1 \le k \le 3t - 2 \ .$$

Substituting this into $f(x)$ and rearranging terms, we finally arrive at

$$f(x) = \sum_{k=0}^{t-1} (-1)^k s_{3k} \left[x^{3t-2-3k} + x^2 g^{(2)}_{t-k-2}(x^3) + g^{(0)}_{t-k-2}(x^3) \right], \qquad (4)$$

where $g^{(0)}_{-1}, g^{(2)}_{-1} := 0$, and, for $d = 0, \ldots, t - 2$,

$$g^{(0)}_d(x^3) := \sum_{i=0}^{d} (-1)^{d+i} A_{3d-3i+4}\, x^{3i},$$

$$g^{(2)}_d(x^3) := \sum_{i=0}^{d} (-1)^{d+i} A_{3d-3i+2}\, x^{3i}$$

are polynomials of degree at most $3d$.

The task is now to show for $t = 3, 5$ that for any choice of $A_k \in \mathbb{F}_q$ ($1 \le k \le 3t - 2$, $3 \nmid k$), $A_1 = 0$, we can find $s_3, \ldots, s_{3t-3} \in \mathbb{F}_q$ such that (4) splits completely over $\mathbb{F}_q$. Note that we can assume at least one of the A_k is non-zero (otherwise choose $s_3, \ldots, s_{3t-3} = 0$). For notational convenience we replace the s_{3k} for odd k and the $A_{3k\pm1}$ for even k with their negatives; thus (4) simplifies to

$$f(x) = \sum_{k=0}^{t-1} s_{3k} \left[x^{3t-2-3k} + x^2 g^{(2)}_{t-k-2}(x^3) + g^{(0)}_{t-k-2}(x^3) \right] \qquad (5)$$

where the explicit forms of the relevant polynomials $g^{(0)}_d$, $g^{(2)}_d$ are:

$$g^{(0)}_0(x^3) = A_4,$$
$$g^{(0)}_1(x^3) = A_4 x^3 + A_7,$$
$$g^{(0)}_2(x^3) = A_4 x^6 + A_7 x^3 + A_{10},$$
$$g^{(0)}_3(x^3) = A_4 x^9 + A_7 x^6 + A_{10} x^3 + A_{13},$$

$$g^{(2)}_0(x^3) = A_2,$$
$$g^{(2)}_1(x^3) = A_2 x^3 + A_5,$$
$$g^{(2)}_2(x^3) = A_2 x^6 + A_5 x^3 + A_8,$$
$$g^{(2)}_3(x^3) = A_2 x^9 + A_5 x^6 + A_8 x^3 + A_{11}.$$

In the next section we state the necessary auxiliary results for our proof of Theorem 1, along with some terminology and notation.

3   Sufficient Criteria

Throughout this section let $f_0, f_1 \in \mathbb{F}_q[x]$ be monic polynomials with $n :=$ deg $f_0 >$ deg $f_1 \geq 2$ and $f_0/f_1 \notin \mathbb{F}_q(x^3)$, and let $F_u(x) := f_0(x) + u f_1(x)$, with u an indeterminate. Denote by $\overline{G}$ the Galois group of F_u over $\overline{\mathbb{F}}_q(u)$ and by G the Galois group of F_u over $\mathbb{F}_q(u)$ (sometimes called the "geometric" and "arithmetic" monodromy group, respectively), viewed as permutation groups on the roots of F_u. Then always $\overline{G} \subseteq G$.

Lemma 1.   *Suppose $\overline{G} = G$. Then there exists an $\alpha \in \mathbb{F}_q$ such that F_α splits completely over $\mathbb{F}_q$ provided $q \geq [(n-2) \cdot n!]^2$.*

Proof. [5], Lemma 5.1 plus remark on p. 325.

   Let $\mathbb{K}$ be a field. A polynomial $g \in \mathbb{K}[x]$ will be called *simple* if all its factors over its splitting field have multiplicity 1 except for exactly one which has multiplicity 2.

   A rational function g/h with $g, h \in \mathbb{K}[x]$ is called *(functionally) decomposable over* $\mathbb{K}$ if there exist rational functions $Q = Q_1/Q_2$, $R = R_1/R_2$ with $Q_1, Q_2, R_1, R_2 \in \mathbb{K}[x]$, $\gcd(Q_1, Q_2) = \gcd(R_1, R_2) = 1$, such that $g(x)/h(x) = Q(R(x))$ and neither Q nor R is a linear fraction. If g/h is not decomposable it is called *indecomposable*.

   Suppose g/h is decomposable with decomposition as above. Write $\omega_i := $ deg Q_i and $\rho_i := $ deg R_i $(i = 1, 2)$. If deg $g >$ deg h, it is always possible to arrange that

$$\frac{g(x)}{h(x)} = c \cdot Q(R(x)) \tag{6}$$

with $c \in \mathbb{K}$, Q_1, Q_2, R_1, R_2 monic, $\omega_1 > 1$, $\rho_1 > 1$, $\omega_1 > \omega_2$, $\rho_1 > \rho_2$. Unless otherwise mentioned, assume all decompositions in this paper have been normalised in this way.

   Writing out (6), we obtain

$$\frac{g(x)}{h(x)} = \frac{R_2^{\omega_1}(x) \cdot Q_1(R_1(x)/R_2(x))}{R_2^{\omega_1 - \omega_2}(x) \cdot [R_2^{\omega_2}(x) \cdot Q_2(R_1(x)/R_2(x))]} \tag{7}$$

where the numerator and the expression in square brackets on the right-hand side are polynomials. Suppose g and h are co-prime, then we can equate numerators and denominators and conclude

$$\text{deg } g = \omega_1 \rho_1, \qquad \text{deg } h = (\omega_1 - \omega_2)\rho_2 + \omega_2 \rho_1 ; \tag{8}$$

in particular, g/h must then be indecomposable whenever deg g is prime or deg $h = 1$.

   We return to Lemma 1 and give sufficient conditions for $\overline{G} = G$. Let $\mathcal{S}_n, \mathcal{A}_n$ denote the symmetric group and alternating group, respectively, of order n.

Lemma 2.   *Suppose F_u satisfies all of the following conditions:*

    (i) *f_0 and f_1 are co-prime,*
    (ii) *f_0/f_1 is indecomposable over $\mathbb{F}_q$,*
    (iii) *F_β is simple for some $\beta \in \overline{\mathbb{F}}_q$.*

Then $\overline{G} = G = \mathcal{S}_n$.

Proof. [5], Lemma 3.1.—In combination with (i), (ii) implies that G is a primitive group and (iii) that $\overline{G}$ contains a transposition. This suffices to give $\overline{G} = \mathcal{S}_n$. $\square$

If F_β'' is identically zero (irrespective of the choice of β) then it is impossible to establish condition (iii). In characteristic 2 an alternative is to require that $\deg f_0 - \deg f_1 = 2$ and f_1 be square-free, which again yields a transposition in $\overline{G}$. Though this remains true in the ternary case, it is not helpful in the situation of Sect. 2, because the degrees of the summands of (5) corresponding with the s_{3k} decrease in steps of three. Instead we use the following criterion, writing Δ_x for the discriminant of $F_u(x)$ as a polynomial in x over $\overline{\mathbb{F}}_q(u)$.

Lemma 3.   *Suppose F_u satisfies all of the following conditions:*

    (i) *f_0 and f_1 are co-prime,*
    (ii) *f_0/f_1 is indecomposable over $\mathbb{F}_q$,*
    (iii) *$\deg f_0 - \deg f_1 = 3$ and f_1 has no factor of multiplicity divisible by 3,*
    (iv) *Δ_x is either a non-square in $\overline{\mathbb{F}}_q(u)$ ("situation S") or a square in $\mathbb{F}_q(u)$ ("situation A").*

Then $\overline{G} = G$, and this group is equal to $\mathcal{S}_n$ (in "situation S") or $\mathcal{A}_n$ (in "situation A").

Proof. As before, (i) and (ii) ensure that G is a primitive group. From (iii) we obtain a 3-cycle in $\overline{G}$ (from ramification "at infinity"; see [5], Sect. 4, with $e_r = e_r(\infty) = 3$ at the beginning of the second paragraph on p. 330). Therefore $\mathcal{A}_n \subseteq \overline{G}$. Now (iv) implies that either $\mathcal{S}_n = \overline{G} \subseteq G \subseteq \mathcal{S}_n$ or $\mathcal{A}_n \subseteq \overline{G} \subseteq G \subseteq \mathcal{A}_n$. $\square$

Our strategy for a proof of Theorem 1 will now be as follows. Express the polynomial $f(x)$ as $F_u(x) = f_0(x) + uf_1(x)$, $u \in \mathbb{F}_q^*$, where f_0 or f_1 depends on a parameter $v \in \mathbb{F}_q$, and then show that v can be chosen in a way that the conditions of Lemma 2 or Lemma 3 are satisfied. This is usually achieved by showing that the number of v, even in $\overline{\mathbb{F}}_q$, for which one of the conditions is NOT satisfied is (much) smaller than q. Then Lemma 1 applies and proves the existence of a suitable choice of u for which $f(x)$ splits completely, as well as (with $n = 3t - 2$) the explicit lower bounds for a sufficient size of q stated in Theorem 1.

4   Designed Distance Eight

This is equivalent to taking $t = 3$, and we have the choice of s_3 and s_6. Put $s_3 := u$ and $s_6 := uv$ with $u \in \mathbb{F}_q^*$, $v \in \mathbb{F}_q$. Then $f(x) = F_u(x) = f_0(x) + u f_1(x)$ with

$$f_0(x) = x^7 + A_2 x^5 + A_4 x^3 + A_5 x^2 + A_7 ,$$
$$f_1(x) = x^4 + A_2 x^2 + v x + A_4 .$$

Distinguish four cases as follows.

(1) A_4 and A_7 are not both zero, A_2 and A_4 are not both zero.
(2) A_4 and A_7 are not both zero, $A_2 = A_4 = 0$.
(3) $A_4 = A_7 = 0$, $A_2 \neq 0$.
(4) $A_4 = A_7 = A_2 = 0$.

We show that in all cases $v \in \mathbb{F}_q$ can be found such that Lemma 3 applies.

Case (1). Suppose v is such that f_0 and f_1 have a common root $\gamma \in \overline{\mathbb{F}}_q$. By assumption $\gamma \neq 0$, so $f_1(\gamma) = 0$ determines $v = -(\gamma^4 + A_2 \gamma^2 + A_4)/\gamma \in \overline{\mathbb{F}}_q$ uniquely. Since there are at most seven non-zero roots of f_0 in $\overline{\mathbb{F}}_q$, there are at most seven values v in $\mathbb{F}_q$ for which f_0 and f_1 are not co-prime. Exclude these from further consideration.

Indecomposability of f_0/f_1 is clear (for all remaining v) because deg $f_0 = 7$ is prime.

To establish condition (iii) of Lemma 3 we show that with few exceptions f_1 is square-free. Suppose f_1 has a repeated root γ in $\overline{\mathbb{F}}_q$. Then

$$f_1(\gamma) = \gamma^4 + A_2 \gamma^2 + v\gamma + A_4 = 0 , \tag{9}$$
$$f_1'(\gamma) = \gamma^3 - A_2 \gamma + v = 0 . \tag{10}$$

Taking $(9) - \gamma \cdot (10)$, one obtains $-A_2 \gamma^2 + A_4 = 0$, which implies $A_2 \neq 0$. Solving for γ and substituting in (10) we see that there are at most two values of v, namely $\pm\sqrt{A_4/A_2}\,(A_2 - A_4/A_2)$, for which f_1 is not square-free and which we exclude.

Finally, Maple was used to compute Δ_x, a polynomial in u of degree ≤ 5. If $A_2 \neq 0$ then the coefficient of u^5, as a polynomial in v, has leading term $-A_2^6 v^5$. This is non-zero provided we avoid $v = 0$, so that in this case Δ_x is a non-square in $\mathbb{F}_q(u)$. Otherwise, if $A_2 = 0$ but $A_5 \neq 0$, we must have $A_4 \neq 0$; then the coefficient of u^5 in Δ_x is $A_4^3 A_5^3 \neq 0$ and Δ_x is again a non-square in $\mathbb{F}_q(u)$. Finally, if $A_2 = A_5 = 0$ (whence again $A_4 \neq 0$), then $\Delta_x = -A_4^3 (A_4 v + A_7)^3 u^3 - A_7^6$, so that if we avoid $v = -A_7/A_4$, then once more Δ_x is a non-square in $\mathbb{F}_q(u)$.

In summary, all four conditions of Lemma 3 hold for all but at most ten values $v \in \mathbb{F}_q$.

Case (2). In this case $A_7 \neq 0$. Choose $v = 0$ (any other choice of v would lead to a root of multiplicity three in f_1). Then $f_0(x) = x^7 + A_5 x^2 + A_7$, $f_1(x) = x^4$ and the conditions (i)–(iii) of Lemma 3 are clearly satisfied.

By Maple, $\Delta_x = A_5^3 A_7^3 u^2 + A_5^5 A_7^2 u + A_7^7 A_7 - A_5^6$. If $A_5 \neq 0$ then the discriminant of Δ_x as a polynomial in u over $\mathbb{F}_q$ is $A_5^3 A_7^9 \neq 0$, hence Δ_x is a non-square in $\overline{\mathbb{F}}_q(u)$. If $A_5 = 0$ then $\Delta_x = -A_7^6$; this is a square in $\mathbb{F}_q(u)$ exactly if -1 is a square in $\mathbb{F}_q$, i.e. exactly if m is even. (This is where the restriction on m is vital.)

Case (3). Here $f_0(x)$ and $f_1(x)$ have always a common factor x. Work instead with $f_0^*(x) := f_0(x)/x = x^6 + A_2 x^4 + A_5 x$ and $f_1^*(x) := f_1(x)/x = x^3 + A_2 x + v$. (Lemma 1 can be applied to a polynomial of lower degree than f_0, and if $F_u^*(x) := f_0^*(x) + u f_1^*(x)$ splits then so does $F_u(x) = x \cdot F_u^*(x)$. Henceforth we omit the asterisks.)

Obviously one has to avoid $v = 0$ to make f_0 and f_1 co-prime. Apart from that we see, adapting the argument from Case (1), that at most five more values of $v \in \mathbb{F}_q$ (given by the non-zero roots of f_0) have to be excluded for co-primality.

Suppose f_0/f_1 is decomposable. From (8), only two types of decomposition can occur: $(\omega_1, \omega_2, \rho_1, \rho_2) = (2, 1, 3, 0)$ or $(3, 0, 2, 1)$. In the first case put $Q_1(x) := x^2 + Ax + B$, $Q_2(x) := x + C$, $R_1(x) := x^3 + Dx^2 + Ex + F$. Comparing $Q_2(R_1(x)) = x^3 + Dx^2 + Ex + F + C$ with $f_1(x) = x^3 + A_2 x + v$ implies $D = 0$ and $E = A_2 \neq 0$. But then from the quadratic coefficients of $Q_1(R_1(x))$ and $f_0(x)$ we find $AD + E^2 - DF = 0$, a contradiction. Hence this type of decomposition cannot occur. Similarly, in the second case we would have to have $f_1(x) = R_2^3(x)$, hence $A_2 = 0$, which is also a contradiction. Therefore f_0/f_1 is indecomposable (regardless of v).

As $f_1'(x) = A_2 \neq 0$, f_1 is square-free for all $v \in \mathbb{F}_q$.

Finally, $\Delta_x = -A_2^3 (A_5 + A_2 v)^3 u^3 - A_5^6$. Avoiding $v = -A_5/A_2$ makes Δ_x a non-square in $\overline{\mathbb{F}}_q(u)$, so that again all conditions of Lemma 3 are satisfied for a suitable choice of v.

Case (4). Again choose $v = 0$ and divide $f_0(x)$ and $f_1(x)$ by their common factor x^2. Then $f_0(x) = x^5 + A_5$ and $f_1(x) = x^2$. These obviously satisfy conditions (i)–(iii) of Lemma 3, and $\Delta_x = -A_5^4$ is a square in $\mathbb{F}_q(u)$ provided m is even. This completes the proof of part (i) of Theorem 1.

5   Designed Distance Fourteen

Now $t = 5$, and we have the choice of s_3, s_6, s_9 and s_{12}.

We distinguish three cases. If $A_2, \dots, A_{10}$ are not all zero, let $j \in \{1, 2, 3\}$ be minimal such that A_{3j-1} and A_{3j+1} are not both zero, and put $A_{3j-1} =: C_2$, $A_{3j+1} =: C_0$. The three cases are

(I) $A_2 = \dots = A_{10} = 0$,
(II) $C_0 g_3^{(2)} \neq C_2 g_3^{(0)}$,
(III) $C_0 g_3^{(2)} = C_2 g_3^{(0)}$.

Case (I). Choose $s_3 := u \in \mathbb{F}_q^*$. (All other s_{3k} are understood to be zero.)

If $A_{13} \neq 0$ then $f_0(x) = x^{13} + A_{11}x^2 + A_{13}$ and $f_1(x) = x^{10}$, whereas if $A_{13} = 0$ (then $A_{11} \neq 0$) we divide out the common factor x^2 and work with $f_0(x) = x^{11} + A_{11}$ and $f_1(x) = x^8$. In both cases conditions (i)–(iii) of Lemma 3 are obviously satisfied, f_0 being each time of prime degree.

To establish condition (iv), we used Maple to study the discriminants. Consider first the situation where $A_{13} \neq 0$. If $A_{11} = 0$ then $\Delta_x = A_{13}^{12}$, a square in $\mathbb{F}_q(u)$. Otherwise, if A_{11} is non-zero, the discriminant of Δ_x as a polynomial in u is $A_{13}^{21}A_{11}^3 \neq 0$, therefore Δ_x is a non-square in $\overline{\mathbb{F}}_q(u)$. Finally, in the situation where $A_{13} = 0$ we have $\Delta_x = A_{11}^{10}$, again a square in $\mathbb{F}_q(u)$. Hence in all cases Lemma 3 applies.

Case (II). In this case (only) we use Lemma 2.

First observe that under the assumption of this case one of the coefficients $A_{13}, A_{10}, \ldots, A_{3j+1} = C_0$ must be non-zero, i.e. $g_3^{(0)} \neq 0$. If $C_0 = 0 = A_{13}$, fix l with $j < l < 4$ such that $A_{3l+1} \neq 0$ and put $w := 1$. Otherwise, if one of C_0, A_{13} is non-zero, put $w := 0$ (then the choice of l will be irrelevant).

Now choose $s_{12-3j} := u \in \mathbb{F}_q^*$, $s_{12-3l} := w$ (ignore this if $w = 0$) and $s_{12} := v \in \mathbb{F}_q$. Then $f(x) = F_u(x) = f_0(x) + uf_1(x)$ with

$$f_0(x) = x^{13} + x^2 g_3^{(2)}(x^3) + g_3^{(0)}(x^3) + \\ + wx^{3l+1} + wx^2 g_{l-1}^{(2)}(x^3) + wg_{l-1}^{(0)}(x^3) + vx,$$

$$f_1(x) = x^{3j+1} + C_2 x^2 + C_0,$$

where the w-terms in $f_0(x)$ are present if and only if $C_0 = A_{13} = 0$.

(i) *Co-primality.* By construction, one of f_0, f_1 has a non-zero constant term (A_{3l+1} being the constant term of $g_{l-1}^{(0)}$), hence x is not a common factor of $f_0(x)$ and $f_1(x)$. Argue as in Case (1) of Sect. 4; since $\deg f_1 = 3j + 1 \leq 10$, at most ten values $v \in \mathbb{F}_q$ have to be excluded to ensure that f_0 and f_1 are co-prime.

(ii) *Indecomposability.* This follows immediately from $\deg f_0 = 13$.

(iii) *Simplicity of F_β for a suitable $\beta \in \mathbb{F}_q$.* We employ three steps.

STEP 1. We show that for all but a bounded number of v there exists $\beta \in \overline{\mathbb{F}}_q$ such that F_β has a repeated factor.

Suppose $v \in \overline{\mathbb{F}}_q$ is such that F_β is square-free for all $\beta \in \overline{\mathbb{F}}_q^*$. Then the system

$$F_u(x) = f_0(x) + uf_1(x) = 0,$$
$$F_u'(x) = f_0'(x) + uf_1'(x) = 0 \tag{11}$$

has no solution $(u, x) = (\beta, \gamma) \in \overline{\mathbb{F}}_q^* \times \overline{\mathbb{F}}_q$.

Now consider $E(x) := f_0'(x)f_1(x) - f_0(x)f_1'(x)$. A root γ of $E(x)$ such that $f_1'(\gamma) \neq 0$ would imply a solution to (11) by putting $\beta := -f_0'(\gamma)/f_1'(\gamma)$. Hence every root of E must also be a root of the fixed polynomial f_1'. As $\deg f_1' = 3j$

and E turns out to be of degree 11 or less, this limits the number of possible choices for E by

$$\binom{11 + 3j}{3j} \le 167960 \ .$$

Since every choice of E admits at most one value of v, this is also the maximal number of values v we have to exclude from $\mathbb{F}_q$ to guarantee the existence of an F_β with a repeated factor.

STEP 2. We show that, except for few v, the polynomial F_β has no factor of multiplicity greater than two.

Assume that $v \in \mathbb{F}_q$ is such that $F_\beta(x)$ has a root γ of multiplicity ≥ 3. Observe first that if $\gamma = 0$ then $0 = F'_\beta(\gamma) = v$, i.e. by avoiding $v = 0$ we can assume that $\gamma \ne 0$ and henceforth divide by γ. Furthermore, γ satisfies both

$$f'_0(x)f_1(x) - f_0(x)f'_1(x) = 0 \tag{12}$$

and

$$f''_0(x)f_1(x) - f_0(x)f''_1(x) = 0 \ . \tag{13}$$

By evaluating their left-hand sides, (12) becomes

$$x^{3j+2}g_3^{(2)}(x^3) + wx^{3j+2}g_{l-1}^{(2)}(x^3) - C_2x^{14} - wC_2x^{3l+2} + C_0x^{12} -$$
$$- C_0xg_3^{(2)}(x^3) + wC_0xg_{l-1}^{(2)}(x^3) - x^{3j}g_3^{(0)}(x^3) - wx^{3j}g_{l-1}^{(0)}(x^3) + \tag{14}$$
$$+ C_2xg_3^{(0)}(x^3) + wC_2xg_{l-1}^{(0)}(x^3) + v(-C_2x^2 + C_0) = 0$$

and (13) becomes

$$-x^{3j+1}g_3^{(2)}(x^3) - C_0g_3^{(2)}(x^3) - wx^{3j+1}g_{l-1}^{(2)}(x^3) - wC_0g_{l-1}^{(2)}(x^3) +$$
$$+ C_2x^{13} + C_2g_3^{(0)}(x^3) + wC_2x^{3l+1} + wC_2g_{l-1}^{(0)}(x^3) + vC_2 = 0 \ . \tag{15}$$

Suppose $C_2 = 0$. Then (15) says that γ is a root of a polynomial of degree at most $7 + 3j$ and independent of v, and (14) says that γ determines a unique v. Therefore we have to exclude at most sixteen values $v \in \mathbb{F}_q$ to make sure that F_β has no triple or higher factor.

Otherwise, if $C_2 \ne 0$, we can eliminate v from (14) and (15) to obtain

$$\frac{1}{C_2x} \cdot f_1(x) \cdot \left[C_0g_3^{(2)}(x^3) - C_2g_3^{(0)}(x^3) + w(C_0g_{l-1}^{(2)}(x^3) - C_2g_{l-1}^{(0)}(x^3)) \right] = 0.$$

The polynomial in square brackets is a fixed polynomial of degree ≤ 6, not identically zero, of which γ must be a root (as $f_1(\gamma) = 0$ would contradict coprimality). By (14) we see again that γ determines a unique v, so that in this case we have to discard no more than six values of v.

STEP 3. We show that for most v the polynomial F_β has no more than one repeated factor and is therefore simple. The following argument bears comparison with [5], pp. 341–342, "Proof of (D)".

We retain the notation $E(x)$ for $f_0'(x)f_1(x) - f_0(x)f_1'(x)$. Assume that γ_1, γ_2 are distinct multiple roots of F_β. Then $F_\beta(\gamma_i) = F_\beta'(\gamma_i) = 0$ $(i = 1, 2)$, and this also implies $E(\gamma_i) = 0$ $(i = 1, 2)$. Now, $E(x)$ is equal to $h_1(x) - vD(x)$ with $D(x) := C_2 x^2 - C_0$ and

$$
\begin{aligned}
h_1(x) :=\ & x^{3j+2} g_3^{(2)}(x^3) + w x^{3j+2} g_{l-1}^{(2)}(x^3) - C_2 x^{14} - w C_2 x^{3l+2} + C_0 x^{12} - \\
& - C_0 x g_3^{(2)}(x^3) + w C_0 x^{3l} - w C_0 x g_{l-1}^{(2)}(x^3) - x^{3j} g_3^{(0)}(x^3) - \\
& - w x^{3j} g_{l-1}^{(0)}(x^3) + C_2 x g_3^{(0)}(x^3) + w C_2 x g_{l-1}^{(0)}(x^3)\ .
\end{aligned}
$$

Solve the equation $E(x) = 0$ for v and put $v = h_1(x)/D(x) =: H_1(x)$. Then substitute this expression for v into f_0 to find $f_0(x)/f_1(x) = h_2(x)/D(x) =: H_2(x)$ with

$$
h_2(x) := x^2 g_3^{(2)}(x^3) - g_3^{(0)}(x^3) + w \left[x^2 g_{l-1}^{(2)}(x^3) - g_{l-1}^{(0)}(x^3) \right]\ .
$$

As in [5], for any rational function $\varphi(x) = \varphi_1(x)/\varphi_2(x) \in \overline{\mathbb{F}}_q(x)$ we write

$$
B_\varphi(X, Y) := \frac{\varphi_1(X)\varphi_2(Y) - \varphi_1(Y)\varphi_2(X)}{X - Y} \in \mathbb{F}_q[X, Y]\ .
$$

One readily verifies that $B_{H_1}(X, Y)$ and $B_{H_2}(X, Y)$ have total degree at most 15 and $15 - 3j$, respectively, and that $(X, Y) = (\gamma_1, \gamma_2)$ is a solution to

$$
B_{H_1}(X, Y) = B_{H_2}(X, Y) = 0
$$

that determines a unique $v = H_1(\gamma_1) = H_1(\gamma_2)$. If B_{H_1} and B_{H_2} are co-prime, it follows now from Bezout's theorem [7] that the number of such pairs (γ_1, γ_2), and hence the number of values v we have to exclude, is bounded by $15 \cdot (15 - 3j) \le 180$. Therefore, what remains to be shown is that B_{H_1} and B_{H_2} indeed satisfy the condition of being co-prime. By Lemma 4 of [3], or by [6], it suffices to show that H_1 and H_2 do not decompose a functions of the same non-linear rational function.

Suppose $C_0 = 0$. Then h_1 has linear coefficient $C_2(A_{13} + wA_{3l+1}) \ne 0$, therefore x (but not x^2) cancels from $H_1(x)$, so that its denominator has degree 1, which implies that H_1 is indecomposable.

Next suppose $C_0 \ne 0$, $C_2 = 0$. Then H_2 is a polynomial of degree $12 - 3j$. For $j = 3$ this is prime. For $j = 1$ the only possible decomposition type in terms of the degrees $(\omega_1, \omega_2, \rho_1, \rho_2)$ is $(3, 0, 3, 0)$, and for $j = 2$ it could be $(3, 0, 2, 0)$ or $(2, 0, 3, 0)$. All of these can be ruled out "by hand" (i.e. assuming an explicit decomposition and comparing coefficients, similarly to Case (3) of Sect. 4). Hence H_2 is indecomposable.

Finally suppose $C_0 C_2 \ne 0$. Assume first that H_2 is in lowest terms. Then for $j = 1$ and $j = 3$ we get indecomposability because $\deg h_2$ is prime, and for $j = 2$ the only possible decomposition types are $(\omega_1, \omega_2, \rho_1, \rho_2) = (4, 0, 2, 1)$ or $(2, 1, 4, 0)$, both of which can again be eliminated by explicit discussion. Next assume a linear factor cancels from H_2. Then we get indecomposability from degree 1 of the denominator. Finally assume that $C_2x^2 - C_0$ divides $h_2(x)$. Note that $w = 0$ (because $C_0 \neq 0$) and

$$h_1(x) = -x^{12}(C_2x^2 - C_0) + x^{3j}h_2(x) + xL(x^3)$$

where $L(x^3) := C_2g_3^{(0)}(x^3) - C_0g_3^{(2)}(x^3)$ is not identically zero by assumption. From

$$x^2L(x^3) = (C_2x^2 - C_0)g_3^{(0)}(x^3) - C_0h_2(x)$$

it is clear that $C_2x^2 - C_0$ divides $L(x^3)$ and hence also $h_1(x)$. Moreover, as $L(x^3)$ is a polynomial in x^3, we must in fact have $(C_2x^2 - C_0)^3 \mid L(x^3)$. For $j = 2, 3$ this already yields a contradiction, since in these cases $\deg L(x^3) \leq 3$. For $j = 1$,

$$L(x^3) = \frac{k}{C_2^3}(C_2x^2 - C_0)^3 \quad \text{with} \ k := C_2A_7 - C_0A_5 \in \mathbb{F}_q^* . \tag{16}$$

Performing the division of $h_2(x)$ by $C_2x^2 - C_0$, we obtain

$$H_2(x) = x^9 + \frac{A_5}{C_2}x^6 - \frac{k}{C_2^2}x^4 + \text{terms of lower degree},$$

which can only decompose as $Q(R(x))$ with $\deg Q = \deg R = 3$. Now $H_1(x) = -x^{12} + x^3H_2(x) + xL(x^3)/(C_2x^2 - C_0)$ is in fact of degree ≤ 9, and the x^7-term has coefficient $-k/C_2^2$. If $H_1(x) = P(R(x))$, with R as above and $\deg P \leq 3$, then clearly the coefficient of x^7 is zero, i.e. $k = 0$ in contradiction to (16). Therefore H_1 cannot decompose as a function of the same function as H_2. This completes the third step of (iii) and thereby Case (II).

Case (III). If one of A_{10}, A_{13} is non-zero put $w := 0$. In the situation $A_{10} = A_{13} = 0$ proceed as follows: put $w := 0$ if also $A_7 = A_4 = 0$ (i.e. $g_3^{(0)} = 0$ identically, which is possible in this case); otherwise choose $l \in \{1, 2\}$ such that $A_{3l+1} \neq 0$ and put $w := 1$.

  Now choose $s_3 := u \in \mathbb{F}_q^*$, $s_{12-3l} := w$ (ignore this if $w = 0$) and $s_{12} := uv$ with $v \in \mathbb{F}_q$. Then $f(x) = F_u(x) = f_0(x) + uf_1(x)$ with

$$f_0(x) = x^{13} + x^2g_3^{(2)}(x^3) + g_3^{(0)}(x^3) + wx^{3l+1} + wx^2g_{l-1}^{(2)}(x^3) + wg_{l-1}^{(0)}(x^3),$$
$$f_1(x) = x^{10} + x^2g_2^{(2)}(x^3) + g_2^{(0)}(x^3) + vx.$$

In the special case $g_3^{(0)} = 0$ cancel a factor x and work with

$$f_0(x) = x^{12} + xg_3^{(2)}(x^3), \qquad f_1(x) = x^9 + xg_2^{(2)}(x^3) + v .$$

In all cases we aim to establish the conditions of Lemma 3.

(i) *Co-primality.* In the situation where $g_3^{(2)} \neq 0$ we have arranged that x is not a common factor of f_0 and f_1. By the usual argument, excluding at most thirteen values v from $\mathbb{F}_q$ ensures co-primality.

In the case $g_3^{(2)} = 0$ we have to avoid $v = 0$ and up to eleven other values $v \in \mathbb{F}_q$.

(ii) *Indecomposability.* If $g_3^{(0)} \neq 0$ we have deg $f_0 = 13$ prime.

The situation for $g_3^{(0)} = 0$ is more complicated. From (8), there are six possible decomposition types $(\omega_1, \omega_2, \rho_1, \rho_2)$, namely $(3,0,4,3)$, $(2,1,6,3)$, $(3,2,4,1)$, $(4,3,3,0)$, $(4,1,3,2)$ and $(6,3,2,1)$. The first four can be ruled out "by hand" as above. For the remaining two equate denominators in (7) to find that $R_2^3(x) \mid f_1(x)$ and also $R_2^3(x) \mid f_1'(x) = A_2 x^6 + A_5 x^3 + A_8 \in \mathbb{F}_q[x^3]$. Hence $R(x)$ divides a fixed quadratic polynomial in $\overline{\mathbb{F}}_q[x]$ and can have at most two roots. Moreover, $R_2(x) \mid f_1(x) - x f_1'(x) = (x + v^{1/9})^9$, so we must have $v = -\gamma^9$ where γ is a root of $R_2(x)$. This leaves us with at most two values of v which need to be excluded to ensure indecomposability.

(iii) *No factor of multiplicity divisible by 3.* We show indeed that f_1 is square-free for all but a small number of v. Clearly, if this is shown for f_1 in its original meaning, this covers also the case where f_1 is reduced by a factor x. Suppose $\gamma \in \overline{\mathbb{F}}_q$ is a double root of f_1. Then it satisfies both

$$f_1(x) = x^{10} + x^2 g_2^{(2)}(x^3) + g_2^{(0)}(x^3) + vx = 0 \qquad (17)$$

and

$$f_1'(x) = x^9 - x g_2^{(2)}(x^3) + v = 0 . \qquad (18)$$

From (18) we get $v = -x^9 + x g_2^{(2)}(x^3)$, which in (17) gives $-x^2 g_2^{(2)}(x^3) + g_2^{(0)}(x^3) = 0$. This means γ is a root of a fixed polynomial of degree at most 8 that is not identically zero. Consequently there are at most eight simultaneous solutions to (17) and (18), each of which determines a unique v. Excluding these values of v makes f_1 square-free.

(iv) *Discriminant.* This is deferred to the next and final section.

6 $\mathcal{S}_n$ or $\mathcal{A}_n$?

The outstanding new feature of the ternary case, in comparison to the binary case [4], is the appearance of the alternating group $\mathcal{A}_n$ as the common Galois group $G = \overline{G}$. We have already seen situations—Case (2) with $A_5 = 0$ and Case (4) in Sect. 4, and Case (I) in Sect. 5 with one of A_{11}, A_{13} zero—where this is genuinely the case (i.e. no matter which of the possible v we choose). On the other hand, the use of Lemma 2 in Case (II), the "most general" case, in Sect. 5 means that here we are automatically always in "situation S". The overall impression is that "situation S" is the standard case, but "situation A" inevitably occurs in some exceptional situations. This will be confirmed by Case (III) of Sect. 5, which we are now going to complete with the analysis of the discriminant Δ_x (criterion (iv) of Lemma 3). Recall that Case (III) imposes relationships between the A_k.

To this end we use the following lemma, an elementary proof of which we include in a forthcoming paper on the general problem. In terms of ramification theory it exploits the relationship between the "different" and derivative, as in [1], Chap. 1, Prop. 6.

Lemma 4.   *Let f_0, f_1 and F_u be as in Sect. 3, $f := f_0/f_1$. Assume that f_0 and f_1 are co-prime. Write the part of $E(x) = f_0'(x)f_1(x) - f_0(x)f_1'(x)$ that is co-prime to f_1 as*

$$\prod_{j=1}^{r}(x - \beta_j)^{d_j} \qquad (\beta_j \in \overline{\mathbb{F}}_q) \ .$$

Then the discriminant Δ_x of $F_u(x)$ is equal to

$$c \cdot \prod_{j=1}^{r}(u - f(\beta_j))^{d_j} \qquad \text{with a constant } c \in \mathbb{F}_q^* \ .$$

Suppose, in our situation, that α is a common root of E and f_1. Then $f_0(\alpha)f_1'(\alpha) = 0$ and, since $f_0(\alpha) \neq 0$, α must be a repeated root of f_1. As we have already arranged in part (iii) that such roots do not exist, we can assume that E is co-prime to f_1. By Lemma 4 this means that all its factors contribute to the degree of Δ_x in u. Hence to prove that Δ_x is a non-square in $\mathbb{F}_q(u)$ it suffices to show that the degree of $E(x)$ is odd.

Consider first the situation where $g_3^{(0)} \neq 0$. Then $C_0 \neq 0$, $g_i^{(2)} = Cg_i^{(0)}$ $(i = 0, \ldots, 3)$ with $C := C_2/C_0$, and

$$E(x) = (Cx^2 - 1) \cdot \left[A_{13}(x^9 + v) + vx^3 g_2^{(0)}(x^3) \right] \ .$$

If $j = 1$ then $E(x)$ is of degree 9 or 11 provided we avoid $v = -A_{13}/C_0$. If $j = 3$ then $A_{10} = C_0 \neq 0$, hence $w = 0$, and $E(x)$ reduces to $(Cx^2 - 1)(A_{13}x^9 + vA_{10}x^3 + vA_{13})$, which is either of degree 9 or 11 (if $A_{13} \neq 0$), or of degree 3 or 5 (if $A_{13} = 0$ and we avoid $v = 0$). It remains $j = 2$. Then either $w = 1$ with $l = 1$, or $w = 0$, and

$$E(x) = (Cx^2 - 1) \cdot \left[A_{13}x^9 + (vA_7 - wA_{10})x^6 + vA_{10}x^3 + (vA_{13} + vwA_7) \right] \ .$$

This is of degree 9 or 11 if $A_{13} \neq 0$.

From now on suppose $A_{13} = 0$ (still in the case $j = 2$). Then

$$E(x) = (Cx^2 - 1) \cdot \left[(vA_7 - wA_{10})x^6 + vA_{10}x^3 + vwA_7 \right]$$

is of degree 8 whenever $v \neq 0$ (note that $A_7 = C_2 \neq 0$ and one of w, A_{10} is zero). In this case we can try to show that Δ_x has a factor of odd multiplicity. Closer inspection reveals four possibilities to examine as follows.

(a) $A_7 \neq 0$, all other A_k are zero.
(b) $A_7, A_{10} \neq 0$, all other A_k are zero.
(c) $A_5, A_7 \neq 0$, all other A_k are zero.
(d) $A_5, A_7, A_{10} \neq 0$ (then also $A_8 = A_5 A_{10}/A_7 \neq 0$), all other A_k are zero.

(a) Here $f_0(x) = x^{13} + x^7 + A_7 x^6 + A_7$, $f_1(x) = x^{10} + A_7 x^3 + vx$. Maple calculates $\Delta_x = v^6 A_7^{12} u^6$, which is a square in $\mathbb{F}_q(u)$—once again we find ourselves in "situation A".

(b) Here $f_0(x) = x^{13} + A_7 x^6 + A_{10} x^3$ and $f_1(x) = x^{10} + A_7 x^3 + vx + A_{10}$. We have $E(x) = -v A_7 x^3 (x^3 + A_{10}/A_7)$; this has triple roots 0 and $\alpha := \sqrt[3]{-A_{10}/A_7}$. As $f(0) = 0$, the root 0 contributes a factor u^3 to Δ_x by Lemma 4. Hence, to show that Δ_x is a non-square in $\overline{\mathbb{F}}_q(u)$, it suffices that $f(\alpha) \neq 0$. Indeed, one finds $f_0(\alpha) = \alpha^{13} \neq 0$.

(c) Here $f_0(x) = (x^6 + 1) f_2(x)$ and $f_1(x) = x^3 f_2(x) + vx$ with $f_2(x) := x^7 + A_5 x^2 + A_7$. $E(x) = v A_5 (x^2 - A_7/A_5)(x^6 + 1)$ has roots $\pm \alpha$, where $\alpha := \sqrt{A_7/A_5}$, and $\pm \beta$, where $\beta := \sqrt{-1}$. One finds $f_0(\beta) = f_0(-\beta) = 0$, so that $\pm \beta$ together contribute u^6 to Δ_x. Therefore, to show that Δ_x is a non-square in $\overline{\mathbb{F}}_q(u)$, it suffices that $f(-\alpha) \neq f(\alpha)$. It is easily checked that the choice $v = A_7^2/A_5$ is permissible (that is, conditions (i)–(iii) of Lemma 3 are satisfied and f_1 is square-free). With this, $f_1(\alpha) = \alpha^{10} = f_1(-\alpha)$ and $f_0(-\alpha) - f_0(\alpha) = \alpha^{13} + \alpha^7 = \alpha^7(A_7^3/A_5^3 + 1)$, so that indeed $f(-\alpha) \neq f(\alpha)$ whenever $A_5 \neq -A_7$. If, however, $A_5 = -A_7$, then the same choice of v becomes $-A_7$, and Maple finds $\Delta_x = v^6 A_7^{12} u^8$, a square in $\mathbb{F}_q(u)$ ("situation A", at least for this particular v).

(d) Here $f_0(x) = x^3 f_2(x)$ and $f_1(x) = f_2(x) + vx$ with $f_2(x) := x^{10} + A_5 x^5 + A_7 x^3 + A_8 x^2 + A_{10}$. $E(x) = v A_5 x^3 (x^2 - A_7/A_5)(x^3 + A_{10}/A_7)$ has roots 0, $\pm \alpha$ with $\alpha := \sqrt{A_7/A_5}$, and $\beta := \sqrt[3]{-A_{10}/A_7}$. One finds $f_0(0) = 0$ and $f_0(\beta) = \beta^{13} \neq 0$, so 0 contributes u^3 to Δ_x, and β contributes a triple factor different from u^3. Therefore Δ_x is a non-square in $\overline{\mathbb{F}}_q(u)$, unless exactly one of $f(\pm \alpha)$ coincides with 0 and the other with $f(\beta)$.

To study the latter possibility, assume without loss of generality that $f(\alpha) = 0$ and $f(-\alpha) = f(\beta)$. One calculates

$$f(\beta) = \frac{\lambda_1}{\lambda_2 + v} \quad \text{with } \lambda_1 := \beta^{12}, \ \lambda_2 := \beta^9,$$

and, using $f(\alpha) = 0$,

$$f(-\alpha) = \frac{\lambda_3}{\lambda_4 + v} \quad \text{with } \lambda_3 := -\alpha^5 A_7, \ \lambda_4 := \alpha^2 A_7 \ .$$

Then $f(\beta) = f(-\alpha)$ if and only if $(\lambda_1 - \lambda_3)v = \lambda_2 \lambda_3 - \lambda_1 \lambda_4$. Clearly, if $\lambda_3 \neq \lambda_1$ this can be avoided by a suitable change of v. It remains to examine the situation for $\lambda_3 = \lambda_1$. Then we have also $\lambda_1 \lambda_4 = \lambda_2 \lambda_3$, which is equivalent to $\beta = -\alpha$, and from $\lambda_3 = \lambda_1$ we further deduce $A_7 = -\alpha^7$. Let ε denote a square root of A_7 and η a square root of A_5 such that $\alpha = \varepsilon/\eta$ (i.e. $f(\varepsilon/\eta) = 0$). Then $\varepsilon^2 = A_7 = -\varepsilon^7/\eta^7$, i.e. $\eta^7 = -\varepsilon^5$, and this relation allows us to write our coefficients as

$$A_5 = \eta^2, \quad A_7 = \varepsilon^2, \quad A_8 = \frac{\varepsilon^3}{\eta} = -\frac{A_5^3}{A_7} \quad \text{and} \quad A_{10} = -\eta^4 = -A_5^2 \ .$$

Now, with the help of Maple, and using $\eta^7 = -\varepsilon^5$, one determines

$$\Delta_x = \big[(\eta^{12}v^4 - \eta^{10}\varepsilon^4 v^3 + \eta^8 \varepsilon^8 v^2)u^4 - $$
$$- (\eta^{14}\varepsilon^2 v^3 + \eta^{12}\varepsilon^6 v^2)u^3 + \eta^{16}\varepsilon^4 v^2 u^2 \big]^2 =$$

$$= \big[(A_5^6 v^4 - A_5^5 A_7^2 v^3 + A_5^4 A_7^4 v^2)u^4 - $$
$$- (A_5^7 A_7 v^3 + A_5^6 A_7^3 v^2)u^3 + A_5^8 A_7^2 v^2 u^2 \big]^2 ,$$

a square in $\mathbb{F}_q(u)$. This is another remarkable sporadic appearance of "situation A". (For an example where this actually occurs, take $A_5 = A_7 = 1$ and $A_8 = A_{10} = -1$.)

Finally, we have to consider the case $g_3^{(0)} = 0$. Then $E(x) = (A_{11}+vA_2)x^9 + vA_5 x^6 + vA_8 x^3 + vA_{11}$. If $j = 1$ avoid $v = -A_{11}/A_2$ to obtain $\deg E = 9$, and if $j = 3$ then the degree of E is 3 or 9, according to whether A_{11} is zero or not. So suppose $j = 2$. Then $\deg E = 9$ if $A_{11} = 0$. For $A_{11} \neq 0$ we have $E(x) = vA_5 x^3(x^3 + A_8/A_5)$. This has always the triple root 0, which, as $f_0(0) = 0$, contributes u^3 to Δ_x. The second triple root of E is $\alpha := \sqrt[3]{-A_8/A_5}$. Only when $A_8 = 0$ do the roots coincide and we find $\Delta_x = v^6 A_5^{12} u^6$ ("situation A"). Otherwise $f_0(\alpha) = \alpha^{12} \neq 0$ and Δ_x is a non-square in $\overline{\mathbb{F}}_q(u)$.

With this the proof of Theorem 1 is complete.

References

1. J. W. S. Cassels and A. Fröhlich (eds.), "Algebraic Number Theory", Academic Press, London, 1967.
2. G. D. Cohen, M. G. Karpovsky, H. F. Mattson Jr. and J. R. Schatz, "Covering Radius—Survey and Recent Results", *IEEE Transactions on Information Theory*, Vol. IT-31 No. 3 (1985), 328–343.
3. S. D. Cohen, "Uniform Distribution of Polynomials over Finite Fields", *J. London Math. Soc. (2)* **6** (1972), 93–102.
4. S. D. Cohen, "The Length of Primitive BCH Codes with Minimal Covering Radius", *Designs, Codes and Cryptography* **10** (1997), 5–16.
5. S. D. Cohen, "Polynomial Factorisation and an Application to Regular Directed Graphs", *Finite Fields and Their Applications* **4** (1998), 316–346.
6. M. D. Fried and R. E. MacRae, "On Curves with Separated Variables", *Math. Ann.* **180** (1969), 220–226.
7. W. Fulton, "Algebraic Curves", Benjamin, New York, 1969.
8. T. Helleseth, "On the Covering Radius of Cyclic Linear Codes and Arithmetic Codes", *Discr. Appl. Math.* **11** (1985), 157–173.
9. Y. Kaipainen, "On the Covering Radius of Long Non-binary BCH Codes", dissertation, University of Turku, 1995.
10. R. Lidl and H. Niederreiter, "Introduction to Finite Fields and Their Applications", Cambridge University Press, 1994.
11. H. Stichtenoth, "Algebraic Function Fields and Codes", Springer, Berlin/Heidelberg, 1993.

The Gray Map on $\mathrm{GR}(p^2, n)$ and Repeated-Root Cyclic Codes

Horacio Tapia-Recillas [*]

Departamento de Matemáticas, Universidad Autónoma Metropolitana-I
09340 México, D.F., MEXICO (htr@xanum.uam.mx)

Abstract. It is shown that the Gray map on $\mathbb{Z}_{p^2}^n$, where p is a prime and n a positive integer, yields the same result as an appropriate extension of the well-known "$(u|u+v)$-construction". It is also shown that, up to a permutation, which is a generalization of Nechaev's permutation, the Gray image of certain $\mathbb{Z}_{p^2}$-codes of length n constructed from $\mathbb{F}_p$-cyclic codes of length n are $\mathbb{F}_p$-cyclic codes of length pn with multiple roots. These results generalize some of those appearing in [21]. Examples are given in order to illustrate the ideas.

1 Introduction

If n is a positive integer and $\mathbb{Z}_4$ is the ring of integers modulo 4, it follows from their definitions that the Gray map on $\mathbb{Z}_4^n$ and the "$(u|u+v)$-construction" yields the same result (cf. §3.1). Furthermore if C_1 and C_2 are binary cyclic codes of length n with generating polynomials g_1 and $g_1 g_2$ respectively, where g_1 and g_2 are coprime divisors of $x^n - 1$, n odd, then, up to a permutation, the "$(u|u+v)$-construction" of these cyclic codes is the same as the repeated-root binary cyclic code of length $2n$ with generator polynomial $g_1^2 g_2$, a problem treated in [21] (see also [2]). The purpose of this note is to show that the same kind of results are valid for the case of the Gray map Φ defined on $\mathbb{Z}_{p^2}^n$, the ring of integers modulo p^2, where p is a prime and an appropriate extension of the "$(u|u+v)$-construction". Let $C_i = <g_1 g_2 \cdots g_i>$, for $i = 1, 2, ..., p$, be $\mathbb{F}_p$-cyclic codes of length n, where the polynomials g_i are monic pairwise coprime divisors of $x^n - 1$, with $(n, p) = 1$. Then, up to a permutation which is a generalization of Nechaev's permutation (cf. §2), it is shown that the Gray map image $\Phi(D)$ of a $\mathbb{Z}_{p^2}$-code D of length n constructed from the cyclic codes C_i's, is a repeated-root $\mathbb{F}_p$-cyclic code of length pn whose generating polynomial is obtained from those polynomials generating the cyclic codes C_i (see §4 for details). Some examples are given in order to illustrate the ideas. The ring $\mathbb{Z}_{p^2}$ plays an important role in other contexts, for instance in the construction of partial and relative difference sets in the Galois ring $\mathrm{GR}(p^2, m)$ and related results (cf. [5], [15]). In the description of the Kerdock code over this Galois ring ([20]) and nonlinear p-ary sequences ([11]).

---

[*] Research partially supported by Red de Criptología (CONACYT-UAM,I), México.

G. Mullen, A. Poli and H. Stichtenoth (Eds.): Fq7 2003, LNCS 2948, pp. 181–196, 2003.
© Springer-Verlag Berlin Heidelberg 2003

The problem of repeated-root binary cyclic codes has been discussed in [21], [2], [10]. If C_1 and C_2 are binary codes the construction $C_2 + 2C_1$ has been used recently in [4] to express the Gray images of (free) cyclic codes over the ring $\mathbb{F}_2 + u\mathbb{F}_2$. Also in [14] the same construction is used to study cyclic self-dual codes over the ring $\mathbb{Z}_4$ and in [3] where the codes C_1 and C_2 are the Reed-Muller codes $RM(r, m)$ and $RM(m - r - 1, m)$ for $0 \leq r \leq \frac{m-1}{3}$, respectively, to study type II codes over the ring $\mathbb{Z}_4$. In [22] the same construction is used to describe some families of $\mathbb{Z}_4$-cyclic codes.

The paper is organized as follows: in the next section a generalization of Nechaev's permutation ([13], [23]) which will be useful in the rest of the paper is given. In Section 3, well-known facts about the "$(u|u+v)$-construction", Gray map and repeated-root cyclic codes in the binary case are recalled. In Section 4, the main results are presented and some examples are provided in the last section.

2   The p-Permutation

For an odd positive integer n let the permutation σ be defined on the set $\{0, 1, 2, ..., 2n-1\}$ as $\sigma = (1, n+1)(3, n+3) \cdots (2i+1, n+2i+1) \cdots (n-2, 2n-2)$, where the elements $0, 2, 4, ..., 2n-1$ are invariant under σ, induces a permutation Π on the cartesian product $\mathbb{F}_2^{2n}$ in the following way. If

$$\mathbf{u} = (u_0, u_1, ..., u_{n-1} | u_n, u_{n+1}, ..., u_{2n-1}) \in \mathbb{F}_2^{2n}$$

then

$$\Pi(\mathbf{u}) = (u_{\sigma(0)}, u_{\sigma(1)}, ..., u_{\sigma(n-1)}, | u_{\sigma(n)}, u_{\sigma(n+1)}, ..., u_{\sigma(2n-1)}).$$

This permutation on $\mathbb{F}_2^{2n}$, called Nechaev's permutation (cf. [13]), has been used by several authors in determining properties of binary and quaternary codes ([23], [9], [16], [18]).

For a prime p and any positive integer n such that $p \leq n$, let N_{np} be the set

$\{0$	1	2	$\cdots$	p	$\cdots$	$n-1$
n	$n+1$	$n+2$	$\cdots$	$n+p$	$\cdots$	$2n-1$
$2n$	$2n+1$	$2n+2$	$\cdots$	$2n+p$	$\cdots$	$3n-1$
$\vdots$	$\vdots$	$\vdots$	$\vdots$	$\vdots$	$\vdots$	$\vdots$

$$(p-1)n \ (p-1)n+1 \ (p-1)n+2 \cdots (p-1)n+p \cdots np-1\}.$$

Observe that this array has n columns numbered $0, 1, 2, 3, ..., n-1$ and p rows numbered $0, 1, 2, ..., p-1$. We define the permutation σ on the set N_{np} as the composition (product), i.e., $\sigma = \sigma_0 \cdots \sigma_{n-1}$ of the following p-cycles σ_j given on each column of the above arrangement as:

- σ_j = identity if $j \equiv 0 \bmod p$
- $\sigma_j = (j, (j)_p n + j, (2j)_p n + j, ..., ((p-1)j)_p n + j)$, for $j = 1, 2, ..., p-1$ where $(s)_p$ means reduction of s modulo p

- $\sigma_m = \sigma_j$ if $m \equiv j \mod p$, for $j = 0, 1, 2, ..., p-1$

This permutation induces a permutation Π on the cartesian product $\mathbb{F}_p^{np}$ in the following way. If

$$\mathbf{u} = (u_0, u_1, ..., u_{n-1} | u_n, u_{n+1}, ..., u_{2n-1} | \cdots | u_{(p-1)n}, u_{(p-1)n+1}, ..., u_{pn-1}) \in \mathbb{F}_p^{np}$$

then the vector $\Pi(\mathbf{u})$ with np coordinates is the concatenation of the consecutive rows of the following arrangement:

$$\Pi(\mathbf{u}) = \begin{pmatrix} u_{\sigma(0)} & u_{\sigma(1)} & \cdots & u_{\sigma(n-1)} \\ u_{\sigma(n)} & u_{\sigma(n+1)} & \cdots & u_{\sigma(2n-1)} \\ \vdots & \vdots & \vdots & \vdots \\ u_{\sigma((p-1)n)} & u_{\sigma((p-1)n+1)} & \cdots & u_{\sigma(np-1)} \end{pmatrix}$$

This permutation on $\mathbb{F}_p^{np}$ will be called the "p-permutation" and it will be used in the next section to describe some properties of the Gray map image of codes defined over the ring $\mathbb{Z}_{p^2}$. The permutation Π has also been considered in [9]. Observe that if $p = 2$ the p-permutation is precisely Nechaev's permutation.

3 Gray Map, "$(u|u+v)$-Construction" and Cyclic Codes: the Binary Case

In this section some facts about the "$(u|u+v)$-construction", the classical Gray map, i.e., over $\mathbb{Z}_4$, and repeated-root binary cyclic codes are recalled. To be more precise, it is shown that if C_1 and C_2 are two binary cyclic codes of length n, $D = C_2 + 2C_1 = \{\mathbf{c_2} + 2\mathbf{c_1} \in \mathbb{Z}_4^n, \mathbf{c_i} \in C_i\}$ and Φ is the Gray map on $\mathbb{Z}_4^n$ then $\Phi(D)$ is the same as the well-known "$(u|u+v)$-construction" on certain subsets of $\mathbb{F}_2^n$ associated to C_1 and C_2. It is also shown that the code obtained by applying the 2-permutation to the "$(u|u+v)$-construction", or equivalently, to the Gray map image $\Phi(D)$ of the code D, is a repeated-root binary cyclic code of length $2n$ whose generator is given in terms of the generators of the codes C_1 and C_2. The problem of repeated-root cyclic codes was treated in [21] (see also [2]). In the next section, this construction is generalized to $\mathbb{F}_p$-codes, a relation with a suitable "$(u|u+v)$-construction" and the generalized Gray map on the Galois ring $GR(p^2, m)$ are given.

3.1 The "$(u|u+v)$-Construction" and the Gray Map

We first recall the "$(u|u+v)$-construction" ([12], page 76). Let U and V be two (non-empty) subsets of the cartesian product R^m, where R is a finite (commutative) ring and m is a positive integer. The "$(u|u+v)$-construction" on U and V is the subset $\Gamma(U, V) = (U|U+V)$ of R^{2m} given by:

$$\{(\mathbf{u}|\mathbf{u}+\mathbf{v}) : \mathbf{u} \in U, \mathbf{v} \in V\} = \{(u_0, ..., u_{m-1} | u_0 + v_0, ..., u_{m-1} + v_{m-1})\}$$

where $\mathbf{u} = (u_0, ..., u_{m-1}) \in U$, $\mathbf{v} = (v_0, ..., v_{m-1}) \in V$ and the bar means concatenation.

Observe that if $R = \mathbb{F}_q$, a finite field with q elements, if U is a $[m, r, \delta]$-$\mathbb{F}_q$ linear code and V is also a $[m, s, \lambda]$-$\mathbb{F}_q$ linear code, then $(U|U + V)$ is a $[2m, r + s, d = \min\{2\delta, \lambda\}]$-$\mathbb{F}_q$ linear code ([12]).

We recall the definition of the Gray map on $\mathbb{Z}_4^m$, where $\mathbb{Z}_4$ is the ring of integers modulo 4 and n is a positive integer (cf. [7]). First observe that any element $a \in \mathbb{Z}_4$ can be expressed as $a = r_0(a) + 2r_1(a)$, its binary expansion, where $r_0(a), r_1(a)$ are in $\mathbb{F}_2$. The Gray map $\Phi : \mathbb{Z}_4 \longrightarrow \mathbb{F}_2^2$ is given by $\Phi(a) = (r_1(a), r_1(a) + r_0(a))$. This map can be extended to $\mathbb{Z}_4^n$ in the natural way (coordinate-wise): if $\mathbf{a} = (a_0, a_1, ..., a_{n-1}) \in \mathbb{Z}_4^n$, then,

$$\Phi : \mathbb{Z}_4^n \longrightarrow \mathbb{F}_2^{2n},$$

is given by:

$$\Phi(\mathbf{a}) = (r_1(a_0), ..., r_1(a_{n-1})|r_1(a_0) \oplus r_0(a_0), ..., r_1(a_{n-1}) \oplus r_0(a_{n-1})).$$

It is well-known that Φ is a bijective isometry with respect to the Lee metric on $\mathbb{Z}_4^n$ and the Hamming metric on $\mathbb{F}_2^{2n}$ ([7]).

Let

$$R_0 = \{(r_0(v_0), r_0(v_1), ..., r_0(v_{n-1})) \in \mathbb{F}_2^n : \mathbf{v} = (v_0, v_1, ..., v_{n-1}) \in \mathbb{Z}_4^n\}$$

and let

$$R_1 = \{(r_1(v_0), r_1(v_1), ..., r_1(v_{n-1})) \in \mathbb{F}_2^n : \mathbf{v} = (v_0, v_1, ..., v_{n-1}) \in \mathbb{Z}_4^n\}.$$

From the above definitions it follows that the "$(u|u+v)$-construction", $(R_1|R_1 + R_0)$ on the sets R_1 and R_0, is precisely the image of the Gray map on $\mathbb{Z}_4^n$, i.e.:

$$\Phi(\mathbb{Z}_4^n) = (R_1|R_1 + R_0).$$

In particular, if C_1 and C_2 are two binary codes of length n, then the code $D = C_2 + 2C_1 = \{\mathbf{a} + 2\mathbf{b} = (a_0 + 2b_0, ..., a_{n-1} + 2b_{n-1}) : \mathbf{a} = (a_0, ..., a_{n-1}) \in C_2, \mathbf{b} = (b_0, ..., b_{n-1}) \in C_1\} \subseteq \mathbb{Z}_4^n$ is such that its Gray map image $\Phi(D)$ is

$$\Phi(D) = (C_1|C_1 + C_2).$$

It follows that $\Phi(D)$ is a $\mathbb{F}_2$-linear code.

3.2   The Gray Map and Cyclic Codes

Let n be an odd positive integer, $x^n - 1 = f_1(x)f_2(x) \cdots f_r(x)$, with each $f_i(x)$ monic irreducible and pairwise coprime. If $g_1(x) = f_1(x) \cdots f_k(x)$, $g_2(x) = f_{k+1}(x) \cdots f_s(x)$, which are coprime, let $C_1 = <g_1(x)>$, $C_2 = <g_1(x)g_2(x)>$ be the ideals in $\mathbb{F}_2[x]/(x^n - 1)$ generated by $g_1(x)$ and $g_1(x)g_2(x)$ respectively, i.e., cyclic codes of length n. In [21] it is shown that the repeated-root cyclic code $C = < g_1(x)^2 g_2(x) > \subset \mathbb{F}_2[x]/(x^{2n} - 1)$ is equivalent, up to a permutation on the coordinates, to the linear code $(C_1 | C_1 + C_2)$. Hence the cyclic code C is equivalent to the Gray map image $\Phi(D)$ of the code $D = C_2 + 2C_1 \subset \mathbb{Z}_4^n$. Observe that in this case $C_2 \subseteq C_1$.

It is easy to see that the permutation that takes the binary linear code $\Phi(D)$ to the linear (cyclic) code C is precisely Nechaev's permutation, introduced in §2. In fact, in [21] it is shown that if $a(x) \in C_1$ and $b(x) \in C_2$ then the polynomial

$$w(x) = (x^n - 1)a(x) + b(x) + (x^n - 1)b_e(x^2)$$

is an element of the ideal $C = < g_1(x)^2 g_2(x) > \subset \mathbb{F}_2[x]/(x^{2n} - 1)$, where $b_e(x^2)$, the even part of $b(x)$, is such that $b(x) = b_e(x^2) + xb_o(x^2)$. If $a(x) = a_0 + a_1x + \cdots + a_{n-1}x^{n-1}$ and $b(x) = b_0 + b_1x + \cdots + b_{n-1}x^{n-1}$, it is easy to see that the vector $\mathbf{w}$ associated to the polynomial $w(x)$ is the concatenation of the rows of the following arrangement:

$$\begin{pmatrix} a_0, & a_1 + b_1, & a_2, & a_3 + b_3, \cdots, & a_{n-1}, \\ a_0 + b_0, & a_1, & a_2 + b_2, & a_3 & , \cdots, a_{n-1} + b_{n-1} \end{pmatrix}$$

Applying Nechaev's permutation to the vector $\mathbf{w}$, that is, the 2-permutation Π as introduced in §2, we obtain:

$$\begin{pmatrix} a_0, & a_1, & a_2, & a_3 & , \cdots, & a_{n-1}, \\ a_0 + b_0, & a_1 + b_1, & a_2 + b_2, & a_3 + b_3, & \cdots, & a_{n-1} + b_{n-1} \end{pmatrix}$$

which is precisely the element $(\mathbf{a} | \mathbf{a} + \mathbf{b})$ of $(C_1 | C_1 + C_2)$, where $\mathbf{a} = (a_0, a_1, \ldots, a_{n-1})$ and $\mathbf{b} = (b_0, b_1, \ldots, b_{n-1})$. Since the Gray map is bijective, by dimension arguments it follows that $\Pi(C) = (C_1 | C_1 + C_2) = \Phi(D)$; i.e.,

$$C = \Pi^{-1}(\Phi(D)).$$

We observe that any binary cyclic code of length $2n$ arises in this way. In fact if $C' = < g'(x) > \subset \mathbb{F}_2[x]/(x^{2n} - 1)$ is an ideal, where $g'(x)$ is a monic divisor of $(x^{2n} - 1)$, then since $x^n - 1 = f_1(x)f_2(x) \cdots f_r(x)$, $f_i(x) \neq f_j(x)$, $0 \leq i, j \leq r$, with each $f_i(x)$ monic irreducible over $\mathbb{F}_2[x]$, we have $x^{2n} - 1 = (x^n - 1)^2 = f_1^2(x)f_2^2(x) \cdots f_r^2(x)$. Thus $g'(x)$ has the form $f_{i_1}^2(x) \cdots f_{i_r}^2(x)f_{j_1}(x) \cdots f_{j_s}(x) = a(x)^2 b(x)$, where $a(x)$ and $b(x)$ are binary monic coprime divisors of $x^n - 1$. Let

$$C_1' = < a(x) >, \ C_2' = < b(x) >,$$

then the above argument shows that

$$\Pi(C') = (C_1' | C_1' + C_2').$$

According to [23], a $\mathbb{Z}_4$-cyclic code M of length n can be thought of as a principal ideal $M = < g(x) >$ of $\mathbb{Z}_4[x]/(x^n - 1)$, where $g(x) = a(x)[b(x) + 2]$

and $a(x)$, $b(x)$ are coprime factors of $x^n - 1$ in $\mathbb{Z}_4[x]$. Then $M_1 = < \tilde{a}(x) >$ and $M_2 = < \tilde{a}(x)\tilde{b}(x) >$, where "$\tilde{y}$" means reduction modulo 2, are binary cyclic codes of length n, which may be called the "projections" of M on $\mathbb{F}_2[x]/(x^n - 1)$. Observe that $M_2 \subset M_1$. It would be interesting to give conditions on the cyclic code M under which it could be obtained from its projections, i.e., from the binary cyclic codes M_1 and M_2 via the $M_2 + 2M_1$ construction as described above. If Π and Φ are as above, it would be also interesting to give conditions on M such that $\Pi^{-1}(\Phi(M)) = < \tilde{a}^2(x)\tilde{b}(x) >$. In [22] the authors deal partially with this question.

4 The Gray Map, "$(u|u + v)$-Construction" and Cyclic Codes over $\mathbb{Z}_{p^2}$

Let p be a prime. In order to avoid confusion in the rest of the paper, "+" will denote the sum operation in the ring $\mathbb{Z}_{p^2}$ and "$\oplus_p$" will denote the sum in the finite field $\mathbb{F}_p$. In this section it is shown that if n is a positive integer such that $(p, n) = 1$, then the Gray map Φ on $\mathbb{Z}_{p^2}^n$ is the same as a suitable generalization of the "$(u|u + v)$-construction". Furthermore if $C_p, ..., C_1$ are $\mathbb{F}_p$-cyclic codes of length n and $D = (C_p \oplus_p \cdots \oplus_p C_2) + pC_1 \subset \mathbb{Z}_{p^2}^n$ then, up to the p-permutation introduced in Section 2, the Gray map image $\Phi(D)$ is the same as a repeated-root $\mathbb{F}_p$-cyclic code C of length pn. The generator of this code is given in terms of the generators of the codes C_i, $i = 1, 2, ..., p$ (for $p = 2$ this problem was treated in [21] and [2]).

4.1 The Gray Map on $\mathbb{Z}_{p^2}^n$ and "$(u|u + v)$-Construction"

We show first that if p is a prime and n a positive integer such that $(p, n) = 1$, the (generalized) Gray map on $\mathbb{Z}_{p^2}^n$ and a version of the "$(u|u+v)$-construction" yields the same result.

Recall that any element $u \in \mathbb{Z}_{p^2}$ has the p-adic expansion: $u = r_0(u) + r_1(u)p$, where $r_i(u) \in \mathbb{F}_p$.

If $\mathbf{u} = (u_0, u_1, ..., u_{n-1}) \in \mathbb{Z}_{p^2}^n$ let:

$$r_0(\mathbf{u}) = (r_0(u_0), r_0(u_1), ..., r_0(u_{n-1})), \quad r_1(\mathbf{u}) = (r_1(u_0), r_1(u_1), ..., r_1(u_{n-1})).$$

Identifying $\mathbb{F}_p^{np}$ with p copies of $\mathbb{F}_p^n$, the (generalized) Gray map

$$\Phi : \mathbb{Z}_{p^2}^n \longrightarrow \mathbb{F}_p^{np}$$

is defined as:

$$\Phi(\mathbf{a}) = (r_1(\mathbf{a}), r_1(\mathbf{a}) \oplus_p (p-1)r_0(\mathbf{a}), r_1(\mathbf{a}) \oplus_p (p-2)r_0(\mathbf{a}), ..., r_1(\mathbf{a}) \oplus_p r_0(\mathbf{a}))$$

where the sum and the products $(p - i)r_0(\mathbf{a})$, with $i = 1, 2, ..., p$, are taken in $\mathbb{F}_p$. The above definition of the Gray map is equivalent to the one given in [6] (see also [9], [17]). In any case the Gray map is injective.

Let the *homogeneous weight*, $wt_{\hom}$, on $\mathbb{Z}_{p^2}$ be defined by

$$
wt_{\hom}(a) = \begin{cases} 0 & \text{if } a = 0, \\ p & \text{if } a \in p\mathbb{Z}_{p^2} \setminus \{0\}, \quad \forall\, a \in \mathbb{Z}_{p^2}, \\ (p-1) & \text{otherwise} \end{cases}
$$

and, for $\mathbf{a} \in \mathbb{Z}_{p^2}^n$, the value $wt_{\hom}(\mathbf{a}) \in \mathbb{Z}$ is taken as the sum of the homogeneous weight of its components. The *homogeneous metric*, $\delta_{\hom}$, is given by $\delta_{\hom}(\mathbf{a}, \mathbf{b}) = wt_{\hom}(\mathbf{a} - \mathbf{b})$ for all $\mathbf{a}, \mathbf{b} \in \mathbb{Z}_{p^2}^n$. Let δ_H denote the usual Hamming distance on $\mathbb{F}_p^{np}$. Then we have (cf. [7], [6], [1], [16]):

Theorem 1. *The Gray map Φ is an isometry between $(\mathbb{Z}_{p^2}^n, \delta_{\hom})$ and $(\mathbb{F}_p^{np}, \delta_H)$.*

For (non-empty) subsets $U_1, ..., U_p$ of $\mathbb{F}_p^n$, a natural generalization of the "$(u|u + v)$-construction" is the following subset $\Gamma(U_1, ..., U_p)$ of $\mathbb{F}_p^{pn}$ given by:

$$
(U_1|U_1 \oplus_p (p-1)[U_2 \oplus_p \cdots \oplus_p U_p]|\cdots|U_1 \oplus_p [U_2 \oplus_p \cdots \oplus_p U_p]
$$

where "$|$" means concatenation. A typical element of this set has the form:
$(\mathbf{u}^{(1)}|\mathbf{u}^{(1)} \oplus_p (p-1)[\mathbf{u}^{(2)} \oplus_p \cdots \oplus_p \mathbf{u}^{(p)}]|\mathbf{u}^{(1)} \oplus_p (p-2)[\mathbf{u}^{(2)} \oplus_p \cdots \oplus \mathbf{u}^{(p)}]|\cdots|$
$\mathbf{u}^{(1)} \oplus_p [\mathbf{u}^{(2)} \oplus_p \cdots \oplus_p \mathbf{u}^{(p)}])$,

where $\mathbf{u}^{(j)} = (u_0^{(j)}, u_1^{(j)}, ..., u_n^{(j)}) \in U_j$, for $j = 1, 2, ..., p$.

Let C_i, $i = 1, 2, ..., p$ be $\mathbb{F}_p$-linear codes of length n and let $D = (C_p \oplus_p C_{p-1} \oplus_p \cdots \oplus_p C_2) + pC_1 \subset \mathbb{Z}_{p^2}^n$. A typical element $\mathbf{v} = (v_0, v_1, ..., v_{n-1}) \in D$ has the form $v_i = (a_i^{(p)} \oplus_p a_i^{(p-1)} \oplus_p \cdots \oplus_p a_i^{(2)}) + p a_i^{(1)}$, where $(a_0^{(j)}, ..., a_{n-1}^{(j)}) \in C_j$, for $j = 1, 2, ..., p$ and $i = 0, 1, ..., n - 1$.

Thus if $\mathbf{a}^{(j)} = (a_0^{(j)}, ..., a_{n-1}^{(j)}) \in C_j$ for $j = 1, 2, ..., p$, we can write

$$
\mathbf{v} = (\mathbf{a}^{(p)} \oplus_p \mathbf{a}^{(p-1)} \oplus_p \cdots \oplus_p \mathbf{a}^{(2)}) + p\mathbf{a}^{(1)}.
$$

If $\mathbf{v} = (v_0, v_1, ..., v_{n-1}) \in D$ is as described above, from the definition of the Gray map it follows that:

$$
\Phi(\mathbf{v}) = (\mathbf{a}^{(1)}|\mathbf{a}^{(1)} \oplus_p (p-1)[\mathbf{a}^{(2)} \oplus_p \cdots \oplus_p \mathbf{a}^{(p)}]
$$
$$
|\mathbf{a}^{(1)} \oplus_p (p-2)[\mathbf{a}^{(2)} \oplus_p \cdots \oplus_p \mathbf{a}^{(p)}]|\cdots|\mathbf{a}^{(1)} \oplus_p [\mathbf{a}^{(2)} \oplus_p \cdots \oplus_p \mathbf{a}^{(p)}]
$$

where "$|$" means concatenation and $\mathbf{a}^{(j)} = (a_0^{(j)}, a_1^{(j)}, ..., a_{n-1}^{(j)})$, for $j = 1, 2, ..., p$.

The Gray map image, $\Phi(D)$, of the $\mathbb{Z}_{p^2}$-code D as given above is precisely the "$(u|u+v)$-construction" $\Gamma(C_1, ..., C_p)$ on the codes C_i, i.e., $\Phi(D) = \Gamma(C_1, ..., C_p)$.

4.2   The Gray Map and Cyclic Codes

In this section, assuming each C_i is an $\mathbb{F}_p$-cyclic code, it is shown that by applying the inverse of the p-permutation Π introduced in §2 to the Gray image $\Phi(D)$ of the $\mathbb{Z}_{p^2}$-code D as defined above, an $\mathbb{F}_p$-repeated-root cyclic code C of length np is obtained. The generator of this cyclic code C is given in terms of the generators of the cyclic codes C_i.

Let p be a prime and let n be a positive integer such that $(p,n) = 1$ with $p \le n$. Let $g_1, g_2, ..., g_p \in \mathbb{F}_p[x]$ be different monic pairwise coprime divisors of $x^n - 1$ and let $C_j = < g_1 g_2 \cdots g_j >$ for $j = 1, 2, ..., p$ be the $\mathbb{F}_p$-cyclic codes of length n generated by the polynomial $g_1 g_2 \cdots g_j$; i.e., they are ideals in the ring $R(p,n) = \mathbb{F}_p[x]/(x^n - 1)$.

It will be shown that the cyclic code C, i.e., the ideal of $R(p, np)$ generated by the polynomial $g_1^p g_2^{p-1} \cdots g_{p-1}^2 g_p$ is precisely $\Pi^{-1}(\Phi(D))$, where $D = (C_p \oplus_p \cdots \oplus_p C_2) + pC_1$ and Π^{-1} is the inverse of the p-permutation Π introduced in §2. In order to prove this, it will be shown that the codeword

$$\Pi^{-1}(\Phi(\mathbf{v})) = \Pi^{-1}\Big[(\mathbf{a^{(1)}}|\mathbf{a^{(1)}} \oplus_p (p-1)[\mathbf{a^{(2)}} \oplus_p \cdots \oplus_p \mathbf{a^{(p)}}]\|$$

$$\mathbf{a^{(1)}} \oplus_p (p-2)[\mathbf{a^{(2)}} \oplus_p \cdots \oplus_p \mathbf{a^{(p)}}]\| \cdots \|\mathbf{a^{(1)}} \oplus_p [\mathbf{a^{(2)}} \oplus_p \cdots \oplus_p \mathbf{a^{(p)}}]\Big]$$

is equal to the vector associated to a polynomial in the ideal $C = < g_1^p g_2^{p-1} \cdots g_{p-1}^2 g_p >$.

We first observe that any element $a(x) = a_0 + a_1 x + a_2 x^2 + \cdots + a_{n-1} x^{n-1}$ of $R(p,n)$ can be written as:

$$a(x) = a^{(0)}(x^p) + x a^{(1)}(x^p) + \cdots + x^{p-1} a^{(p-1)}(x^p)$$

where

$$a^{(j)}(x^p) = a_j + a_{j+p} x^p + \cdots + a_{j+pt} x^{pt} + \cdots + a_{j+ps} x^{ps}$$

for $j = 0, 1, 2, ..., p-1$ and $j + ps \le \deg a(x)$.

Let n be as above, let $n \equiv r \bmod p$, with $1 \le r \le p-1$ and let γ be a positive integer such that $r\gamma \equiv -1 \bmod p$, i.e., $\gamma \equiv -r^{-1} \bmod p$. Associated to the polynomial $a(x)$, let

$$\bar{a}(x) = \sum_{j=0}^{p-1} a^{(j)}(x^p) x^{j+n(j\gamma)_p}$$

be an element of $\mathbb{F}_p[x]$, where $(j\gamma)_p$ is the reduction of $j\gamma$ modulo p.

Since all non-constant monomials of $\bar{a}(x)$ have exponents divisible by p then $\bar{a}(x) = (\alpha(x))^p$ for some $\alpha(x) \in \mathbb{F}_p[x]$.

Proposition 1. *With the notation as above,*

$$\bar{a}(x) \equiv a(x) \bmod (x^n - 1).$$

Proof. The proof follows immediately by observing that

$$\bar{a}(x) - a(x) = \sum_{j=0}^{p-1} (a^{(j)}(x^p)x^j)(x^{n(j\gamma)_p} - 1)$$

and

$$x^{nm} - 1 = (x^n - 1)(x^{(m-1)n} + x^{(m-2)n} + \cdots + 1)$$

for any integer $m \geq 1$.

From the previous Proposition it follows that if $f(x)$ is any irreducible factor of $x^n - 1$ and $f(x)|a(x)$, then $f(x)^p|\bar{a}(x)$.

If the expression for $\bar{a}(x)$ given above is written as $\bar{a}(x) = \sum_{i=0}^{p-1} A_i(x)x^{n_i}$ where $n_0 < n_1 < \cdots < n_{p-1}$ (increasing powers of x), it can be seen that for any $i = 0, 1, ..., p - 2$ the difference $n_i - n_{i+1}$ is a multiple of p. If $n \equiv 1 \bmod p$, $n_i - n_{i+1} = kp$ for the same positive integer k and for all $i = 0, 1, ..., p - 2$, i.e., the difference $n_i - n_{i+1}$ is a constant multiple of p. In this case it is easier to control the powers of x in the operations which involve the expression for $\bar{a}(x)$. For simplicity, from now on it will be assumed that $n \equiv 1 \bmod p$. In this case $\gamma = -1 \bmod p$, i.e., we can take $\gamma = p - 1$. From the arguments given below it is easy to see that if $n \equiv r \bmod p$, with the appropriate modifications the same results are obtained. Let C_i for $i = 1, 2, ..., p$ and C be the $\mathbb{F}_p$-cyclic codes introduced above.

If $a_i(x) \in C_i$, for $i = 1, 2, ..., p$, let

$$a_i(x) = a_i^{(0)}(x^p) + xa_i^{(1)}(x^p) + \cdots + x^{p-1}a_i^{(p-1)}(x^p)$$

and consider the corresponding polynomials

$$\bar{a}_i(x) = \sum_{j=0}^{p-1} [a_i^{(j)}(x^p)x^j]x^{n(j\gamma)_p}$$

as described above. Let

$$w(x) = (x^n - 1)^{p-1}\bar{a}_1(x) + (x^n - 1)^{p-2}x^n \sum_{j=2}^{p} \bar{a}_j(x).$$

Proposition 2. *With the notation as above, the polynomial $w(x)$ is an element of the ideal $C = < g_1^p g_2^{p-1} \cdots g_{p-1}^2 g_p >$.*

Proof. First observe that since g_1 divides $x^n - 1$ then g_1^{p-1} divides $(x^n - 1)^{p-1}$ and since $a_1 \in C_1$, by Proposition 1, g_1 divides $\bar{a}_1$. So, g_1^p divides the first term of $w(x)$. Also, $g_1^{p-2}|(x^n - 1)^{p-2}$ and from the fact that $a_i \in C_i$ it follows that $g_1|\bar{a}_i$, implying that g_1^p divides each one of the other terms of $w(x)$. Thus g_1^p divides $w(x)$. Since $g_j|(x^n - 1)$ and $a_j \in C_j = < g_1 g_2 \cdots g_j >$, by Proposition 1, a similar argument as above shows that for $j = 2, 3, ..., p - 1$, g_j^{p-j+1} divides each term of $w(x)$, proving the assertion.

The next step is to show that the polynomial $w(x)$, or equivalently, the vector $\mathbf{w}$ of $\mathbb{F}_p^{np}$ associated to the polynomial $w(x)$, has the form $\Pi^{-1}(\Phi(\mathbf{v}))$ for some $\mathbf{v} \in D$.

In order to do so we find the vector $\mathbf{w}$ corresponding to the polynomial $w(x)$. First observe that over the field $\mathbb{F}_p$ the following identity holds:

$$\frac{(y-1)^p}{y-1} = y^{p-1} + y^{p-2} + \cdots + y + 1 = (y-1)\sum_{j=1}^{p-1} jy^{p-j-1}.$$

In particular we have:

$$(x^n - 1)^{p-2} = \sum_{j=1}^{p-1} jx^{n(p-j-1)}.$$

For any $b(x) = b_0 + b_1 x + \cdots + b_{n-1}x^{n-1} \in R(p, n)$, let $B(\bar{b}, n)$ be the arrangement with p rows and n columns, where if $s \equiv j \bmod p$, and $0 \leq s \leq n-1$, its s−th column is the transpose of the following p-tuple:

$$\begin{pmatrix} \mathbf{s} & \mathbf{s+n} & \cdots & \mathbf{s+rn} & \cdots & \mathbf{s+(p-1)n} \\ (p-j)b_j & (p-j-1)b_j & \cdots & (p-j-r)b_j & \cdots & (p-j-(p-1))b_j \end{pmatrix}$$

where the boldface entries indicate the position in the arrangement and the terms $(p-t)$ are reduced modulo p.

Let $\bar{b}(x)$ be the corresponding polynomial associated to $b(x)$ as defined above. Then, by reducing modulo $x^{pn} - 1$, it is easily seen that the vector $\mathbf{w}(b)$ corresponding to the polynomial $\bar{b}(x)x^n(x^n - 1)^{p-2}$ is the concatenation of the rows of the arrangement $B(\bar{b}, n)$.

If $w(x) = (x^n - 1)^{p-1}\bar{a}_1(x) + (x^n - 1)^{p-2}x^n \sum_{j=2}^{p} \bar{a}_j(x)$, it follows from the above observation that the vector associated to the term $(x^n - 1)^{p-2}x^n \sum_{j=2}^{p} \bar{a}_j(x)$ is the sum of the vectors $\mathbf{w}(a_2), \ldots, \mathbf{w}(a_p)$.

It is also easily seen that if $a_1(x) = a_0^{(1)} + a_1^{(1)}x + \cdots + a_r^{(1)}x^r + \cdots + a_{n-1}^{(1)}x^{n-1}$, the vector $\mathbf{w}(a_1)$ associated to the polynomial $(x^n - 1)^{p-1}\bar{a}_1(x)$, that is, the first term in the expression of $w(x)$, is the vector of length pn divided into p blocks of length n, each block having the form:

$$(a_0^{(1)}, a_1^{(1)}, \ldots, a_r^{(1)}, \ldots, a_{n-1}^{(1)}).$$

Summarizing, the vector $\mathbf{w}$ associated to the polynomial $w(x)$ is the vector

$$\mathbf{w}(a_1) + \mathbf{w}(a_2) + \cdots + \mathbf{w}(a_p).$$

For $0 \leq j \leq p-1$ the action of the p-permutation Π on the j-column of the arrangement $B(\bar{b}, n)$ is just shifting-down j-places, and if $t \equiv j \bmod p$, the action of Π on the t-column is the same as the action on the j-column. Thus we conclude that

$$\Pi(\mathbf{w}) = \Phi(\mathbf{v})$$

where $\mathbf{v} = (\mathbf{a}^{(p)} \oplus_p \mathbf{a}^{(p-1)} \oplus_p \cdots \oplus_p \mathbf{a}^{(2)}) + p\mathbf{a}^{(1)}$ with $\mathbf{a}^{(j)} = (a_0^{(j)}, a_1^{(j)}, ..., a_{n-1}^{(j)})$ and $a_1^{(j)}$ are the coefficients of the polynomial $a^{(j)}(x) \in C_j$, for $j = 1, 2, ..., p$.

If $\rho_i = \deg(g_i)$ and $m = \sum_{i=1}^{p}(p-i+1)\rho_i$, it is easy to see that the cardinality of D and the cardinality of C are the same and both are equal to p^{np-m}. From the fact that the Gray map Π is injective we conclude that $\Pi(C) = \Phi(D)$, i.e.,

$$C = \Pi^{-1}(\Phi(D)).$$

5   Examples

In order to illustrate the results presented in the previous section, some examples are provided. The first example is given with certain detail.

Example 1. In this example we take $p = 3$ and $n = 4$.

The Gray map on $\mathbb{Z}_{3^2}^4$ and "$(u|u + v)$-construction"

First we show that the generalized Gray map on $\mathbb{Z}_{3^2}^4$ and the generalization of the "$(u|u + v)$-construction" given previously yields the same result.

If U_1, U_2, U_3 are (non-empty) subsets of $\mathbb{F}_3^4$ the generalization of the "$(u|u + v)$-construction" on these subsets is the subset $\Gamma(U_1, U_2, U_3)$ of $(\mathbb{F}_3^4)^3 = \mathbb{F}_3^{12}$ given as:

$$\Gamma(U_1, U_2, U_3) = \{(\mathbf{u_1}|\mathbf{u_1} + 2(\mathbf{u_2} + \mathbf{u_3})|\mathbf{u_1} + \mathbf{u_2} + \mathbf{u_3}), \ \mathbf{u_i} \in U_i\}$$

(where the bar "$|$" means concatenation).

Recall that any element $u \in \mathbb{Z}_{3^2}$ can be expressed as $u = r_0(u) + 3r_1(u)$ with $r_i(u) \in \mathbb{F}_3$. If $\mathbf{u} = (u_0, \ldots, u_3) \in \mathbb{Z}_3^4$ let $r_0(\mathbf{u}) = (r_0(u_0), \ldots, r_0(u_3))$ and $r_1(\mathbf{u}) = (r_1(u_0), ..., r_1(u_3))$. Identifying $\mathbb{F}_3^{12}$ with 3 copies of $\mathbb{F}_3^4$, the Gray map $\Phi : \mathbb{Z}_{3^2}^4 \longrightarrow \mathbb{F}_3^{12}$ is defined as:

$$\Phi(\mathbf{u}) = (r_1(\mathbf{u})|r_1(\mathbf{u}) \oplus_3 2r_0(\mathbf{u})|r_1(\mathbf{u}) \oplus_3 r_0(\mathbf{u})).$$

Let

$$V = \{\mathbf{v} = (v_0, v_1, v_2, v_3) \in \mathbb{Z}_{3^2}^4 \ : v_i = (c_i \oplus_3 b_i) + 3a_i; \ a_i, b_i, c_i \in \mathbb{F}_3\}.$$

From the above definition of the Gray map on $\mathbb{Z}_{3^2}^4$ it follows that:

$$\Phi(\mathbf{v}) = (\mathbf{a}|\mathbf{a} \oplus_3 2(\mathbf{b} \oplus_3 \mathbf{c})|\mathbf{a} \oplus_3 (\mathbf{b} \oplus_3 \mathbf{c}))$$

where $\mathbf{a} = (a_0, a_1, a_2, a_3)$, $\mathbf{b} = (b_0, b_1, b_2, b_3)$, $\mathbf{c} = (c_0, c_1, c_2, c_3)$ are in $\mathbb{F}_3^4$. Hence $\Phi(V) = \Gamma(R_1, R_2, R_3)$, where $R_i = \mathbb{Z}_{3^2}^4$.

If C_1, C_2, C_3 are ternary linear codes of length $n = 4$, let $D = (C_3 \oplus_3 C_2) + 3C_1 \subseteq \mathbb{Z}_{3^2}^4$. A typical element $\mathbf{v} = (v_0, v_1, v_2, v_3) \in D$ has the form $v_i = (c_i \oplus_3 b_i) + 3a_i$, i.e., $r_0(v_i) = c_i \oplus_3 b_i$ and $r_1(v_i) = a_i$ where $\mathbf{a} = (a_0, a_1, a_2, a_3) \in C_1$, $\mathbf{b} = (b_0, b_1, b_2, b_3) \in C_2$ and $\mathbf{c} = (c_0, c_1, c_2, c_3) \in C_3$ with $c_i, b_i, a_i \in \mathbb{F}_3$, for $i = 0, 1, 2, 3$. Therefore, $\mathbf{v} = (\mathbf{c} \oplus_3 \mathbf{b}) + 3\mathbf{a}$.

From the above observations it follows that $\Phi(D)$ is the "$(u|u+v)$-construction" $\Gamma(C_1, C_2, C_3)$ on the codes C_i, that is,

$$\Phi(D) = \Gamma(C_1, C_2, C_3).$$

Thus the Gray map on $\mathbb{Z}_{3^3}^4$ and the generalized "$(u|u+v)$-construction" as introduced above give the same result.

The Gray map and cyclic codes

Let $R(3, 4) = \mathbb{F}_3[x]/(x^4 - 1)$ and let $x^4 - 1 = g_1 g_2 g_3$ where $g_1 = (x - 1)$, $g_2 = (x - 2)$, $g_3 = (x^2 + 1)$ are irreducible elements of $\mathbb{F}_3[x]$. Let $C_1 = <g_1>$, $C_2 = <g_1 g_2>$, $C_3 = <g_1 g_2 g_3>$ be the $\mathbb{F}_3$-cyclic codes of length 4 generated by the corresponding polynomials, i.e., they are ideals in the ring $R(3, 4)$.

It will be shown that the ideal C of $R(3, 12) = \mathbb{F}_3[x]/(x^{12} - 1)$ generated by the polynomial $g_1^3 g_2^2 g_3$ is equal, up to the inverse of the 3-permutation Π as introduced in Section 2, to the Gray map image $\Phi(D)$ of the code $D = (C_3 \oplus_3 C_2) + 3C_1$, i.e., $C = <g_1^3 g_2^2 g_3> = \Pi^{-1}(\Phi(D))$.

In order to do so it will first be shown that for any element $\mathbf{v} = (\mathbf{c} \oplus_3 \mathbf{b}) + 3\mathbf{a} \in D$ the codeword $\Pi^{-1}(\Phi(\mathbf{v})) = \Pi^{-1}(\mathbf{a}|\mathbf{a} \oplus_3 2(\mathbf{b} \oplus_3 \mathbf{c})|\mathbf{a} \oplus_3 (\mathbf{b} \oplus_3 \mathbf{c}))$ corresponds to an element of the ideal $<g_1^3 g_2^2 g_3>$.

Any polynomial $a(x) = a_0 + a_1 x + a_2 x^2 + a_3 x^3 \in R(3, 4)$ can be written as (remember that $p = 3$ and $n = 4$):

$$a(x) = a^{(0)}(x^3) + x a^{(1)}(x^3) + x^2 a^{(2)}(x^3)$$

where

$$a^{(0)}(x^3) = a_0 + a_3 x^3, \quad a^{(1)}(x^3) = a_1, \quad a^{(2)}(x^3) = a_2.$$

Since $n = 4 \equiv 1 \bmod 3$, let

$$\bar{a}(x) = a^{(0)}(x^3) + x^{4+2} a^{(2)}(x^3) + x^{2 \cdot 4 + 1} a^{(1)}(x^3) = (a_0 + a_3 x^3) + a_2 x^6 + a_1 x^9.$$

Observe that $\bar{a}(x) \equiv a(x) \bmod (x^4 - 1)$ and each non-constant monomial of $\bar{a}(x)$ has degree multiple of 3, and $\bar{a}(x) = \alpha(x)^3$ for some polynomial $\alpha(x) \in R(3, 4)$. In particular if $f(x)$ is an irreducible factor of $(x^4 - 1)$ which divides $a(x)$ then $f(x)^3$ divides $\bar{a}(x)$.

Let $a(x) = a_0 + a_1 x + a_2 x^2 + a_3 x^3 \in C_1$, $b(x) = b_0 + b_1 x + b_2 x^2 + b_3 x^3 \in C_2$ and $c(x) = c_0 + c_1 x + c_2 x^2 + c_3 x^3 \in C_3$ and consider the corresponding polynomials $\bar{a}(x), \bar{b}(x), \bar{c}(x)$ as defined above. Let

$$w(x) = (x^4 - 1)^2 \bar{a}(x) + x^4(x^4 - 1)[\bar{b}(x) + \bar{c}(x)].$$

We claim that the vector $\mathbf{w}$ associated to the polynomial $w(x)$ is equivalent, up to the inverse of the 3-permutation, to the codeword

$$\Phi(\mathbf{v}) = (\mathbf{a}|\mathbf{a} \ominus_3 2(\mathbf{b} \ominus_3 \mathbf{c})|\mathbf{a} \ominus_3 (\mathbf{b} \ominus_3 \mathbf{c}))$$

where $\mathbf{a} = (a_0, a_1, a_2, a_3)$, $\mathbf{b} = (b_0, b_1, b_2, b_3)$, $\mathbf{c} = (c_0, c_1, c_2, c_3)$ are in $\mathbb{F}_3^4$.

First observe that reducing modulo $x^{12} - 1$:

$$(x^4 - 1)^2 \overline{a}(x) = \overline{a}(x) + x^4 \overline{a}(x) + x^8 \overline{a}(x) =$$
$$a_0 + a_1 x + a_2 x^2 + a_3 x^3 + a_0 x^4 + a_1 x^5 + a_2 x^6$$
$$+ a_3 x^7 + a_0 x^8 + a_1 x^9 + a_2 x^{10} + a_3 x^{11}$$

and the corresponding vector is

$$(a_0, a_1, a_2, a_3 | a_0, a_1, a_2, a_3 | a_0, a_1, a_2, a_3).$$

Also, it is easy to see that

$$x^4(x^4 - 1)\overline{b}(x) = -b_1 x + b_2 x^2 - b_0 x^4 + b_1 x^5 - b_3 x^7 b_0 x^8 - b_2 x^{10} + b_3 x^{11}$$

corresponds to the vector

$$(0, -b_1, b_2, 0| - b_0, b_1, 0, -b_3 | b_0, 0, -b_2, b_3)$$

and

$$x^4(x^4 - 1)\overline{c}(x) = -c_1 x + c_2 x^2 - c_0 x^4 + c_1 x^5 - c_3 x^7 c_0 x^8 - c_2 x^{10} + c_3 x^{11}$$

is associated to the vector

$$(0, -c_1, c_2, 0| - c_0, c_1, 0, -c_3 | c_0, 0, -c_2, c_3).$$

Therefore the vector $\mathbf{w}$ associated to the polynomial $w(x)$ is the successive concatenation of the rows of the following arrangement

$$\begin{pmatrix} a_0 & a_1 - b_1 - c_1 & a_2 + b_2 + c_2 & a_3 \\ a_0 - b_0 - c_0 & a_1 + b_1 + c_1 & a_2 & a_3 - b_3 - c_3 \\ a_0 + b_0 + c_0 & a_1 & a_2 - b_2 - c_2 & a_3 + b_3 + c_3 \end{pmatrix}$$

Applying the 3-permutation Π as introduced in §2 to the vector $\mathbf{w}$, we obtain

$$\Pi(\mathbf{w}) = \begin{pmatrix} a_0 & a_1 & a_2 & a_3 \\ a_0 - b_0 - c_0 & a_1 - b_1 - c_1 & a_2 - b_2 - c_2 & a_3 - b_3 - c_3 \\ a_0 + b_0 + c_0 & a_1 + b_1 + c_1 & a_2 + b_2 + c_2 & a_3 + b_3 + c_3 \end{pmatrix}$$

which is precisely the element

$$\Phi(\mathbf{v}) = (\mathbf{a}|\mathbf{a} \ominus_3 2(\mathbf{b} \ominus_3 \mathbf{c})|\mathbf{a} \ominus_3 \mathbf{b} \ominus_3 \mathbf{c})$$

where $\mathbf{v} = (\mathbf{c} \oplus_3 \mathbf{b}) + 3\mathbf{a}$.

From the fact that $a(x) \in C_1$, $b(x) \in C_2$, $c(x) \in C_3$, for each polynomial $\gamma(x) \in \mathbb{F}_3[x]$, $\overline{\gamma}(x) \equiv \gamma(x)$ mod $(x^4 - 1)$ and $\overline{\gamma}(x)$ is the third power of a polynomial, it follows that $w(x)$ is an element of the ideal $C = <g_1^3 g_2^2 g_3>$. If $\rho_i = \deg(g_i)$, $i = 1, 2, 3$ and $t = 3\rho_1 + 2\rho_2 + \rho_3$, the cardinality of C and D are the same and equal to 3^{12-t}. Since the Gray map is injective we conclude that:

$$C = \Pi^{-1}(\Phi(D)).$$

Example 2. In this case we take $p = 5$ and $n = 11$.

Let $R(5, 11) = \mathbb{F}_5[x]/(x^{11} - 1)$. Any element $a(x) = a_0 + a_1 x + \cdots + a_9 x^9 + a_{10} x^{10}$ of $R(5, 11)$ can be written as

$$a(x) = a^{(0)}(x^5) + a^{(1)}(x^5)x + a^{(2)}(x^5)x^2 + a^{(3)}(x^5)x^3 + a^{(4)}(x^5)x^4$$

where

$$a^{(0)}(x^5) = a_0 + a_5 x^5 + a_{10} x^{10}, \ a^{(1)}(x^5) = a_1 + a_6 x^5, \ a^{(2)}(x^5) = a_2 + a_7 x^5,$$
$$a^{(3)}(x^5) = a_3 + a_8 x^5, \ a^{(4)}(x^5) = a_4 + a_9 x^5.$$

Since $n \equiv 1$ mod p, let

$$\overline{a}(x) = a^{(0)}(x^5) + [a^{(1)}(x^5)x]x^{11(4)} + [a^{(2)}(x^5)x^2]x^{11(3)}$$
$$+ [a^{(3)}(x^5)x^3]x^{11(2)} + [a^{(4)}(x^5)x^4]x^{11(1)}.$$

Let $g_1, g_2, g_3, g_4, g_5 \in \mathbb{F}_5[x]$ be different pairwise coprime divisors of the polynomial $x^{11} - 1$ and let $C_i = <g_1 g_2 \cdots g_i>$ for $i = 1, 2, ..., 5$ be the ideals of $R(5, 11)$ generated by the corresponding polynomials. For $a_i(x) = a_0^{(i)} + a_1^{(i)} x + \cdots + a_{10}^{(i)} x^{10} \in C_i$ let

$$w(x) = \overline{a}_1(x)(x^{11} - 1)^4 + [\overline{a}_2(x) + \cdots + \overline{a}_5(x)](x^{11} - 1)^3 x^{11}.$$

The arrangement $B(a^{(j)}, 11)$ for $j = 2, ..., 5$ has the form:

$$
\begin{pmatrix}
0 & 1 & 2 & 3 & 4 & 5 & 6 & 7 & 8 & 9 & 10 \\
0 & 4a_1^{(j)} & 3a_2^{(j)} & 2a_3^{(j)} & a_4^{(j)} & 0 & 4a_6^{(j)} & 3a_7^{(j)} & 2a_8^{(j)} & a_9^{(j)} & 0 \\
11 & 12 & 13 & 14 & 15 & 16 & 17 & 18 & 19 & 20 & 21 \\
4a_0^{(j)} & 3a_1^{(j)} & 2a_2^{(j)} & a_3^{(j)} & 0 & 4a_5^{(j)} & 3a_6^{(j)} & 2a_7^{(j)} & a_8^{(j)} & 0 & 4a_{10}^{(j)} \\
22 & 23 & 24 & 25 & 26 & 27 & 28 & 29 & 30 & 31 & 32 \\
3a_0^{(j)} & 2a_1^{(j)} & a_2^{(j)} & 0 & 4a_4^{(j)} & 3a_5^{(j)} & 2a_6^{(j)} & a_7^{(j)} & 0 & 4a_9^{(j)} & 3a_{10}^{(j)} \\
33 & 34 & 35 & 36 & 37 & 38 & 39 & 40 & 41 & 42 & 43 \\
2a_0^{(j)} & a_1^{(j)} & 0 & 4a_3^{(j)} & 3a_4^{(j)} & 2a_5^{(j)} & a_6^{(j)} & 0 & 4a_8^{(j)} & 3a_9^{(j)} & 2a_{10}^{(j)} \\
44 & 45 & 46 & 47 & 48 & 49 & 50 & 51 & 52 & 53 & 54 \\
a_0^{(j)} & 0 & 4a_2^{(j)} & 3a_3^{(j)} & 2a_4^{(j)} & a_5^{(j)} & 0 & 4a_7^{(j)} & 3a_8^{(j)} & 2a_9^{(j)} & a_{10}^{(j)}
\end{pmatrix}
$$

The vector $\mathbf{w}(a_j)$ associated to the polynomial $\bar{a}_j(x)x^{11}(x^{11}-1)^3$ is the (successive) concatenation of the rows of the arrangement $B(a^{(j)}, 11)$. Recall that the boldface entries indicate the coordinate position.

The vector $\mathbf{w}(a_1)$ associated to the polynomial $\bar{a}_1(x)(x^{11}-1)^4$ has length 55 and is divided into 5 blocks of length 11. The entries of each one of these blocks are the coefficients of the polynomial $a_1(x)$, i.e., each block has the form $(a_0^{(1)}, a_1^{(1)}, ..., a_{10}^{(1)})$.

Summarizing, the vector associated to the polynomial $w(x)$ as defined above is:

$$\mathbf{w} = \mathbf{w}(a_1) + \mathbf{w}(a_2) + \mathbf{w}(a_3) + \mathbf{w}(a_4).$$

If Π is the 5-permutation as introduced in §2, it is easily seen that $\Pi(\mathbf{w}) = \Phi(\mathbf{a})$ where $\mathbf{a} = (\mathbf{a}^{(5)} \oplus_5 \mathbf{a}^{(4)} \oplus_5 \mathbf{a}^{(3)} \oplus_5 \mathbf{a}^{(2)}) + 5\mathbf{a}^{(1)}$ and $\mathbf{a}^{(j)} = (a_0^{(j)}, a_1^{(j)}, ..., a_{n-1}^{(j)})$ are the coefficients of the polynomial $a^{(j)}(x) \in C_j$, for $j = 1, 2, ..., 5$.

As it was shown, the polynomial $w(x)$ is an element of the ideal, that is, of the cyclic code $C = < g_1^5 g_2^4 g_3^3 g_4^2 g_5 >$ in the ring $R(5, 55) = \mathbb{F}_5[x]/(x^{55}-1)$. In this case the cardinalities of D and C are both equal to $5^{55-(5\rho_1+4\rho_2+3\rho_3+2\rho_4+\rho_5)}$, where $\rho_i = \deg(g_i)$. Since the Gray map Φ is injective we conclude that

$$C = \Pi^{-1}(\Phi(D)).$$

Conclusion. For any prime p the Gray map can be defined on the $\mathbb{Z}_{p^2}$-module $\mathbb{Z}_{p^2}^n$, and, following the ideas presented in [21], it is shown that the Gray map image of a code built from $\mathbb{F}_p$-cyclic codes C_i of length n, is a cyclic code of length pn with multiple roots, whose generator is given in terms of the generators of the codes C_i's. Some examples were provided to illustrate the ideas. It should be noted that the construction of the code $D = (C_p \oplus_p \cdots \oplus_p C_2) + pC_1$ from the codes C_i's could give more information. For instance, if $p = 2$ and $C_1 = C_2^\perp$ it would be interesting to describe the $\mathbb{Z}_4$-code $D = C_2 + 2C_2^\perp$ and give some of its properties.

Acknowledgment. The author would like to thank the referees for their comments which greatly improved the presentation of this paper.

References

1. Carlet C.: $\mathbb{Z}_{2^k}$-linear Codes. IEEE Trans. Inform. Theory, vol.44, (1998) 1543-1547.
2. Castagnoli G., Massey J.L., Schoeller P.A., von Seemann N.: On repeated-Root Cyclic Codes. IEEE Trans. Inform. Theory, vol. 37, No. 2, (1991) 337-342.
3. Bonnecaze A., Solé P., Bachoc C., Mourrain B.: Type II Codes over $\mathbb{Z}_4$. IEEE Trans. Inform. Theory, vol. 43, No. 3, (1997) 969-976.
4. Bonnecaze A., Udaya P.: Cyclic Codes and Self-Dual Codes over $\mathbb{F}_2 + u\mathbb{F}_2$. IEEE Trans. Inform. Theory, vol. 45, No. 4, (1999) 1250-1254.
5. Chen Y.Q., Ray-Chaudhuri D.K., Xiang Q.: Construction of Partial Difference Sets and Relative Difference Sets Using a Galois Ring II. J. of Combinatorial Theory, Series A **76**, (1996) 179-196.

6. Greferath M., Schmidt S.E.: Gray Isometries for Finite Chain Rings and a Nonlinear Ternary $(36, 3^{12}, 15)$ Code. IEEE Trans. Inform. Theory, vol. 45, No. 7, (1999) 2522-2524.
7. Hammons Jr A.R., Kumar P.V., Calderbank A.R., Sloane N.J.A., Solé P.: The $\mathbb{Z}_4$-linearity of Kerdock, Preparata, Goethals, and related codes. IEEE Trans. Inform. Theory, vol. 40, No. 2, (1994) 301-319.
8. Heise W., Honold T., Nechaev A.A.: Weighted modules and representations of codes. Proc. ACCT 6 (Pskov, Russia, 1998), 123-129.
9. Lin S., Blackford J.T.: $\mathbb{Z}_{p^{k+1}}$-Linear Codes. IEEE Trans. Inform. Theory, vol.48, No. 9, (2002) 2592-2605.
10. Lin S., Solé P.: Duadic Codes over $\mathbb{F}_2 + u\mathbb{F}_2$. AAECC **12**, (2001) 365-379.
11. Lin S., Solé P.: Nonlinear p-ary sequences. AAECC **14**, (2003) 117-125.
12. MacWilliams F.J., Sloane N. J. A.: *The Theory of Error-Correcting Codes.* Amsterdam, The Netherlands: North-Holland, (1977).
13. Nechaev A.A.: The Kerdock code in a cyclic form. Math. Appl. vol. 1, (1991) 365-384, (English translation of Diskret. Mat., 1989).
14. Pless V., Solé P., Qian Z.: Cyclic Self-Dual $\mathbb{Z}_4$-Codes. Finite Fields and Their Applications, **3**, (1997) 48-69.
15. Tapia-Recillas H.: Difference sets on the Galois ring $GR(p^2, m)$ and their Gray map image. Submitted for publication.
16. Tapia-Recillas H., Vega G.: A Generalization of Negacyclic Codes. Proc. International Workshop on Coding and Cryptography, WCC 2001, (D. Augot, Ed.), Paris, France, (2001) 519-529.
17. Tapia-Recillas H., G. Vega: On the $\mathbb{F}_p$-Linearity of the Generalized Gray Map Image of a $\mathbb{Z}_{p^k}$-Linear Code. Finite Fields with Applications to Coding Theory, Cryptography and Related Areas, Proc. Int. Conference on Finite Fields and Applications (Fq6), (G. L. Mullen, H. Stichtenoth, H. Tapia-Recillas, eds.), Springer Verlag, ISBN 3-540-43961-7, (2002) 306-312.
18. Tapia-Recillas H., Vega G.: Some constacyclic codes over $\mathbb{Z}_{2^k}$ and binary quasi-cyclic codes. Discrete Applied Mathematics, 128, (2003) 305-316.
19. Tapia-Recillas H., Vega G. On the $\mathbb{Z}_{2^k}$-Linear and Quaternary Codes. To appear in SIAM Journal on Discrete Mathematics.
20. van Asch B., van Tilborg H.C.A.: Two families of nearly-linear codes over $\mathbb{Z}_p$, p odd. AAECC **11**, (2001) 313-329.
21. van Lint J.: Repeated-Root Cyclic Codes. IEEE Trans. Inform. Theory, vol. 37, No. 2, (1991) 343-345.
22. Vega G., Wolfmann J.: Some families of $\mathbb{Z}_4$-cyclic codes. Preprint.
23. Wolfmann J.: Binary Images of Cyclic Codes over $\mathbb{Z}_4$. IEEE, Trans. Inform. Theory, vol. 47, No.5, (2001) 1773-1779.

Primitive Polynomials over Small Fields

Stephen D. Cohen

Department of Mathematics
University of Glasgow
Glasgow G12 8QW
Scotland.
E-mail: sdc@maths.gla.ac.uk

1 Introduction

Given a positive integer n, any root of a *primitive* polynomial $x^n + a_1 x^{n-1} + \cdots + a_n$ over the finite field $\mathbb{F}_q$ of q elements (where q is a power of a prime p) is a *primitive* (generating) element of the extension $\mathbb{F}_{q^n}$, by definition having multiplicative order $q^n - 1$. For many purposes it is valuable to be assured of the existence of a primitive polynomial with a proportion of its coefficients prescribed (e.g., with many of the coefficients zero). Results are known guaranteeing the existence of a primitive polynomial of degree n over $\mathbb{F}_q$ with a fixed number m of the "first" coefficients $a_1, \ldots, a_m$ prescribed, where $1 \le m \le 3$. See [1], [11], [7], [9], [10], [6], [3], [14] : some relevant facts are summarised at the end of this Introduction. Further results apply, for *sufficiently large q* (dependent on n), for a varying number of prescribed coefficients: thus, it is known that up to the first $\lfloor \frac{n}{2} \rfloor$ coefficients can be specified, see [8], [15]. Moreover, whereas the natural approach to such problems (working within the fields themselves) means introducing a restriction that the characteristic p should exceed m, recently Fan and Han [4], [5], [6] have shown how to eliminate such restrictions in problems of this nature by working p-adically and in Galois rings based upon $\mathbb{F}_q$. Nevertheless, as formulated thus far, results of the latter type — wherein a proportion of the coefficients are specified — have little validity when $\mathbb{F}_q$ (i.e., q) is small. In particular, they say nothing in the important case of the binary field $\mathbb{F}_2$. In this connection there is the result of Shparlinski [16], this time significant for *sufficiently large n* (in an unspecified manner), which established the existence of a primitive polynomial over $\mathbb{F}_2$ of weight $\frac{n}{4} + o(n)$.

The purposes of this paper are to give a streamlined exposition of the Fan-Han method and to derive some unconditional results effective even for small fields. Key features are the application of estimates of Winnie Li [13] for mixed character sums with polynomial arguments over Galois rings and a sieving technique. The first main conclusion is that, for any n, there exists a primitive binary polynomial with one quarter of its coefficients prescribed.

Theorem 1. *Given arbitrary positive integers n and $m \le \frac{n}{4}$, there exists a primitive binary polynomial $f(x) = x^n + a_1 x^{n-1} + \cdots + a_{n-1} x + 1 \in \mathbb{F}_2[x]$ with either the first m coefficients $a_1, \ldots, a_m$ or the last m coefficients $a_{n-m}, \ldots, a_{n-1}$ prescribed in advance (as 0's or 1's).*

G. Mullen, A. Poli and H. Stichtenoth (Eds.): Fq7 2003, LNCS 2948, pp. 197–214, 2003.
© Springer-Verlag Berlin Heidelberg 2003

Through consideration of the reciprocal polynomial $f^*(x) := x^n f(1/x)$ of $f(x)$, it suffices to prove Theorem 1 when it is the *first* m coefficients that are prescribed.

For ternary polynomials $(q = 3)$, we prove unconditionally that one can specify up to one third of the first or last coefficients of a primitive polynomial.

Theorem 2. *Given arbitrary positive integers n and $m \leq \frac{n}{3}$, there exists a primitive ternary polynomial $f(x) = x^n + a_1 x^{n-1} + \cdots + a_{n-1} x + (-1)^{n-1} \in \mathbb{F}_3[x]$ with either the first m coefficients $a_1, \ldots, a_m$ or the last m coefficients $a_{n-m}, \ldots, a_{n-1}$ prescribed in advance.*

The constant term in a primitive ternary polynomial is necessarily $(-1)^{n-1}$. Again by consideration of the (monic) reciprocal polynomial $f^*(x) := (-1)^{n-1} x^n f(1/x)$, it suffices to prove Theorem 2 when it is the *first* m coefficients that are prescribed.

As q increases one could specify a larger proportion (up to one half), but we are content with the general result which follows. The single exception (with $q = 4$, $n = 3$, $m = 1$) was recorded already in [1].

Theorem 3. *Given a prime power $q > 3$ and arbitrary positive integers n and $m \leq \frac{n}{3}$, there exists a primitive polynomial $x^n + a_1 x^{n-1} + \cdots + a_n \in \mathbb{F}_q[x]$ with the first m coefficients $a_1, \ldots, a_m$ prescribed in advance, with the exception that there is no primitive cubic over $\mathbb{F}_4$ with zero first coefficient.*

The declared emphasis of this study is on "small" fields, because it is these that come nearest to testing the above theorems. Therefore we provide a complete analysis for fields $\mathbb{F}_q$ with $q \leq 5$. Of course, the techniques are valid for all values of q and n and we indicate how the proof flows in general. The difficulties are more organisational than actual. Nevertheless, for values of $q \leq 13$, there are some cases which have to established by direct computation: for example, $n = 12$ when $q = 13$. Note that there is a reasonable expectation that, with some further calculation, Theorem 1 may be "upgraded" to yield the equivalent result to Theorem 2.

To complete this introduction, we explain how Theorems 1, 2 and 3 follow from known results in most cases for $n < 12$. First, the author's study [1] of primitive polynomials with prescribed trace yields the results for $n \leq 7$ $(q = 2)$ and $n \leq 5$ $(q > 2)$. Next, Fan and Han [6] have shown that, for arbitrary q, there is a primitive polynomial of every degree $n \geq 8$, with the first *three* coefficients prescribed (though most of the details of the calculations in the cases that have to be settled by direct computation have been suppressed). For q odd, these calculations are confirmed in detail in [14], where, with a few possible exceptions, all degrees $n \geq 7$ are dealt with. This yields Theorem 1 for $n \leq 15$ and Theorems 2 and 3 for $8 \leq n \leq 11$. Earlier, for odd q, Han had shown the existence of primitive polynomials with two prescribed coefficients for $n \geq 7$ and, more recently, Cohen and Mills [3] have clarified this working and extended it, in great detail, to cover the cases $n = 5, 6$. Thus, our conclusions are established for $n \leq 11$ when q is odd. For $q \geq 4$ even, it is not clear that Theorem 3 has been covered when $n = 6$ or 7, though from [7] and [9], it seems there is a primitive polynomial with arbitrary *second* coefficient. In conclusion, when q is odd we can assume forthwith that $n \geq 12$. When q (≥ 4) is even, we shall also make brief reference to the cases of $n = 6$, 7.

2   Symmetric Functions and Power Sums in p-adic Rings

For a general field F, let $f(x)$ be an irreducible monic polynomial of degree n in $F[x]$ with (complete set of) roots $\xi_1, \ldots, \xi_n$ in a suitable splitting field. Then (up to sign), the coefficients are the symmetric functions of its roots: more precisely, $f(x) = x^n - \sigma_1 x^{n-1} + \sigma_2 x^{n-2} - \cdots + (-1)^n \sigma_n$, where σ_i is the i-th symmetric function. The basic idea is to connect the values of the symmetric functions with those of sums of powers of the roots, which are easier to handle. Hence, for each $i = 1, 2, 3, \ldots$, define $s_i := \sum_{j=1}^n \xi_j^i$: by symmetry, for each i, $s_i \in F$. By repeated application of Newton's identities, which take the form

$$r\sigma_r = \sigma_{r-1}s_1 - \sigma_{r-2}s_2 + \cdots + (-1)^{r-1}s_r, \quad r = 1, \ldots, n,$$

for a given $m \leq n$, evidently the values of $s_1, \ldots, s_m$ are determined (uniquely) by those of $\sigma_1, \ldots, \sigma_m$. On the other hand, although Newton's identities (as stated) hold in fields of arbitrary characteristic, the converse statement, namely, that the values of $\sigma_1, \ldots, \sigma_m$ are determined by those of $s_1, \ldots, s_m$, holds only in fields whose characteristic is zero or exceeds m. The conclusion is that, in order to develop this principle for application to an arbitrary finite field $\mathbb{F}_q$, it is necessary to transfer the argument to a ring of higher (prime-power) characteristic (a Galois ring), or even characteristic 0 (a p-adic ring or field). This is the motivation behind the discussion which follows. In it, $q = p^k$, p prime, is a given prime power and n is a given positive integer.

Let $\mathbb{Q}_p$ be the p-adic field (i.e., the completion of the rational field with respect to the usual p-adic metric). Its ring of integers is, of course, closed under division by integers ($\in \mathbb{Z}$) indivisible by p. Also, let K_n be the splitting field (in C_p, the usual completion of the algebraic closure of $\mathbb{Q}_p$) of the polynomial $x^{q^n} - x$ over $\mathbb{Q}_p$. Define Γ_n to be the set of roots of this polynomial — the Teichmüller points of K_n. Clearly, its non-zero elements form a cyclic group of order $q^n - 1$. Indeed, K_n is the unique unramified extension of K_1 of degree n. Let R_n denote the ring of integers of K_n. Then $\Gamma_n \subseteq R_n = \{\sum_{i=0}^{\infty} \gamma_i p^i, \gamma_i \in \Gamma_n\}$. Moreover, R_n is a local ring with unique maximal ideal pR_n and $R_n/pR_n \cong \mathbb{F}_{q^n}$.

Now, let e be a positive integer. Define $\Gamma_{n,e}$ to be the set of classes of elements of Γ_n that are congruent mod p^e, i.e., γ_1 and γ_2 are in the same class if $\gamma_1 - \gamma_2 \in p^e R_n$. In this context, retain the notation γ for the class containing γ. Then $\gamma^{q^n} = \gamma$ for $\gamma \in \Gamma_{n,e}$. Passing to classes mod p^e of elements of R_n yields a ring (*Galois ring*) $R_{n,e} = \{\sum_{i=0}^{e-1} \gamma_i p^i, \gamma_i \in \Gamma_{n,e}\} \cong R_n/p^e R_n$, so that $R_{n,e}$ has cardinality q^{ne}. Note that, for each $e \geq 1$, $R_{n,e}/pR_{n,e} \cong \mathbb{F}_{q^n}$ also. Observe too that $R_{n,1} = \Gamma_{n,1}$, which can be identified with $\mathbb{F}_{q^n}$. Conversely, each $\gamma \in \Gamma_{n,1}$ yields a unique *lift* (also denoted by γ) to every $\Gamma_{n,e}$ and ultimately to Γ_n itself.

Of course, an element of (multiplicative) order r (a divisor of $q^n - 1$) in $\Gamma_{n,1}$ lifts to an element of the same order in each $\Gamma_{n,e}$ and in Γ_n. In particular, a primitive element (a generator or element of order $q^n - 1$) lifts to a primitive element in each case.

Next, we consider objects relating to the extension $\mathbb{F}_{q^n}/\mathbb{F}_q$. Of course K_1 is a subfield of K_n, with $\Gamma_{1,e} \subseteq \Gamma_{n,e}$, and $R_{n,1}$ a subring of $R_{n,e}$. Similar relationships apply to the Galois rings. Further, note that the Galois group of K_n/K_1 is isomorphic to that of $\mathbb{F}_{q^n}/\mathbb{F}_q$, being cyclic of order n and generated by the Frobenius automorphism τ_n, where $\tau_n(\gamma) = \gamma^q$, $\gamma \in \Gamma_n$. More generally, on R_n, $\tau_n(\sum_{i=0}^{\infty} \gamma_i p^i) = \sum_{i=0}^{\infty} \gamma_i^{\,q} p^i$ (where each $\gamma_i \in \Gamma_n$). This induces a ring homomorphism τ_n on $R_{n,e}$ such that $\tau_n(\sum_{i=0}^{e-1} \gamma_i p^i) = \sum_{i=0}^{e-1} \gamma_i^{\,q} p^i$ (where now each $\gamma_i \in \Gamma_{n,e}$).

Recall that over $\mathbb{F}_q$ (and so over $R_{1,1}$)), $x^{q^n} - x$ is the product of all monic irreducible polynomials of degree a divisor of n. A typical monic irreducible polynomial $f(x)$ of degree d (a divisor of n) in $R_{1,1}[x]$ has the form

$$f(x) = (x - \gamma)(x - \gamma^q) \cdots (x - \gamma^{q^{d-1}}) = x^d - \sigma_1 x^{d-1} + \cdots + (-1)^d \sigma_d, \quad (2.1)$$

where $\gamma \in \Gamma_{n,1}$ and each $\sigma_j \in \Gamma_{1,1} = R_{1,1}$. The polynomial f *lifts* to a (unique) irreducible polynomial of degree d over each $R_{1,e}$ and over R_1 having the same form, except that γ is the corresponding lifted element of $\Gamma_{1,e}$ or Γ_1. But note that, in general, the coefficients σ_j in (2.1) lie in $R_{1,e}$ (or R_1), but may not be in $\Gamma_{1,e}$ (or Γ_1). From the above, the *order* of the polynomial f (which equals the order of any of its roots) or any of its lifts has the same value (a divisor of $q^n - 1$). In particular, f is *primitive* if it is irreducible of degree n and has order $q^n - 1$: this holds if and only if any of its lifts are primitive.

For any $\gamma \in \Gamma_n$, define its trace (over R_1) as $T_n(\gamma) := \gamma + \tau_n(\gamma) + \cdots + \tau_n^{n-1}(\gamma) = \gamma + \gamma^q + \cdots + \gamma^{q^{n-1}} \in R_1$. Further, if τ is defined as the generator of $\mathrm{Gal}(K_n/\mathbb{Q}_p)$ such that $\tau^k = \tau_n$, where $q = p^k$, then evidently $T_n(\gamma^p)(= T_n(\tau(\gamma))) = \tau(T_n(\gamma))$, $\gamma \in \Gamma_n$. It follows that $T_n(\gamma^{p^j}) = (T_n(\gamma))^{p^j}$, $j = 0, 1, 2, 3, \ldots$. A trace function T_n with similar properties is induced on $\Gamma_{n,e}$.

Next, let $\gamma \in \Gamma_n$ be the root of a lifted irreducible polynomial $f(x) \in R_1[x]$. Later we impose conditions to ensure γ is *primitive*: for the moment it suffices that f has degree n. Thus, (2.1) holds with $d = n$. Here σ_i denotes the i-th symmetric function of the roots $\gamma, \gamma^q, \ldots, \gamma^{q^{n-1}}$. Employing the trace notation, we have that s_i, the sum of the i-th powers of the roots of f, is given by $s_i = T_n(\gamma^i) \in R_1$. Of course, each s_i depends only on f and not on the specific root γ: moreover, all this translates to the expansion of f as a polynomial in $R_{1,e}[x]$. Take $m \le n$. Whereas, as we indicated earlier, knowledge of $s_1, \ldots, s_m$ (for $f[x] \in \mathbb{F}_q[x]$ or $R_{1,1}[x]$) is generally insufficient to determine $\sigma_1, \ldots, \sigma_m$, the latter can be obtained from suitable information about $s_1, \ldots, s_m$ for the lift of f to R_n or indeed to $R_{n,e}$ for sufficiently large e. This is the key to the major advance of [4], [5] .

We proceed to work with a lifted irreducible polynomial f of degree n in $R_1[x]$ and eventually its reduction to $R_{1,e}$, with e to be chosen. From now on, we shall reserve the letter t for a positive integer indivisible by p. Note from above that, for any such t, the value of s_{tp^i} for any $i \geq 0$ is already determined by s_t, and is given by $s_t^{(i)} := \tau^i(s_t)$. So to specify $\{s_1, \ldots, s_m\}$, say, it suffices to know the values of $\{s_t : t \leq m\}$. Next, for any t, write $s_t = \sum_{j=0}^{\infty} g_{t,j} p^j$, $g_{t,j} \in \Gamma_1$, whence $s_t^{(i)} = \sum_{j=0}^{\infty} g_{t,j}^{p^i} p^j$. Since each positive integer l has a unique expression in the form $l = tp^j$, then any "s-component" $g_{t,j}$ is uniquely associated with the integer tp^j. The following lemma is the replacement for Newton's identities in the general context. In its statement, $a \equiv b \bmod p$ for $a, b \in R_1$ is interpreted to mean that $a = b$ as members of $R_{1,1}$.

Lemma 1. *Let $\gamma \in \Gamma_n$ be the root of an irreducible polynomial $f(x) \in R_1[x]$ of degree n as above. Suppose $m \leq n$. Write $m = Tp^J$, $p \nmid T$, $J \geq 0$. Then*

$$\sigma_m \equiv (-1)^{m-1} g_{T,J}^{p^J} + P_{m-1} \bmod p,$$

where P_{m-1} is a polynomial function over $\mathbb{Z}$ of the members of $\{g_{t,j} : tp^j < m\}$.

Proof. The reciprocal polynomial to f is

$$f^*(x) := 1 - \sigma_1 x + \sigma_2 x^2 + \cdots + (-1)^n \sigma_n x^n = \prod_{j=0}^{n-1} (1 - \gamma^{q^j} x) \in R_1[x].$$

Now $f^*(x) \in R_1[x]$ can be regarded as an (infinite) formal power series over K_1. Invoking the expansion of f^* in terms of the exponential power series function $\exp x$ over K_1 (see [12], Ch IV), we have a formal identity

$$f^*(x) = \exp\left(-\sum_{r=1}^{\infty} \frac{T_n(\gamma^r) x^r}{r}\right) = \exp\left(-\sum_{r=1}^{\infty} \frac{s_r x^r}{r}\right)$$

$$= \exp\left(-\sum_{\substack{t=1 \\ p \nmid t}}^{\infty} \sum_{i=0}^{\infty} \frac{s_t^{(i)} x^{tp^i}}{tp^i}\right) = \prod_{\substack{t=1 \\ p \nmid t}}^{\infty} \exp\left(-\sum_{i=0}^{\infty} \frac{s_t^{(i)} x^{tp^i}}{tp^i}\right)$$

$$= \prod_{\substack{t=1 \\ p \nmid t}}^{\infty} \prod_{j=0}^{\infty} \prod_{i=0}^{\infty} \exp\left(-\frac{g_{t,j}^{p^i} p^{j-i} x^{tp^i}}{t}\right). \tag{2.2}$$

It is automatic that the net coefficient of any power of x in the expression (2.2) is in R_1: all contributions to the coefficient which apparently lie in $K_1 \setminus R_1$ must cancel.

Now $(-1)^m \sigma_m$ is the coefficient of x^m in (2.2) and derives solely from the expansion of terms with $tp^i \leq m$. Further, for the value of $\sigma_m \bmod p$ we need restrict consideration to terms with $i \leq j$. (This is because, if $l := j - i \geq 1$, then in the formal power series expansion of $\exp(cp^l x)$, $c \in R_1$, the coefficient

$\dfrac{c^r p^{lr}}{r!}$ of x^r lies in pR_1, since the power of p dividing $r!$ is $\left\lfloor \dfrac{r}{p} \right\rfloor + \left\lfloor \dfrac{r}{p^2} \right\rfloor + \cdots < r$.)
The remaining terms involve only s-components $g_{t,j}$ with $tp^j \leq m$. In particular, the expansion $(\bmod\, p)$ of the term in (2.2) with $t = T$, $j = J$ begins
$$\exp\left(-\frac{g_{T,J}^{p^J}}{T} x^m + \cdots\right) = 1 - \frac{g_{T,J}^{p^J}}{T} x^m + \cdots. \text{ This series also appears in the full}$$
expansion of (2.2): all other contributions $(\bmod\, p)$ to the coefficient of x^m effectively involve (a $\mathbb{Z}$-polynomial in) s-components $g_{t,j}$ with $tp^j \leq m - 1$. This completes the proof.

Given $m \leq n$ as in Lemma 1, define, for each $t \leq m$, the integer $e_t \geq 1$ by the inequalities $tp^{e_t - 1} \leq m < tp^{e_t}$. Set $e := e_1$; thus $e_t \leq e$ for all $t \leq m$. The application which follows relates to lifts of irreducible polynomials to $R_{1,e}$ for this choice of e.

Corollary 4. *Suppose $m \leq n$ and $A \subseteq R_{1,e}[x]$ is a set of lifted irreducible polynomials of degree n with the property that, given any prescribed set $\{a_t \in R_{1,e_t} : t \leq m\}$, there exists a polynomial in A for which $p^{e-e_t} s_t = p^{e-e_t} a_t$ for all $t \leq m$. Then, there exists an irreducible polynomial $f(x) \in R_{1,1}[x]$ whose lift is in A and whose first m coefficients are arbitrary specified values in $R_{1,1}$.*

Proof. Call the value of tp^j of an s-component its *rank*. The given assumptions can be summarised by saying that there is a polynomial in A with its s-components of rank $\leq m$ prescribed. Since there is a unique s-component of each rank, the total number of prescribed s-components is exactly m. Equivalently, $\sum_{t \leq m} e_t = m$.

Suppose that two members of A are such that their s-components of rank $\leq m$ do not all coincide $(\bmod\, p)$, and let $l = t_l p^{j_l}$, where $1 \leq l \leq m$, be the smallest rank where there is disagreement. Then the corresponding values of $g_{t_l, j_l}^{p^{j_l}} \bmod p$ are different, and, by Lemma 1, the corresponding values of $\sigma_l \bmod p$ also are different. Thus, there is bijection between the set of q^m possible values of the s-components of polynomials in A of rank $\leq m$ and the q^m possible values of the first m coefficients $(\bmod\, p)$ of such polynomials. The result follows.

3   Estimates of Primitive Elements with Specified Traces

From now on, assume the prime power $q = p^k$, the integers $m \leq n$ and e and, for each $t \leq m$, the further integers e_t are as in Corollary 4. By that result, we have to show that there exists a primitive $\gamma \in \Gamma_{n,e}$ for which the traces $p^{e-e_t} s_t = T_n(p^{e-e_t} \gamma^t) = p^{e-e_t} a_t \in R_{1,e}$ for an arbitrary given set $\{a_t \in R_{1,e_t} : t \leq m\}$.

We proceed to describe a standard expression for the characteristic function for a primitive element (generator) of a cyclic group in terms its multiplicative characters χ, starting with $\Gamma_{n,e}^* \cong \mathbb{F}_{q^n}^*$. Let $\widehat{\Gamma}_{n,e}^* (\cong \mathbb{F}_{q^n}^*)$ denote the group of multiplicative characters of $\Gamma_{n,e}$. With $\chi \in \widehat{\Gamma}_{n,e}^*$ is an associated character over the set of non-zero Teichmüller points Γ_n^* that extends to a character with conductor 1 over K_n itself, see [13], Section 5.

Indeed, we consider a more general set than the one of merely primitive elements. Let Q be any divisor of $q^n - 1$. An element γ of $\Gamma_{n,e}^*$ is called *Q-free* if $\gamma = \alpha^d$, $\alpha \in \Gamma_{n,e}^*$, $d \mid Q$ implies $d = 1$. In particular, γ is primitive if and only if it is $q^n - 1$-free.

For any $d \mid q^n - 1$, write χ_d for a typical character in $\widehat{\Gamma}_{n,e}^*$ of order d. Thus χ_1 is the trivial character. We employ a useful "integral" notation for weighted sums; namely, for $d \mid q^n - 1$, set

$$\int_{d \mid Q} \chi_d := \sum_{d \mid Q} \frac{\mu(d)}{\phi(d)} \sum_{(d)} \chi_d,$$

where ϕ and μ denote the functions of Euler and Möbius respectively and the inner sum runs over all $\phi(d)$ characters of order d. (Observe that only square-free divisors d have any influence.) Then the characteristic function for the subset of Q-free elements of $\Gamma_{n,e}$ is

$$\theta(Q) \int_{d \mid Q} \chi_d(\gamma), \quad \gamma \in \Gamma_{n,e}, \tag{3.1}$$

where $\theta(Q) := \frac{\phi(Q)}{Q} = \prod_{l \mid Q,\, l \text{ prime}} (1 - l^{-1})$.

To deal with the trace conditions, we also need the additive characters of $R_{n,e}$. The canonical additive character $\psi_{(n)}$ is defined by

$$\psi_{(n)}(\xi) = \exp\left(\frac{2\pi i T_{nk}(\xi)}{p^e} \right), \quad \xi \in R_{n,e}.$$

We write ψ for $\psi_{(1)}$ the canonical character on $R_{1,e}$. Thus, for $\xi \in R_{n,e}$, $\psi_{(n)}(\xi) = \psi(T_n(\xi))$. Every additive character on $R_{1,e}$ has the form ψ_α, for some $\alpha \in R_{1,e}$, where $\psi_\alpha(\xi) = \psi(\alpha\xi)$ for all $\xi \in R_{1,e}$. A key property (summing values of ψ over a subring $R_{1,e'}$, $e' \le e$) is that, for $\xi \in R_{1,e}$,

$$\sum_{\alpha \in R_{i,e'}} \psi(p^{e-e'}\alpha\xi) = \begin{cases} q^{e'} & \text{if } \xi \equiv 0 \mod p^{e'} \\ 0 & \text{otherwise.} \end{cases} \tag{3.2}$$

Let $a_t \in R_{1,e_t}$. It follows from (3.2) that the characteristic function of the set of elements γ in $\Gamma_{n,e}$ for which $p^{e-e_t} s_t(\gamma^t)$ $(= T_n(p^{e-e_t}\gamma^t))$ assumes the value $p^{e-e_t} a_t \in R_{1,e}$ is

$$\frac{1}{q^{e_t}} \sum_{\alpha \in R_{1,e_t}} \psi(p^{e-e_t}\alpha(T_n(\gamma^t) - a_t)). \tag{3.3}$$

Assume that, as in Corollary 4, $\{a_t \in R_{1,e_t} : t \le m\}$ are given, and that $Q \mid q^n - 1$. Define $N(Q)$ to be number of Q-free elements in $\Gamma_{n,e}$ with $p^{e-e_t} s_t = p^{e-e_t} a_t \in R_{1,e}$, as at the head of this section. Employing the characteristic functions (3.1) and (3.3), we shall express $N(Q)$ in terms of (mixed additive and multiplicative) character sums with polynomial arguments. To formulate the underlying estimate (taken from [13], e.g., Corollaries 3.1 and 6.1), note that a polynomial $h(x) \in R_{n,e}[x]$ has a p-adic expansion of the form $h(x) = h_0(x) + h_1(x)p + \cdots + h_{e-1}(x)p^{e-1}$, where, for $j = 0, 1, \ldots e-1$, h_j is a polynomial (of degree d_j, say) with coefficients in $\Gamma_{n,e}$, Then the *weighted degree* D_h of h is defined by $D_h := \max(d_0 p^{e-1}, d_1 p^{e-2}, \ldots, d_{e-1})$.

Lemma 2. *Suppose that $h(x)$ in $R_{n,e}[x]$ contains no monomial of degree divisible by p. Then*

$$\left| \sum_{\gamma \in \Gamma_{n,e}} \psi_{(n)}(h(\gamma)) \right| \leq (D_h - 1)q^{\frac{n}{2}}.$$

Further, if χ is a non-trivial multiplicative character of $\Gamma_{n,e}^*$, then

$$\left| \sum_{\gamma \in \Gamma_{n,e}} \psi_{(n)}(h(\gamma))\chi(\gamma) \right| \leq D_h q^{\frac{n}{2}}.$$

The key estimate is the following extension of inequality (6) of [5]. In it, for any positive integer l, we use $\theta(l)$ to denote the ratio $\frac{\phi(l)}{l}$, and $W(l) = 2^{\omega(l)}$ for the number of square-free divisors of l, where $\omega(l)$ is the number of distinct primes dividing l.

Proposition 5. *Suppose that $m < n/2$ and e is as used in Corollary 4. Assume that $\{a_t \in R_{1,e_t} : t \leq m\}$ are given. Suppose that $Q | q^n - 1$. Then, the number of Q-free elements in $R_{n,e}$ with prescribed $p^{e-e_t} s_t \in R_{1,e}$ (as described above) satisfies the bound*

$$N(Q) \geq \theta(Q)\{q^{n-m} - (1 - q^{-m})(mW(Q) - 1)q^{\frac{n}{2}}\} \qquad (3.4)$$
$$> \theta(Q)\{q^{n-m} - W(Q)mq^{\frac{n}{2}}\}. \qquad (3.5)$$

Proof. By (3.1) and (3.3) and the fact (noted in the proof of Corollary 4) that $\sum_{t \leq m} e_t = m$, we have

$$q^m N(Q) = \theta(Q) \sum_{\gamma \in \Gamma_{n,e}} \int_{d|Q} \chi_d(\gamma) \prod_{t \leq m} \left\{ \sum_{\alpha_t \in R_{1,e_t}} \psi(p^{e-e_t}\alpha_t(T_n(\gamma^t) - a_t)) \right\}$$

$$= \theta(Q) \int_{d|Q} \sum_{\{\alpha_t \in R_{1,e_t} : t \leq m\}} \psi(-p^{e-e_t}\alpha_t a_t) S(\psi_{(n)}, \chi_d), \qquad (3.6)$$

where

$$S(\psi_{(n)}, \chi_d) := \sum_{\gamma \in \Gamma_{n,e}} \psi_{(n)}(h(\gamma))\chi_d(\gamma),$$

and

$$h(x) = \sum_{t \leq m} p^{e-e_t}\alpha_t x^t,$$

a polynomial in $R_{1,e}[x]$.

Now evidently, when h is the zero polynomial , i.e., $\alpha_t (\in R_{1,e_t}) = 0$ for all $t \leq m$, then $S(\psi_{(n)}, \chi_d) = q^n - 1$ when $d = 1$ and is 0 for $d > 1$. For all other sets $\{\alpha_t \in R_{1,e_t} : t \leq m\}$, by the convention on t, h satisfies the restriction of Lemma 2. Moreover , the weighted degree of the typical monomial $p^{e-e_t} \alpha_t x^t \in R_{1,e}$ of h is at most $tp^{e_t-1} \leq m$. Accordingly, $D_h \leq m$. Thus, by Lemma 2, the absolute value of the contribution to the right side of (3.6) of the terms with $h \neq 0$ is bounded above by

$$(q^m - 1)q^{\frac{n}{2}}\{m - 1 + (W(Q) - 1)m\} = q^{\frac{n}{2}}(q^m - 1)(mW(Q) - 1),$$

because there are $\phi(d)$ characters χ_d for each $d| Q$. We conclude that (3.4) holds and (3.5) follows.

We remark that improvements in the bound (3.5) are possible — evidently, some of the polynomials h have less than maximal weighted degree, and also more careful summing over one of the α_t is feasible — but (3.5) suffices for our purposes here. Taking $Q = q^n - 1$, we deduce the following conditional version of Theorems 1, 2 and 3.

Corollary 6. *Given a prime power q and arbitrary positive integers n and $m < \frac{n}{2}$, there exists a primitive polynomial $x^n + a_1x^{n-1} + \cdots + a_n \in \mathbb{F}_q[x]$ with the first m coefficients $a_1, \ldots, a_m$ prescribed in advance whenever*

$$q^{\frac{n}{2}-m} > mW(q^n - 1). \tag{3.7}$$

4   Primitive Binary Polynomials

We now suppose $q = 2$ and $m = \lfloor \frac{n}{4} \rfloor$. Then (3.7) certainly holds whenever

$$2^{\frac{n}{4}-\omega} > \frac{n}{4}, \quad \omega := \omega(2^n - 1). \tag{4.1}$$

Of course, a slightly weaker condition suffices when n is indivisible by 4. For larger n it is convenient to employ the following numerical fact.

Lemma 3. *Suppose that $\omega \geq 25$. Then $\omega \leq \frac{n}{5}$.*

Proof. Let $P(r)$ denote the product of the first r *odd* primes. Then, by calculation (MAPLE), the inequality

$$P(r) > 2^{5r}, \quad r \geq 25, \tag{4.2}$$

holds for $r = 25$, the 25-th odd prime being 101. Inequality (4.2) is then evident by induction on r, since higher further primes exceed $2^5 = 32$.

Granted $\omega \geq 25$, suppose that actually $\omega > \frac{n}{5}$. From (4.2),

$$2^n - 1 < 2^{5\omega} - 1 < P(\omega) \leq 2^n - 1,$$

a contradiction. The result follows.

Since the function $2^{x/20} - \frac{x}{4}$ is increasing for $x > 60$ and positive for $x = 90$, it follows from (4.1) that, provided $\omega \leq \frac{n}{5}$ and $n \geq 90$, then Theorem 1 holds for this value of n. Indeed, by Lemma 3, the inequality $\omega \leq \frac{n}{5}$ does hold whenever $\omega \geq 25$ (so that automatically $n \geq 125$). We conclude that Theorem 1 is established except when $\omega \leq 24$ and $n < 5\omega \leq 120$.

In fact, for $n < 120$, $\omega \leq 12$ with equality when $n = 72, 84, 96, 108$. With this bound for ω, it suffices that $2^{n/4} > 1024n$, which holds for $n \geq 65$. Similarly, for $60 \leq n \leq 64$, $\omega \leq 11$ and for $49 < n < 59$, $\omega \leq 8$ and (4.1) holds in these ranges. But the inequality is false for values of n such as 48 ($\omega = 9$), 40 ($\omega = 7$) and 36 ($\omega = 8$).

To deal with most smaller values of n, we apply a sieving technique (effective for all prime power values of q) which yields a criterion which, when $s = 1$, reduces to Corollary 6.

Lemma 4. *Given a prime power q and arbitrary positive integers n, write the product of the distinct primes in $q^n - 1$ as $lp_1 \ldots p_s$, for some divisor l and distinct primes $p_1, \ldots, p_s$. For any $m < \frac{n}{2}$, there exists a primitive polynomial $x^n + a_1 x^{n-1} + \cdots + a_n \in \mathbb{F}_q[x]$ with the first m coefficients $a_1, \ldots, a_m$ prescribed in advance whenever*

$$q^{\frac{n}{2}-m} > mW(l)\left(\frac{s-1}{\delta} + 2\right), \tag{4.3}$$

where $\delta := 1 - \sum_{i=1}^{s} \frac{1}{p_i}$.

Proof. As noted already only the actual primes dividing $q^n - 1$ are significant. Observe also that $W(lp_i) = 2W(l)$ and $\theta(lp_i) = \theta(l)(1 - \frac{1}{p_i})$, $i = 1, \ldots, s$. With the notation of Proposition 5, an elementary sieving argument (see [3]) yields

$$N(q^n - 1) \geq \left(\sum_{i=1}^{s} N(lp_i)\right) - (s-1)N(l)$$

$$= \delta N(l) + \left(\sum_{i=1}^{s}\left[N(lp_i) - \left(1 - \frac{1}{p_i}\right)N(l)\right]\right)$$

$$\geq \theta(l)q^{\frac{n}{2}}\left\{\delta(q^{\frac{n}{2}-m} - mW(l)) - \sum_{i=1}^{s} m\left(1 - \frac{1}{p_i}\right)W(l)\right\},$$

since

$$\left|N(lp_i) - \left(1 - \frac{1}{p_i}\right)N(l)\right| \leq \theta(l)mq^{\frac{n}{2}}(W(lp_i) - W(l)),$$

by the argument of Proposition 5 applied to terms involving characters of order dp_i, where $d \mid l$. Because $W(lp_i) = 2W(l)$, the result follows.

In the binary case, for $n \leq 48$, we attempt to apply Lemma 4 with $q = 2$, $m = \lfloor \frac{n}{4} \rfloor$ and (usually) $l = 3$. Table 1 shows that the condition is satisfied for a sample of (the most delicate) values of $n \geq 21$ (where the displayed decimals have been truncated). Here (and later) we use Δ to denote the quantity $W(l)\left(\dfrac{s-1}{\delta}+2\right)$ appearing in the right side of (4.3) for the appropriate values of the parameters.

The conclusion from such working is that for $21 \leq n \leq 48$, only the value $n = 24$ fails the condition of Lemma 4 with $l = 3$. Here $m = 6$, $p_1 \ldots p_s = 5 \cdot 7 \cdot 13 \cdot 17 \cdot 241$, $\delta = .5172$ and the left and right sides of the condition are 64 and 116.7, respectively. There are similar failures for $n = 18$ and 16. The result holds easily by Corollary 6 for $n = 19, 17$ since $2^n - 1$ is a Mersenne prime. Recall from Section 1, that we can assume $n \geq 16$.

n	m	l	$p_1 \ldots p_s$	δ	Δ	$q^{\frac{n}{2}-m}$
48	12	3	$5 \cdot 7 \cdot 13 \cdot 17 \cdot 97 \cdot 241 \cdot 257 \cdot 673$	.5015	382.9	4096
40	10	3	$5 \cdot 11 \cdot 17 \cdot 31 \cdot 41 \cdot 61681$	.5936	208.4	1024
36	9	3	$5 \cdot 7 \cdot 13 \cdot 19 \cdot 37 \cdot 73 \cdot 109$	.4776	263.7	512
30	7	3	$7 \cdot 11 \cdot 31 \cdot 151 \cdot 331$	.7243	105.8	256
28	7	3	$5 \cdot 29 \cdot 43 \cdot 113 \cdot 127$	.7255	112.8	128
25	6	1	$31 \cdot 601 \cdot 1801$	.9655	24.5	90.5
22	5	3	$23 \cdot 89 \cdot 683$	.9438	41.1	64
21	5	1	$7 \cdot 127 \cdot 337$	.8463	21.8	45.2

Table 1: Sieving table for $q = 2$

The final step in the binary case was to use MAPLE 6 to print out 64 primitive binary polynomials of degree 24 with each possible choice of its first 6 coefficients. For $n = 20$, 32 primitive polynomials with all choices of the first 5 coefficients, and for $n = 18$ or 16, 16 polynomials with all choices of the first 4 coefficients were similarly obtained. To summarise the outcome we shall, in this paper, use the term "weight r" to describe a polynomial with (at most) r non-zero coefficients amongst those of (non-specified, non-constant) monomials x^i, $1 \leq i < n - m$. Here, in every case it sufficed to look at weight 2 polynomials. This completes the proof of Theorem 1.

5   Primitive Ternary Polynomials

We next suppose $q = 3$ and $m = \lfloor \frac{n}{3} \rfloor$. Then (3.7) certainly holds whenever

$$3^{\frac{n}{6}}/2^\omega > \frac{n}{3}, \quad \omega := \omega(3^n - 1). \tag{5.1}$$

Of course, a slightly weaker condition suffices when n is indivisible by 3. The following lemma is the counterpart of are Lemma 3.

Lemma 5. *Suppose that* $\omega \geq 65$. *Then* $\omega \leq \frac{n}{4.17}$.

Proof. This proceeds as for Lemma 3, but based (instead of on (4.2)) on the inequality

$$P(r) > 3^{4.17r}, \quad r \geq 65,$$

where now $P(r)$ denotes the product of the first r primes ($\neq 3$). Note that the 65-th such prime is 317, which exceeds $3^{4.17}$.

By (5.1) and Lemma 5, Theorem 2 holds unless $\omega \leq 64$ and $n < 4.17\omega < 267$ provided $3^{n/6}/2^{n/4.17} > \frac{n}{3}$. This last inequality holds for all $n \geq 266$. In summary, we may suppose $n \leq 266$.

For $72 \leq n \leq 266$, it is not necessary to find the complete factorisation of $3^n - 1$. For MAPLE has an "easy" factorisation option which quickly finds all prime factors $< 10^5$ (and many larger prime factors). From this it follows that, in place of Lemma 5, we have the stronger fact that $\omega \leq \frac{n}{5}$ for $48 < n \leq 266$. Since $3^{n/6}/2^{n/5} > \frac{n}{3}$ for $n \geq 72$, we conclude that Theorem 2 holds any value of $n \geq 72$.

For most values of n in the range $14 \leq n \leq 71$, application of Lemma 4 (using $l = 2$) is successful, as indicated in the Table 2 which features a selection of values in the range (including all the most delicate cases).

n	m	$p_1 \ldots p_s$	δ	Δ	$q^{\frac{n}{2}-m}$
48	16	$5 \cdot 7 \cdot 13 \cdot 17 \cdot 41 \cdot 73 \cdot 97 \cdot 577 \cdot 769 \cdot 6481$	.4628	752.7	6561
36	12	$5 \cdot 7 \cdot 13 \cdot 19 \cdot 37 \cdot 73 \cdot 757 \cdot 530713$	.4855	396.0	729
30	10	$5 \cdot 7 \cdot 11 \cdot 13 \cdot 31 \cdot 37 \cdot 61 \cdot 271 \cdot 4561$	.6367	228.4	243
27	9	$13 \cdot 109 \cdot 151 \cdot 433 \cdot 757 \cdot 8209$	.9101	115.1	140.2
25	8	$11 \cdot 29 \cdot 8951 \cdot 391151$	.9089	67.2	140.2
22	7	$23 \cdot 67 \cdot 661 \cdot 3851$	.9398	72.6	81
20	6	$5 \cdot 11 \cdot 61 \cdot 1181$	.6918	76.0	81
19	6	$1597 \cdot 363889$	.9993	36.0	46.7
17	5	$1871 \cdot 34511$	.9943	30.0	46.7
14	4	$547 \cdot 1093$	.9972	24.0	27

Table 2: Sieving table for $q = 3$

Since we may take $n \geq 12$, the only values of n not covered so far are those in the set $\mathcal{S} := \{24, 21, 18, 16, 15, 13, 12\}$. MAPLE 6 was used to find relevant primitive polynomials for each $n \in \mathcal{S}$ and each choice of first $\lfloor \frac{n}{3} \rfloor$ coefficients. In every case a weight 2 primitive polynomial sufficed, except that, when $n = 12$, one weight 4 polynomial ($x^{12} + x^{10} - x^8 + x^3 - x^2 - x - 1$) was needed. This completes the proof of Theorem 2.

6   Primitive Polynomials over $\mathbb{F}_4$

Now we consider the first case of Theorem 3 and take $q = 4$ and $m = \lfloor \frac{n}{3} \rfloor$.
Then (3.7) certainly holds whenever

$$2^{\frac{n}{3} - \omega} > \frac{n}{3}, \quad \omega := \omega(4^n - 1). \tag{6.1}$$

Of course, a slightly weaker condition suffices when n is indivisible by 3. The following lemma now tales the place of Lemma 3.

Lemma 6. *Suppose that $\omega \geq 63$. Then $\omega \leq \frac{n}{3.29}$.*

Proof. This time use the numerical inequality

$$P(r) > 4^{3.29r}, \quad r \geq 63,$$

where here $P(r)$ denotes the product of the first r odd primes. Note that the 63-rd such prime is 311, which exceeds $4^{3.29}$.

By (6.1) and Lemma 6, Theorem 3 holds with $q = 4$, $\omega \leq 62$ and $n < 3.29\omega < 204$ provided $2^{\frac{n}{3} - \frac{n}{3.29}} > \frac{n}{3}$. This last inequality holds for all $n \geq 208$. In summary, we may suppose $n \leq 207$.

Using MAPLE's "easy" factorisation option, we find that, in place of Lemma 6, we have the stronger fact that $\omega \leq \frac{n}{4}$ for $48 < n \leq 207$. Since $2^{n/12} > \frac{n}{3}$ for $n \geq 48$, we conclude that Theorem 2 holds for any value of $n \geq 49$. Again, for most values of n in the range $13 \leq n \leq 48$, application of Lemma 4 (using $l = 3$) is successful. Table 3 features a selection of values in the range (including all the most delicate cases) for which success occurs.

n	m	$p_1 \ldots p_s$	δ	Δ	$q^{\frac{n}{2} - m}$
36	12	$5 \cdot 7 \cdot 13 \cdot 17 \cdot 19 \cdot 37 \cdot 73 \cdot 109 \cdot 241 \cdot 433 \cdot 38737$	.4123	692.9	4096
30	10	$5 \cdot 7 \cdot 11 \cdot 13 \cdot 31 \cdot 41 \cdot 61 \cdot 151 \cdot 331 \cdot 1321$	.4058	483.4	1024
24	8	$5 \cdot 7 \cdot 13 \cdot 17 \cdot 97 \cdot 241 \cdot 257 \cdot 673$	.5015	255.3	256
20	6	$5 \cdot 11 \cdot 17 \cdot 31 \cdot 41 \cdot 61681$	.5936	125.0	256
16	5	$5 \cdot 17 \cdot 257 \cdot 65537$	.7372	60.6	64
14	4	$5 \cdot 29 \cdot 43 \cdot 113 \cdot 127$	.7255	60.1	64

Table 3: Sieving table for $q = 4$

In Table 3, observe that inequality (4) is just satisfied when $n = 24$. From Section 1, there remain the values $n = 18, 15, 12$ and 6 to verify directly. Because the base field $\mathbb{F}_4$ is non-prime we adopted a different strategy in using MAPLE. We selected a primitive element γ of $\mathbb{F}_{4^n} = \mathbb{F}_{2^{2n}}$ by specifying it as a root of a fixed primitive polynomial of degree $2n$ over $\mathbb{F}_2$. Then, we took powers

$\gamma^i, i \geq 1$, $(i, 4^n - 1) = 1$ with i odd, and calculated the m-tuple of elements in $\mathbb{F}_4$ comprising the first m symmetric functions of the conjugates of γ^i over $\mathbb{F}_4$ (effectively, the first m coefficients of the minimal polynomial of γ^i). Every time we found such an m-tuple we also calculated the conjugate m-tuple over $\mathbb{F}_4$, since this is associated with the conjugate primitive polynomial, having γ^{2i} as a root. This process ceased once a primitive polynomial associated with each of the 4^m choices of first m coefficients (and their conjugates) had been identified. The (odd) value of i reached by this stage is denoted by i_m. The results are displayed in Table 4. In particular, the computation for the case in which $n = 18$, $m = 6$ took several days to complete and and was the longest undertaken.

n	m	4^m	$f_\gamma(x)$	i_m
6	2	16	$x^{12} + x^4 + x + 1$	163
12	4	256	$x^{24} + x^4 + x^3 + x + 1$	4601
15	5	1024	$x^{30} + x^6 + x^4 + x + 1$	13201
18	6	4096	$x^{36} + x^{11} + 1$	88147

Table 4: Values of i_m for $q = 4$.

7   Primitive Polynomials over $\mathbb{F}_5$

Next we take $q = 5$ and continue to denote $\lfloor \frac{n}{3} \rfloor$ by m. Certainly, holds whenever

$$5^{\frac{n}{6}}/2^\omega > \frac{n}{3}, \quad \omega := \omega(5^n - 1). \tag{7.1}$$

The following lemma is now relevant.

Lemma 7. *Now let $P(r)$ denote the product of the first r primes (excluding 5). Then*

$$P(r) > 5^{2.828r}, \quad r \geq 63.$$

Lemma 8. *Suppose that $\omega \geq 63$. Then $\omega \leq \frac{n}{2.828}$.*

Proof. This time use the numerical inequality

$$P(r) > 5^{2.828r}, \quad r \geq 63.$$

where here $P(r)$ denotes the product of the first r primes ($\neq 5$). Note that the 63-rd such prime is 311, which exceeds $5^{2.828}$.

By (7.1) and Lemma 8, Theorem 3 with $q = 5$, $\omega \leq 62$ and $n < 2.828\omega < 176$ provided, $2^{\frac{n}{3} - \frac{n}{3.29}} > \frac{n}{3}$. This last inequality holds for all $n \geq 176$. In summary, we may suppose $n \leq 175$.

By MAPLE's "easy" factorisation option, we find that, in place of Lemma 8, we have the stronger fact that $\omega \leq \frac{2n}{7}$ for $36 < n \leq 175$ (with equality when $n = 42$). Since $5^{n/12}/2^{2n/7} > \dfrac{n}{3}$ for $n \geq 36$, we conclude that Theorem 2 holds for any value of $n \geq 36$. Again, for most values of n in the range $13 \leq n \leq 35$, application of Lemma 4 (using $l = 2$) is successful. Table 5 features a selection of values in the range (including all the most delicate cases) for which success occurs.

n	m	$p_1 \ldots p_s$	δ	Δ	$q^{\frac{n}{2}-m}$
30	10	$3 \cdot 7 \cdot 11 \cdot 31 \cdot 61 \cdot 71 \cdot 181 \cdot 521 \cdot 1741 \cdot 7621$	.3620	537.2	3125
24	8	$3 \cdot 7 \cdot 13 \cdot 31 \cdot 313 \cdot 601 \cdot 390001$	.4097	266.2	625
20	6	$3 \cdot 11 \cdot 13 \cdot 41 \cdot 71 \cdot 521 \cdot 9161$	.4583	181.0	625
16	5	$3 \cdot 13 \cdot 17 \cdot 313 \cdot 114897$	.5276	95.8	125
14	4	$3 \cdot 29 \cdot 449 \cdot 19531 \cdot 127$	.6299	54.1	125

Table 5: Sieving table for $q = 5$

Indeed, the only values of $n \geq 12$ (as we may assume) not covered so far are those in $\mathcal{S} := \{18, 15, 12\}$. Finally, MAPLE was used to find relevant weight 2 primitive polynomials for each $n \in \mathcal{S}$ and each choice of first $\lfloor \frac{n}{3} \rfloor$ coefficients as in Section 5.

8 Primitive Polynomials over Larger Fields

From now on, assume $q \geq 7$. As will be evident from the preceding sections, the most delicate cases occur when n is small (for example, $n = 12$). As remarked already, the difficulties for larger n are largely organisational in character. As noted in Section 1, we can assume $n \geq 12$ (except when $q (> 4)$ is even, in which case we should also check $n = 6$ and 7).

We start by showing that Corollary 6 applies when $\omega := \omega(q^n - 1)$ is sufficiently large. In practice, this means $\omega \geq 1547$.

Lemma 9. *Suppose that the positive integer M is such that $\omega(M) \geq 1547$. Then*

$$2^{\omega(M)} < M^{\frac{1}{12}}.$$

Proof. The 1547-th prime is 12983. By calculation, the product $P := \displaystyle\prod_{l \leq 12983} \frac{2}{l^{1/12}}$
(over all primes $l \leq 12983$) is less than 0.91. Since $12983^{1/12} > 2.2 > 2$, it is evident that

$$\frac{2^{\omega(M)}}{M^{1/12}} \leq \prod_{l \mid M} \frac{2}{l^{1/12}} \leq P < 1,$$

and the lemma follows.

To apply Lemma 9, suppose that $\omega := \omega(q^n-1) \geq 1547$. Then (3.7) is satisfied whenever $q^{n/12} > \frac{n}{3}$. By considering its logarithm, the function $q^{x/12} - \frac{x}{3}$ is increasing for $x \geq 12$ and positive for $x = 12$ (indeed positive for $x = 6$) provided $q > 4$. So Theorem 3 holds in this circumstance.

We can therefore assume $\omega \leq 1546$ (and $n < 12 \cdot 1546 = 18552$). To reduce rapidly the range of possible values for ω, temporarily set $w_1 = 1546$. Apply the sieving inequality as follows. Let l be the product of the smallest r (distinct) primes in $q^n - 1$, where, in the first instance, r is taken to have the value 10. (This means of course that it is assumed that $r \leq \omega \leq w_1$.) Thus, in the notation of Lemma 4, $s = \omega - r \leq s_0 = w_1 - r$ (initially $s_0 = 1536$) and

$$\delta \geq \delta_0 := 1 - \sum_{i=r+1}^{w_1} \frac{1}{l_i},$$ where l_i is the i-th prime. In particular, with the specific

values above $\delta_0 = 0.02267$ (truncating). Set $\Delta_0 := 2^r \left(\frac{s_0 - 1}{\delta_0} + 2 \right)$. Then ,

clearly (4.3) is satisfied whenever

$$q > \left(\frac{n}{3} \Delta_0 \right)^{\frac{6}{n}}. \tag{8.1}$$

Again by differentiation, it is evident that $q^{x/6} - \frac{x}{3} \Delta_0$ is increasing for $n \geq 12$ ($q \geq 2$). Therefore, set $q_0 := 2\sqrt{\Delta_0}$ (i.e., the right side of (8.1) when $n = 12$). Indeed, with the initial values shown we have $q_0 = 16653.6$. We conclude that (4.3) holds (under the above conditions) whenever $q \geq q_0$ and we may henceforth assume that $q \leq \lfloor q_0 \rfloor$. Now suppose that, for an appropriate choice of w_2 (where, in the first instance, we select $w_2 = 120$), we have $w_2 \leq \omega \leq w_1$. Then, with $P(j)$ denoting the product of the first j primes we have

$$n \geq \left\lceil \frac{\log P(w_1)}{\log q} \right\rceil \geq \left\lceil \frac{\log P(w_1)}{\log \lfloor q_0 \rfloor} \right\rceil =: n_0.$$

As the final act of this first round evaluate q^*, defined as the right side of (8.1) when $n = n_0$. Indeed, $q^* = 6.836$. It follows that (4.3) holds (and so Theorem 3 is valid) whenever $q > q^*$ and $\omega \geq w_2$. In particular, the above figures establish Theorem 3 whenever $\omega \geq w_2 = 120$.

w_1	r	s_0	δ_0	q_0	q_1	w_2	n_0	q^*
1546	10	1536	.02267	16653.6		120	66	6.836
119	5	114	.12910	335.1		53	40	6.854
52	4	48	.20068	122.9		40	33	6.914
39	3	36	.12170	96.2	89	26	20	18.05
39	3	36	.12170		17	26	32	6.66
25	3	22	.23051	54.5	53	20	16	22.36
25	3	22	.23051		19	20	21	11.53
25	3	22	.23051		11	20	26	7.57
25	3	22	.23051		7	20	32	5.38
19	2	17	.10455	49.8				

Table 6: Sieving table for large q

The outcome of Table 6 is that Theorem 3 is established, except possibly for q one of the 19 prime powers with $7 \leq q \leq 49$ and $\omega \leq 19$. Indeed, at this last stage (with $r = 2$), by amending δ_0 to take into account the fact that $q^n - 1$ is indivisible by the characteristic p, we can assume that $\omega \leq 15$ and $q \leq 31$. Even then, if we take $n \geq 22$ when $q = 7$ and $n \geq 18$ (instead of the calculated value of n_0), we can guarantee that (4.3) holds. For the pairs, (q, n), $7 \leq q \leq 31$, $n \leq 21$ or 17 that remain, by obtaining the factorisation of $q^n - 1$, we find that most satisfy either the basic criterion (6) or the sieve inequality (4.3) with $l = 2$ or 3, according as q is odd or even, respectively. For example, the latter is satisfied when $q = 8$ and $n = 6$. The pairs (q, n), $n \geq 12$ (allowing also $n = 6$ when q is even) that survive to be checked directly are thereby found to be $(q, 12)$ for $7 \leq q \leq 13$. Here, for the prime values $q = 7, 11, 13$, for each choice of first 4 coefficients (28561 choices in the final case), we hunted successfully (as in Sections 4, 5 and 7) until a primitive polynomial of degree 12 was found with weight 2 (for $q = 7$) and weight 1 for $q = 11$ and 13 (except for a single weight 2 case $x^{12} - 4x^{10} - 5x^8 + x^2 + 2x + 2$ when $q = 11$). For $q = 9$, we took powers γ^i, $(i, 9^{12} - 1) = 1$ with i odd, of a primitive element γ of $\mathbb{F}_{3^{24}}$ (a root of $x^{24} + x^{13} + x^5 - 1$), until all 4-tuples of first 4 coefficients and their conjugates had been accounted for (as in Section 6). In similar notation to that used there, we obtained $i_m = 136103$. Finally, for $q = 8$, the same primitive element of $\mathbb{F}_{2^{24}}$ was used as in Section 6. On this occasion, every time a new 4-tuple of first coefficients is obtained, *two* further conjugates are also accounted for. In this way, we obtained $i_m = 91901$. Interestingly, some 4-tuples in $\mathbb{F}_2$ are the last to emerge in the search, namely $(0, 1, 0, 0)$ $(i = 56423)$; $(1, 1, 1, 0)$ $(i = 61267)$; $(1, 0, 1, 1)$ $(i = 75151)$; $(0, 0, 0, 1)$ $(i = 91901)$. This completes the proof of Theorem 3.

References

1. S. D. Cohen, *Primitive elements and polynomials with arbitrary trace*, Discrete Math. **83** (1990), 1–7.
2. S. D. Cohen, *Primitive elements and polynomials: existence results*, Finite fields, coding theory, and advances in communications and computing 43–55, Lecture Notes in Pure and Appl. Math., 141, Dekker, New York, 1993.
3. S.D. Cohen and D Mills, *Primitive polynomials with first and second coefficients prescribed*, Finite Fields Appl., **9** (2003), 334–350.
4. S-Q. Fan and W-B. Han, *p-adic formal series and primitive polynomials over finite fields*, Proc. Amer. Math. Soc., **132** (2004), 15–31.
5. S-Q. Fan and W-B. Han, *p-adic formal series and Cohen's problem*, Glasgow Math. J., to appear.
6. S-Q. Fan and W-B. Han, *Character sums over Galois rings and primitive polynomials over finite fields*, Finite Fields Appl., to appear.
7. W-B. Han, *The coefficients of primitive polynomials over finite fields* Math. Comp. **65** (1996), 331–340.
8. W-B. Han, *On Cohen's problem*, Chinacrypt '96, Academic Press (China), 231–235, 1996 (Chinese).

9. W-B. Han, *On two exponential sums and their applications*, Finite Fields Appl. **3** (1997), 115–130.

10. W-B. Han, *The distribution of the coefficients of primitive polynomials over finite fields*, Cryptography and computational number theory 43–57, Progr. Comput. Sci. Appl. Logic, 20, Birkhuser, Basel, 2001.

11. D. Jungnickel and S. A. Vanstone, *On primitive polynomials over finite fields*, J. Algebra **124** (1989), 337–353.

12. N. Koblitz p-adic Numbers, p-adic Analysis, and Zeta-Functions, Springer, New Tork, 1984.

13. W- C. W. Li, *Character sums over p-adic fields*, J. Number Theory **74** (1999), 181–229

14. D. Mills, *Existence of primitive polynomials with three coefficients prescribed*, JP J. Algebra Number Theory Appl., to appear.

15. D-B. Ren, *On the coefficients of primitive polynomials over finite fields*, Sichuan Daxue Xuebao **38**, 33–36.

16. I. E. Shparlinski, *On primitive polynomials*, Prob. Peredachi Inform. **23** (1988), 100–103 (Russian).

Vectorial Functions and Covering Sequences

Claude Carlet[1] and Emmanuel Prouff[2]

[1] Claude Carlet, INRIA, projet CODES, BP 105 - 78153, Le Chesnay Cedex, France;
also GREYC-Caen and University of Paris 8
claude.carlet@inria.fr

[2] INRIA Projet CODES and University of Paris 11, Laboratoire de Recherche en
Informatique, 15 rue Georges Clemenceau, 91405 Orsay Cedex, France
prouff@info.unicaen.fr

Abstract. The design of large classes of highly nonlinear resilient vectorial functions (mappings from $\mathbb{F}_2^n$ into $\mathbb{F}_2^m$, also called S-boxes) is needed for iterated block ciphers and for pseudo-random generators with multiple output. In this paper, we recall the diverse known constructions of such S-boxes, and we show that those which provide good candidate functions are, in fact, all in the same class. This class corresponds to a generalization of a well known construction due to Maiorana and MacFarland. We study in detail this construction and we specify it to obtain good S-boxes. In a second part, we generalize to S-boxes the notion of covering sequence. We show that this generalization has the same properties as for Boolean functions, and that it has nice additional properties of stability. We study how this notion can be used to design attacks, and we explain why some functions, including the elements of the new class, cannot be involved in the construction of iterated block ciphers.

1 Introduction

Cryptographic encryption schemes are divided into two main classes: blocks ciphers and stream ciphers. Block ciphers, as DES or AES, are the compositions of several rounds. Each round involves vectorial functions from the binary vector space $\mathbb{F}_2^n$ into the vector space $\mathbb{F}_2^m$, also called S-boxes or (n, m)-functions (the parameters n and m of (n, m)-functions used in block ciphers are often chosen to be equal). To protect these cryptosystems from attacks (such as linear [31] or differential attacks [2]) these (n, m)-functions must have, in general, a high algebraic degree and a high nonlinearity.

Pseudo-random generators in stream ciphers involve a Boolean function to combine the outputs of several linear feedback shift registers or to filter the contents of a single one. To speed up the encryption and decryption of these stream ciphers, one can try to replace respectively the Boolean combining - or filtering - function by a vectorial one. To prevent some kind of attacks called fast correlation attacks against stream ciphers, the functions involved, Boolean or vectorial, must have a high algebraic degree and a high nonlinearity. Moreover, in the particular case of stream ciphers with combining generator, they must also satisfy

G. Mullen, A. Poli and H. Stichtenoth (Eds.): Fq7 2003, LNCS 2948, pp. 215–248, 2003.
© Springer-Verlag Berlin Heidelberg 2003

another property called *high resiliency*. In addition to fast correlation attacks, algebraic attacks, based on the low degree approximation/decomposition, have been introduced (see [15, 21]) against all these systems involving either Boolean or vectorial functions as cryptographic primitives. Among the classical criteria (algebraic degree, nonlinearity and resiliency) only the degree seems related to the resistance of a cryptosystem against algebraic attacks; and it is still necessary to define all new relevant criteria.

The construction of highly nonlinear balanced vectorial functions (resp., the construction of highly nonlinear resilient vectorial functions) is needed to design secure block ciphers or stream ciphers with filtering multi-output generator (resp. secure stream ciphers with combination multi-output generator). Moreover, the class of functions constructed for any given nonlinearity and any given resiliency order, must be sufficiently large to allow cryptographers to choose functions satisfying additional criteria more specific to the implementation.

We will show that all known constructions of highly nonlinear (n, m)-functions (resilient or not) can be obtained as special cases of a single general construction. This construction is a generalization to the vectorial case of the Maiorana-MacFarland construction. We recall the known facts about this construction, giving a general construction including all the previous constructions introduced in [23, 25, 29, 34]. We show that, for $m > n/2$, this construction does not allow to design directly resilient highly nonlinear (n, m)-functions, and we give a way to use vectorial Maiorana-MacFarland functions to design, by concatenation, such (n, m)-functions with parameters satisfying $m > n/2$. This is the only known primary construction of a large set of resilient highly nonlinear (n, m)-functions with $m > n/2$.

In a second part, we generalize to vectorial functions the notion of covering sequences introduced in [11] and we explain how some covering sequences could be used to attack iterated block ciphers. We show that Maiorana-MacFarland functions extended by concatenation admit covering sequences which can be used to attack iterated block ciphers involving them in the round functions, when the round key is introduced by addition (which is the most usual way of introducing it).

2   Notation and Preliminaries

We will have to distinguish in the whole paper between the additions of integers in $\mathbb{R}$, denoted by $+$ and $\sum_i$, and the additions mod 2, denoted by $\oplus$ and $\bigoplus_i$. For simplicity and because there will be no ambiguity, we will denote by $+$ the addition of vectors of $\mathbb{F}_2^n$ (words) and of elements of fields $\mathbb{F}_{2^n}$ with $n > 1$.

We call (n, m)-function any mapping F from $\mathbb{F}_2^n$ into $\mathbb{F}_2^m$. If m equals 1, then the function is called Boolean and we will denote by $\mathcal{B}_n$ the set of all Boolean functions defined on $\mathbb{F}_2^n$.

Denoting the all-zero vector by $\mathbf{0}$, we call support of F and we denote by *Supp F* the set $\{x \in \mathbb{F}_2^n / F(x) \neq \mathbf{0}\}$. Let $\#E$ denote the cardinality of any set E. An (n, m)-function F is said to be balanced if every element of $y \in \mathbb{F}_2^m$ admits the same number 2^{n-m} of pre-images by F, that is $\#\{x \in \mathbb{F}_2^n; F(x) = y\} = 2^{n-m}$.

To every (n, m)-function F, we associate the m-tuple $(f_1, \cdots, f_m)$ of Boolean functions on $\mathbb{F}_2^n$, called *the coordinate functions of F*, such that we have $F(x) = (f_1(x), \cdots, f_m(x))$.

Every (n, m)-function F admits a unique representation as a polynomial over $\mathbb{F}_2^m$ in n binary variables of the form:

$$F(x_1, \cdots, x_n) = \bigoplus_{I \subseteq \{1, \cdots, n\}} a_I \prod_{i \in I} x_i, \; a_I \in \mathbb{F}_2^m.$$

This representation is called the *algebraic normal form* (A.N.F.) of F. We will call *(algebraic) degree* of F and denote by $\deg F$ the degree of its A.N.F. It equals the maximum algebraic degree of its coordinate functions or more generally of the linear combinations of its coordinate functions. But a low minimum degree of the nonzero linear combinations of the coordinate functions of F can be used to cryptanalyse a system. So we also define a second notion of degree which will be called *minimum degree*, defined by

$$\deg_m(F) = \min_{v \in \mathbb{F}_2^{m*}} \deg(v \cdot F).$$

To make easier the study of the properties of F, we classically introduce the *sign* function χ_F of F defined as $\chi_F(x, v) = (-1)^{v \cdot F(x)}$ (if F is Boolean, the *sign* function is defined by $\chi_F(x) = (-1)^{F(x)}$).

For any numerical function φ on $\mathbb{F}_2^n$, the *discrete Fourier transform* of φ, denoted by $\hat{\varphi}$, is a bijective transformation defined by $\widehat{\varphi}(u) = \sum_{x \in \mathbb{F}_2^n} \varphi(x)(-1)^{u \cdot x}$, $u \in \mathbb{F}_2^n$. It satisfies the following relation

$$\hat{\hat{\varphi}} = 2^n \varphi. \tag{1}$$

The Fourier transform of the sign function of a Boolean function f (that we will call *Walsh transform* of f) is the integer valued function $\widehat{\chi_f}$, defined on $\mathbb{F}_2^n$ by the formula:

$$\widehat{\chi_f}(u) = \sum_{x \in \mathbb{F}_2^n} (-1)^{f(x)+u \cdot x} = \sum_{x \in \mathbb{F}_2^n} (-1)^{f(x) \oplus u \cdot x}. \tag{2}$$

More generally, if F is an (n, m)-function, then its Walsh transform, $\widehat{\chi_F}$, is defined on $\mathbb{F}_2^n \times \mathbb{F}_2^m$ by the formula:

$$\widehat{\chi_F}(u, v) = \widehat{\chi_{v \cdot F}}(u) = \sum_{x \in \mathbb{F}_2^n} (-1)^{v \cdot F(x)+u \cdot x}. \tag{3}$$

Every numerical function φ satisfies Parseval's relation

$$\sum_{u\in\mathbb{F}_2^n} \widehat{\varphi}^2(u) = 2^n \sum_{u\in\mathbb{F}_2^n} \varphi^2(u), \tag{4}$$

and in the case of the Walsh transform of an (n,m)-function F, this relation becomes:

$$\forall v \in \mathbb{F}_2^m, \quad \sum_{u\in\mathbb{F}_2^n} \widehat{\chi_F}^2(u,v) = 2^{2n}. \tag{5}$$

We recall now the definition of the convolutional product between two numerical functions φ and ψ on $\mathbb{F}_2^n$. It is denoted by $\varphi \otimes \psi$ and defined on $\mathbb{F}_2^n$ by:

$$(\varphi \otimes \psi)(x) = \sum_{a\in\mathbb{F}_2^n} \varphi(a)\,\psi(x+a). \tag{6}$$

A well-known fact is that the Fourier transform of $\varphi \otimes \psi$ equals the product of $\widehat{\varphi}$ and $\widehat{\psi}$, that is:

$$\widehat{\varphi \otimes \psi} = \widehat{\varphi} \times \widehat{\psi}. \tag{7}$$

Another useful tool for studying a function F is the notion of derivative. The derivative of F with respect to a vector $a \in \mathbb{F}_2^n$ is the (n,m)-function $D_a F : x \mapsto F(x) + F(x+a)$. The derivatives play an important role in cryptography, related to the differential attack [2].

The nonlinearity of a function F is one of the parameters which quantify, from the viewpoint of the Hamming distance, the level of *confusion* put in the system by the function (another such parameter is the degree). The nonlinearity of a vectorial function F is defined as the minimum Hamming distance between the nonzero linear combinations of the coordinate functions of F and the set of all affine functions. Cryptographic functions used in stream or block ciphers must have high nonlinearities to prevent these systems from linear attacks (see [1, 6, 16, 22, 31, 33]).

For every (n,m)-function F, the nonlinearity N_F and the Walsh transform $\widehat{\chi_F}$ satisfy the relation:

$$N_F = 2^{n-1} - \frac{1}{2} \max_{u\in\mathbb{F}_2^n, v\in\mathbb{F}_2^{m*}} |\widehat{\chi_F}(u,v)|. \tag{8}$$

Because of Parseval's relation (5), N_F is upper bounded by $2^{n-1} - 2^{n/2-1}$ for every (n,m)-function F. If n is even and $m \le \frac{n}{2}$, then this bound is tight (see [34]). The functions achieving it are called *bent*. Chabaud and Vaudenay proved in [12] that the nonlinearity N_F is also upper bounded by $2^{n-1} - \frac{1}{2}\sqrt{3 \times 2^n - 2 - 2\frac{(2^n-1)(2^{n-1}-1)}{2^m-1}}$. This bound equals $2^{n-1} - 2^{n/2-1}$ if and only if $m = n - 1$ and it is better than $2^{n-1} - 2^{n/2-1}$ if and only if $m \ge n$. If $m = n$ (the only case of tightness), then Chabaud Vaudenay's bound implies that the maximum possible nonlinearity of any (n, n)-function is upper bounded by $2^{n-1} - 2^{\frac{n-1}{2}}$. The functions achieving this nonlinearity are called *almost bent* and exist only when n is odd. In the other cases (when $m = n$ and n is even or when $m < n < 2m$), the maximum values achieved by the nonlinearity are unknown.

We note that (n, m)-functions are used in fault-tolerant distributed computing, quantum cryptography key distribution and random sequence generation for stream cipherss. To resist divide-and-conquer attacks, these functions have to be balanced and to stay balanced if any t of the inputs are fixed (where t is an integer as large as possible). This property is called *t-resiliency*, and we call *resiliency order* of a balanced function the maximum value of such t. Resiliency can be characterized by means of the Fourier transform :

Proposition 1. *[4, 42] An (n, m)-function F is t-resilient if and only if its Walsh transform satisfies $\widehat{\chi_F}(u, v) = 0$ for every pair $(u, v) \in \mathbb{F}_2^n \times \mathbb{F}_2^m$ such that $v \neq \mathbf{0}$ and $w_H(u) \leq t$, where w_H denotes the Hamming weight.*

Remark 1.
1. A function is balanced if and only if it is 0-resilient, that is: $\widehat{\chi_F}(\mathbf{0}, v) = 0$ for any non-zero vector $v \in \mathbb{F}_2^m$,
2. If G is a t-resilient (n, m)-function and P is a permutation on $\mathbb{F}_2^m$, then the resiliency order of the (n, m)-function $F = P \circ G$ is at least t.

According to Proposition 1, an (n, m)-function F is t-resilient if and only if every nonzero linear combination of its coordinates, denoted by $v \cdot F$ where v belongs to $\mathbb{F}_2^{m*}$, is t-resilient. On the other hand, as observed by Camion and Canteaut in [4], an (n, m)-function F is t-resilient if and only if for every balanced Boolean function $g \in \mathcal{B}_m$, the Boolean function $g \circ F$ is also t-resilient.

We will call (n, m, t)-function an (n, m)-function which is t-resilient.

In the sequel we will need the next well known proposition, introduced by Rothaus in 1976 (see [37]) to define a simple iterative construction of bent functions and which has been generalized some years after to design highly nonlinear resilient functions.

Proposition 2. *Let G and H be respectively an (r, m, t_1)-function and an $(n - r, m, t_2)$-function. Then the (n, m)-function F defined by $F(x, y) = G(x) + H(y)$ is $(t_1 + t_2 + 1)$-resilient and its nonlinearity equals*

$$2^{n-1} - \frac{1}{2} \max_{(u,u') \in \mathbb{F}_2^r \times \mathbb{F}_2^{n-r}, v \in \mathbb{F}_2^{m*}} |\widehat{\chi_G}(u, v)\widehat{\chi_H}(u', v)|,$$

which is lower bounded by $2^{n-1} - \frac{1}{2}(2^r - 2N_G)(2^{n-r} - 2N_H)$.

Proof. For every pair (u, u') of $\mathbb{F}_2^r \times \mathbb{F}_2^{n-r}$ and for every nonzero vector v of $\mathbb{F}_2^m$, we have:

$$\widehat{\chi_F}((u, u'), v) = \sum_{(x,y)\in\mathbb{F}_2^r\times\mathbb{F}_2^{n-r}}(-1)^{v\cdot G(x)+v\cdot H(y)+u\cdot x+u'\cdot y}$$

$$= \left(\sum_{x\in\mathbb{F}_2^r}(-1)^{v\cdot G(x)+u\cdot x}\right)\left(\sum_{y\in\mathbb{F}_2^{n-r}}(-1)^{v\cdot H(y)+u'\cdot y}\right)$$

$$= \widehat{\chi_G}(u, v)\widehat{\chi_H}(u', v).$$

We deduce that $\widehat{\chi_F}((u, u'), v)$ is nonzero if and only if $\widehat{\chi_G}(u, v)$ and $\widehat{\chi_H}(u', v)$ are nonzero: this implies that F is $(t_1 + t_2 + 1)$-resilient. By applying Equation (8), we obtain the nonlinearity of F.

3   Maiorana-MacFarland Functions

3.1   Boolean Case

In [5], Camion, Carlet, Charpin and Sendrier introduce a modification of the construction of Maiorana and MacFarland of bent functions (given in [19, 32]) whose elements, viewed as binary vectors of length 2^n, are the concatenations of affine functions.

Definition 1. *[5] The class $\mathcal{M}_n$ is the set of Boolean functions f of the form:*

$$f_{\phi,h}(x, y) = x \cdot \phi(y) \oplus h(y) \quad, \tag{9}$$

where r and s are any positive integers such that $r + s = n$, ϕ is any function from $\mathbb{F}_2^s$ into $\mathbb{F}_2^r$ and h is any Boolean function on $\mathbb{F}_2^s$.

A function $f_{\phi,h} \in \mathcal{M}_n$ is t-resilient if and only if, for every vector $u \in \mathbb{F}_2^r$ of Hamming weight lower than or equal to t, either the set $\phi^{-1}(u)$ is empty or the function $y \in \mathbb{F}_2^s \mapsto h(y) \oplus u' \cdot y$ is balanced on $\phi^{-1}(u)$ for every vector $u' \in \mathbb{F}_2^s$ of Hamming weight at most $t - w_H(u)$. Indeed, as shown in [5], for every $u \in \mathbb{F}_2^r$ and every $u' \in \mathbb{F}_2^s$, we have $\widehat{\chi_{f_{\phi,h}}}(u, u') = 2^r \sum_{y\in\phi^{-1}(u)}(-1)^{h(y)+u'\cdot y}$ since every (affine) function $x \mapsto f_{\phi,h}(x, y) \oplus u \cdot x \oplus u' \cdot y$ either is constant or is balanced and contributes then for 0 in the sum $\sum_{x\in\mathbb{F}_2^r, y\in\mathbb{F}_2^s}(-1)^{f_{\phi,h}(x,y)\oplus u\cdot x\oplus u'\cdot y}$. A sufficient condition for $f_{\phi,h}$ being t-resilient is that every element in $\phi(\mathbb{F}_2^s)$ has Hamming weight strictly greater than t. This condition does not involve the function h. Another sufficient condition is that every element in $\phi(\mathbb{F}_2^s)$ has weight at least t and that, if it has weight t, then it admits two inverse images y_1 and y_2 by ϕ, and that $h(y_1) = 0$, $h(y_2) = 1$. The nonlinearity of a Boolean function $f_{\phi,h} : \mathbb{F}_2^r \times \mathbb{F}_2^s \mapsto \mathbb{F}_2$ in $\mathcal{M}_n$ equals $2^{n-1} - 2^{r-1} \max_{(u,u')\in\mathbb{F}_2^r\times\mathbb{F}_2^s} \left|\sum_{y\in\phi^{-1}(u)}(-1)^{h(y)+u'\cdot y}\right|$.

When ϕ is injective (resp. takes exactly two times each value of $Im\ \phi$), then this nonlinearity equals $2^{n-1} - 2^{r-1}$ (resp. $2^{n-1} - 2^r$). Also, as proved in [9, 39], the nonlinearity of every Boolean function $f_{\phi,h}$ defined as in Equation (9) satisfies

$$2^{n-1} - 2^{r-1} \max_{u \in \mathbb{F}_2^r} \left(\#\phi^{-1}(u) \right) \leq N_{f_{\phi,h}} \leq$$

$$2^{n-1} - 2^{r-1} \left\lceil \sqrt{\max_{u \in \mathbb{F}_2^r} \#\phi^{-1}(u)} \right\rceil. \quad (10)$$

If every element in $\phi(\mathbb{F}_2^s)$ has Hamming weight strictly greater than k ($f_{\phi,h}$ is then t-resilient with $t \geq k$), then $N_{f_{\phi,h}}$ satisfies also the following relation [9]

$$N_{f_{\phi,h}} \leq 2^{n-1} - 2^{r-1} \left\lceil \frac{2^{s/2}}{\sqrt{\sum_{i=k+1}^{r} \binom{r}{i}}} \right\rceil. \quad (11)$$

The functions $f_{\phi,h}$ such that the Hamming weight $w_H(\phi(y))$ is strictly greater than t for every vector $y \in \mathbb{F}_2^s$ can be viewed as the concatenations of t-resilient affine Boolean functions. More generally, the concatenation of any t-resilient functions produces also t-resilient functions (this property has been used for instance in [38]).

3.2 Multiple-Output Case

The design of resilient vectorial functions by generalizing the construction of Maiorana and MacFarland is natural. One can find a first reference of such a construction in a paper by Nyberg [34] published in 1991. This technique has been used later by Kurosawa et al. [28], Johansson and Pasalic [25], Pasalic and Maitra [29] or more recently (presented as a new technique) by Gupta and Sarkar [23] to produce functions having high resiliency and high nonlinearity. However, a general study of vectorial Maiorana-MacFarland functions has never been done. The aim of this section is to recall what can be said about these functions by simply extending the properties of the Maiorana-MacFarland Boolean functions to the multi-output case. As a direct consequence of this study, we will show that all the constructions of vectorial Maiorana-MacFarland functions presented in the references cited above belong to a unique class.

The construction of t-resilient linear functions is easy: Stinson [40] considered the equivalence between linear resilient functions and what he called large sets of orthogonal arrays and the works of Delsarte [17, 18], in which is studied the relationship between orthogonal arrays and codes, can then be used to straightforwardly establish the connection between linear t-resilient functions and linear codes. The main result of these characterizations is that there exists a linear (n, m, t)-function if and only if there exists a set of 2^m disjoint binary arrays of dimensions $2^{n-m} \times n$, such that in any t columns of the arrays, every one of the 2^t elements of $\mathbb{F}_2^t$ occurs in exaclty 2^{n-m-t} rows and no two rows are identical or equivalently if and only if there exists a linear $[n, m, t+1]$ code (i.e. a subspace of $\mathbb{F}_2^n$ of dimension m whose nonzero elements have Hamming weights greater than or equal to $t + 1$).

We will focus more particularly in this paper on the definition of resilient linear functions from linear codes. Let us first recall that the generator matrix of a linear $[n, k, d]$ code C is the matrix G such that every element x of C can be written in the form $u \times G$, where u is a k-dimensional vector. In [14], Chor et al. state the following proposition, which is a direct consequence of the works of Delsarte [17, 18] and of the characterization of linear t-resilient functions with orthogonal arrays established by Camion et al. [5] for the Boolean case and by Stinson [40] for the vectorial one.

Proposition 3. *Let G be a generating matrix for an $[n, k, d]$ linear code C. Define $L : \mathbb{F}_2^n \mapsto \mathbb{F}_2^k$ by the rule $L(x) = x \times G^T$, where G^T is the transpose of G. Then L is an $(n, k, d-1)$-function.*

Remark 2. Proposition 3 is still trivially true if L is affine instead of linear, that is $L(x) = x \times G^t + a$, where a is a vector of $\mathbb{F}_2^k$.

Notice that the construction of t-resilient functions in Proposition 3 can be generalized by considering some nonlinear codes of length n (subsets of $\mathbb{F}_2^n$) whose dual distance $d^\perp$ is greater than or equal to $t + 1$ (see [41, 8]). The *dual distance of a code C of length n* is the smallest positive integer i such that the coefficient of the monomial $X^{n-i}Y^i$ in the polynomial $\sum_{x,y \in C} (X + Y)^{n - w_H(x+y)} (X - Y)^{w_H(x+y)}$ is nonzero (when the code is linear, the dual distance is equal to the minimum Hamming distance of the dual code).

Since one has, with Proposition 3, a simple way to design t-resilient vectorial affine functions, it is natural to generalize the construction of Maiorana-MacFarland by concatenating affine vectorial functions as we did for Boolean functions.

Definition 2. *The class $\mathcal{M}_{n,m}$ is the set of (n, m)-functions F which can be written in the form:*

$$F(x, y) = x \times \begin{pmatrix} \varphi_{11}(y) & \cdots & \varphi_{1m}(y) \\ \vdots & \ddots & \vdots \\ \varphi_{r1}(y) & \cdots & \varphi_{rm}(y) \end{pmatrix} + H(y), \quad (x, y) \in \mathbb{F}_2^r \times \mathbb{F}_2^s \quad (12)$$

where r and s are two positive integers satisfying $r + s = n$, H is any (s, m)-function and, for every index $i \leq r$ and every index $j \leq m$, φ_{ij} is a Boolean function on $\mathbb{F}_2^s$.

We will call *Maiorana-MacFarland's vectorial function* any function which can be written as in Equation (12). We recalled above that the concatenation of t-resilient functions is still t-resilient. Hence, due to Proposition 3, if the transpose matrix of the matrix involved in Relation (12) is the generator matrix of a linear

$[r, m, d]$-code for every vector y ranging over $\mathbb{F}_2^s$, then the (n, m)-function F is $(d-1)$-resilient.

Any function F belonging to $\mathcal{M}_{n,m}$ can be written in the form:

$$F(x, y) = \left(\bigoplus_{i=1}^{r} x_i \varphi_{i1}(y) \oplus h_1(y), \ldots, \bigoplus_{i=1}^{r} x_i \varphi_{im}(y) \oplus h_m(y) \right) \quad (13)$$

where $H = (h_1, \ldots, h_m)$.

After denoting, for every $i \leq m$, by ϕ_i the (s, r)-function which admits the Boolean functions $\varphi_{1i}, \ldots, \varphi_{ri}$ for coordinate functions, we can rewrite Retation (13) as :

$$F(x, y) = (x \cdot \phi_1(y) \oplus h_1(y), \ldots, x \cdot \phi_m(y) \oplus h_m(y)) . \quad (14)$$

Remark 3. Relation (14) is a simple generalization of the construction of bent (n, m)-functions from Boolean Maiorana-MacFarland's functions which has been proposed by Nyberg in [34] and restated by Chabaud and Vaudenay in [12].

As a direct consequence of Proposition 3 for instance, we have:

Corollary 1. *Let n, m, r and s be three integers such that $n = r + s$. Let F be an (n, m)-function in $\mathcal{M}_{n,m}$ such that, for every $y \in \mathbb{F}_2^s$, the family $(\phi_i(y))_{i \leq m}$ is a basis of an m-dimensional subspace of $\mathbb{F}_2^r$ having $t+1$ for minimum Hamming weight, then F is at least t-resilient.*

Remark 4. Notice that a t-resilient (n, m)-function $F \in \mathcal{M}_{n,m}$ cannot be designed applying Corollary 1 when m is strictly larger than r, since in this case, it is impossible to design m vectors $\phi_1(y)$, ..., $\phi_m(y)$ of $\mathbb{F}_2^r$ which are linearly independent.

According to the facts about the Walsh transform of the Boolean Maiorana-MacFarland functions recalled in Section 3.1, the coefficients of the Walsh transform of the (n, m)-functions F belonging to $\mathcal{M}_{n,m}$ (where $n = r + s$) take the form

$$\widehat{\chi_F}((u, u'), v) = 2^r \sum_{y \in E_{u,v}} (-1)^{v \cdot H(y) + u' \cdot y}, \quad (15)$$

where (u, u') is a pair in $\mathbb{F}_2^r \times \mathbb{F}_2^s$, v is an element of $\mathbb{F}_2^{m*}$ and $E_{u,v}$ denotes the set $\{y \in \mathbb{F}_2^s; \sum_{i=1}^{m} v_i \phi_i(y) = u\}$.

Remark 5. Let f_1, ..., f_m denote the m coordinate functions of an $(r + s, m)$-function F defined as in Relation (14) (i.e. $f_i(x, y) = x \cdot \phi_i(y) \oplus h_i(y)$, $i = 1, \ldots, m$). By using the known properties of Boolean Maiorana-MacFarland functions, we obtain straightforwardly that F is t-resilient if and only if, for every $v \in \mathbb{F}_2^{m*}$ and for every pair $(u, u') \in \mathbb{F}_2^r \times \mathbb{F}_2^s$ of Hamming weight lower than or equal to t, one of the two following conditions is satisfied:

1. the set $E_{u,v}$ is empty,

2. the Boolean function $y \in \mathbb{F}_2^s \mapsto v \cdot H(y) \oplus u' \cdot y$ is balanced on $E_{u,v}$.

However, because the vector v is involved in these two conditions, this characterization is not easy to use for designing resilient Maiorana MacFarland's vectorial functions.

Due to Relations (8) and (15), the nonlinearity N_F of any function $F \in \mathcal{M}_{n,m}$ (where $n = r + s$) defined as in Relation (14) satisfies

$$N_F = 2^{n-1} - 2^{r-1} \max_{(u,u') \in \mathbb{F}_2^r \times \mathbb{F}_2^s, v \in \mathbb{F}_2^{m*}} \left| \sum_{y \in E_{u,v}} (-1)^{v \cdot H(y) + u' \cdot y} \right| . \quad (16)$$

Because of the vector $v \in \mathbb{F}_2^{m*}$ involved in Relation (16), the construction of a highly nonlinear Maiorana-MacFarland vectorial function is more difficult than in the Boolean case. In the next proposition, we give the upper bound and the lower bound on the nonlinearity of Maiorana-MacFarland functions, which come directly from Relations (10) and (11).

Proposition 4. Let F be an (n, m)-function defined as in (14). Then, the nonlinearity N_F of F satisfies

$$2^{n-1} - 2^{r-1} \max_{u \in \mathbb{F}_2^r, v \in \mathbb{F}_2^{m*}} (\#E_{u,v}) \leq N_F$$

$$\leq 2^{n-1} - 2^{r-1} \left\lceil \sqrt{\max_{u \in \mathbb{F}_2^r, v \in \mathbb{F}_2^{m*}} \#E_{u,v}} \right\rceil . \quad (17)$$

Assume that, for every element y, the space spanned by the vectors $\phi_1(y)$, ..., $\phi_m(y)$ admits m for dimension and has a minimum Hamming weight strictly greater than k (so that F is t-resilient with $t \geq k$). Then, we have

$$N_F \leq 2^{n-1} - 2^{r-1} \left\lceil \frac{2^{s/2}}{\sqrt{\sum_{i=k+1}^r \binom{r}{i}}} \right\rceil . \quad (18)$$

Proof. By definition of the nonlinearity of F, we have N_F equal to $\min_{v \in \mathbb{F}_2^{m*}} N_{v \cdot F}$. Hence, we have $N_F = N_{v_0 \cdot F}$ for some nonzero v_0. We deduce $N_F \geq 2^{n-1} - 2^{r-1} \max_{u \in \mathbb{F}_2^r, v \in \mathbb{F}_2^{m*}} (\#E_{u,v})$, according to the lower bound in Relation (10). If the inequality $N_F \leq 2^{n-1} - 2^{r-1} \left\lceil \sqrt{\max_{u \in \mathbb{F}_2^r, v \in \mathbb{F}_2^{m*}} \#E_{u,v}} \right\rceil$ was not satisfied, there would exist a nonzero vector v such that $N_{v \cdot F} < N_F$, according to the upper bound in Relation (10); a contradiction. Inequality (18) is a direct consequence of Inequality (11).

In the following proposition, we specify the functions ϕ_i in order to obtain functions whose nonlinearity can be easily computed.

Proposition 5. Let F be an (n, m)-function defined as in Equation (14).

1. If, for every vector $v \in \mathbb{F}_2^{m*}$, the (s, r)-function $y \mapsto \sum_{i \leq m} v_i \phi_i(y)$ is injective, then F admits $2^{n-1} - 2^{r-1}$ for nonlinearity.
2. If, for every vector $v \in \mathbb{F}_2^{m*}$, the (s, r)-function $y \mapsto \sum_{i \leq m} v_i \phi_i(y)$ takes exactly two times each value of the image set (let us say in the sequel that such function is two-to-one), then F admits $2^{n-1} - 2^r$ for nonlinearity.

Proof. If all the functions $y \mapsto \sum_{i \leq m} v_i \phi_i(y)$ are injective (resp. take exactly two times each value of the image set) when v is a nonzero vector of $\mathbb{F}_2^m$, then all the sets $E_{u,v}$ as defined in Equation (15) are empty or reduced to a singleton (resp. to a pair). We deduce that the maximum achieved by $|\widehat{\chi_F}|$ on $\mathbb{F}_2^n \times \mathbb{F}_2^{m*}$ equals 2^r (resp. 2^{r+1}) and, according to Equation (8), we conclude that N_F equals $2^{n-1} - 2^{r-1}$ (resp. $2^{n-1} - 2^r$).

Remark 6. The injectivity (resp. the two-to-one property) of the (s, r)-functions $y \mapsto \sum_{i \leq m} v_i \phi_i(y)$ $(v \neq \mathbf{0})$ implies $s \leq r$ (resp. $s \leq r + 1$).

The aim of the following Lemma, proposed by Johansson and Pasalic, is to give a way to specify the vectorial functions $\phi_1, \ldots, \phi_m$ to ensure that their nonzero linear combinations satisfy the first hypothesis of Proposition 5.

Lemma 1. [25] Let C be a binary linear $[r, m, t+1]$ code $(m \leq r)$ and let $c_1, \ldots, c_m$ be a basis of C. Let α be a primitive element in the finite field $\mathbb{F}_{2^m}$ and let L_1 be the linear function from $\mathbb{F}_{2^m}$ into C defined by $L_1\left(\sum_{i=1}^m a_i \alpha^{i-1}\right) = \sum_{i=1}^m a_i c_i$. Define $m - 1$ funtions L_i, $i = 2, \ldots, m$, such that for every $0 \leq k \leq 2^m - 2$, $L_i(\alpha^k) = L_1(\alpha^{k+i-1})$. Then, for every vector $v \in \mathbb{F}_2^{m*}$, the function $z \in \mathbb{F}_{2^m} \mapsto \sum_{i=1}^m v_i L_i(z)$ is a bijection from $\mathbb{F}_{2^m}$ into C.

Proof. Since $c_1, \ldots, c_m$ is a basis, L_1 is a bijection. For every vector $v \in \mathbb{F}_2^m$ and every element z of $\mathbb{F}_{2^m}$, we have

$$\sum_{i=1}^m v_i L_i(z) = L_1\left(\left(\sum_{i=1}^m v_i \alpha^{i-1}\right) z\right) \ .$$

The vector v being nonzero, the element $\sum_{i=1}^m v_i \alpha^{i-1}$ is nonzero. Hence, the function $z \in \mathbb{F}_{2^m} \mapsto \sum_{i=1}^m v_i L_i(z)$ is a bijection.

Remark 7. Note that, more generally, if $\beta_1, \ldots, \beta_m$ is a basis of $\mathbb{F}_{2^m}$, and if L_0 is a linear isomorphism between $\mathbb{F}_{2^m}$ and C, then the functions $L_i(z) = L_0(\beta_i z)$, $i = 1, \ldots, m$, have also the property that, for every vector $v \in \mathbb{F}_2^{m*}$, the function $z \in \mathbb{F}_{2^m} \mapsto \sum_{i=1}^m v_i L_i(z)$ is a bijection from $\mathbb{F}_{2^m}$ into C.

Since we have $L_1(0) = L_2(0) = \ldots = L_m(0) = \mathbf{0}$, the functions $L_1, \ldots, L_m$ do not satisfy the hypothesis of Corollary 1 (i.e. the vectors $L_1(z), \ldots, L_m(z)$ are not linearly independent for every $z \in \mathbb{F}_{2^m}$). A solution to derive a family of vectorial functions satisfying the hypothesis of Corollary 1 and one of the two conditions of Proposition 5 is then to right-compose the functions L_i with a same injective - or two-to-one - function π from $\mathbb{F}_2^s$ into $\mathbb{F}_{2^m}^*$. Then, for every nonzero vector $v \in \mathbb{F}_2^{m*}$, the function $y \in \mathbb{F}_2^s \mapsto \sum_{i=1}^m v_i L_i[\pi(y)]$ is injective - or two-to-one -

from $\mathbb{F}_2^s$ into C^*.

For any m-tuple $(L_1, ..., L_m)$ of linear functions from $\mathbb{F}_{2^m}$ into an $[r, m, t+1]$ code defined as in Lemma 1, and for any function π from $\mathbb{F}_2^s$ into $\mathbb{F}_{2^m}^*$, Relation (12) can be rewritten

$$(x, y) \in \mathbb{F}_2^r \times \mathbb{F}_2^s \mapsto x \times \begin{pmatrix} \ell_{11} \circ \pi(y) & \cdots & \ell_{1m} \circ \pi(y) \\ \vdots & \ddots & \vdots \\ \ell_{r1} \circ \pi(y) & \cdots & \ell_{rm} \circ \pi(y) \end{pmatrix} + H(y), \qquad (19)$$

where H is any (s, m)-function and ℓ_{ij} denotes the j-th coordinate function of L_i.

As we show in the following proposition (which is a more general presentation of the results obtained by Johansson and Pasalic in [25], who considered only the case π injective), the nonlinearity and the resiliency of (n, m)-functions defined as in (19) can be easily computed for every function π from $\mathbb{F}_2^s$ into $\mathbb{F}_{2^m}^*$. It can be high if π is injective or two-to-one.

Proposition 6. *Let C be a linear $[r, m, t+1]$ code and let $L_1, ..., L_m$ be m functions defined as in Lemma 1 with respect to C. Let s be any integer, let π be any function from $\mathbb{F}_2^s$ into $\mathbb{F}_{2^m}^*$ and let H be any (s, m)-function. Then, the $(r+s, m)$-function F defined with respect to the functions $L_1, ..., L_m, \pi$ and H as in (19) admits at least t for resiliency order and its nonlinearity satisfies:*

$$N_F = 2^{r+s-1} - 2^{r-1} \max_{z \in \mathbb{F}_{2^m}^*, u' \in \times \mathbb{F}_2^s, v \in \mathbb{F}_2^{m^*}} \left| \sum_{y \in \pi^{-1}(z)} (-1)^{v \cdot H(y) + u' \cdot y} \right|. \qquad (20)$$

If π is injective, then s must be strictly lower than m and N_F equals $2^{r+s-1} - 2^{r-1}$.
If π is two-to-one, then s must be lower than or equal to m and N_F equals $2^{r+s-1} - 2^r$.

Proof. For every nonzero vector $v \in \mathbb{F}_2^m$ and every vector $y \in \mathbb{F}_2^s$, the linear combination $\sum_{i=1}^m v_i [L_i \circ \pi](y)$ belongs to the $[r, m, t+1]$ code C and is nonzero, since $\pi(y)$ never equals zero. We deduce that the vectors $L_1[\pi(y)], ..., L_m[\pi(y)]$ form a basis of C, which implies, according to Corollary 1, that F is at least t-resilient.
On the other hand, let α be a primitive element of $\mathbb{F}_{2^m}$ and let z denote the vector $\pi(y)$. For every nonzero vector $v \in \mathbb{F}_2^m$ and every vector $u \in C^*$ the equation

$$\sum_{i=1}^m v_i L_i(z) = L_1 \left(\sum_{i=1}^m v_i z \alpha^{i-1} \right) = u$$

admits exactly one solution z_0 in $\mathbb{F}_{2^m}^*$, since L_1 is a bijection from $\mathbb{F}_{2^m}^*$ into C^*. Hence, we deduce that $E_{u,v} = \{y \in \mathbb{F}_2^s; \sum_{i=1}^m v_i L_i[\pi(y)] = u\}$ equals $\pi^{-1}(z_0)$

and then, according to Relation (16), we obtain (20). As a direct consequence of Proposition 5, we deduce the nonlinearity of F for the case π injective or two-to-one.

Let F be an $(r + s, m)$-function defined as in (19) with an injective (resp. two-to-one) function π. Since s must be lower than or equal to r in Proposition 6, we deduce that the smaller is the difference $r - s = 2r - n \geq 0$, the larger is the nonlinearity, $2^{n-1} - 2^{r-1}$ (resp. $2^{n-1} - 2^r$), of F. Hence, if we want to construct functions based on Lemma 1 and having the best tradeoff between nonlinearity and resiliency, then we have at the same time to maximize the distance $t + 1$ of the $[r, m, t + 1]$ code and to minimize the value of r (which is greater than or equal to $n/2$). In what follows we give a first construction of resilient highly nonlinear mappings (already described but in more particular ways in [12, 23, 25, 29, 34]; Johansson and Pasalic [25] or Maitra and Pasalic [29] only consider π as an injective function, whereas Nyberg [34], as Chabaud and Vaudenay in [12], focus on the design of bent functions and Gupta and Sarkar consider only one very specific function π as shown in Proposition 7).

Construction 1
Given two integers m and r ($m \leq r$), construct an $[r, m, t+1]$-code C such that t is as large as possible (Brouwer gives in [3] a precise overview of the known codes and of the known bounds on the parameters of these codes). Then, define m linear functions L_1, ..., L_m from $\mathbb{F}_{2^m}$ into C as in Lemma 1, or as in the remark which follows it. Choose an integer s strictly lower than m (resp. lower than or equal to m) and define an injective (resp. two-to-one) function π from $\mathbb{F}_2^s$ into $\mathbb{F}_{2^m}^*$. Choose any (s, m)-function H and denote $r + s$ by n, then the (n, m)-function F whose coordinate functions are defined by $f_i(x, y) = x \cdot [L_i \circ \pi](y) \oplus h_i(y)$ is t-resilient and admits $2^{n-1} - 2^{r-1}$ (resp. $2^{n-1} - 2^r$) for nonlinearity.

Remark 8. In Construction 1, the parameters r, m and s must satisfy either the relation $s < m \leq r$ if π is injective or the relation $s \leq m \leq r$ if π is two-to-one. And the smaller is $r \geq n/2$, the larger is the nonlinearity.

Johansson and Pasalic use in [25] a second construction based on Lemma 1 and involving a family of *nonintersecting codes*, that is a family of codes having the same parameters (same length, same dimension and same minimum distance) and whose pairwise intersection is always reduced to the null vector. In what follows, we give a formalization of this construction and we improve it by considering not only injective functions but also two-to-one ones.

Construction 2
Let $(C_j)_{1 \leq j \leq N}$ be a family of N nonintersecting $[r, m, t + 1]$ codes. For every code C_j, define a family $\left(L_i^{[j]}\right)_{1 \leq i \leq m}$ of linear functions as in Lemma 1. Let s be the greatest integer lower than or equal to $\log_2(N \times (2^m - 1))$ (resp. to $\log_2(N \times (2^m - 1)) + 1$). Let $(E_i)_{1 \leq i \leq N}$ be a partition of $\mathbb{F}_2^s$ such that $\#E_i \leq 2^m - 1$, for $i = 1, ..., N$ (due to the choice of s, such partitions can always be

defined). Define a function π from $\mathbb{F}_2^s$ into $\mathbb{F}_{2^m}^*$ whose restriction to each E_i, $i = 1, ..., N$, is injective (resp. two-to-one). Choose any (s, m)-function H and let n denote the sum $r+s$, then the (n, m)-function F whose coordinate functions are defined by

$$f_i(x, y) = x \cdot \left(\sum_{j=1}^N \delta_{E_j}(y) \times [L_i^{[j]} \circ \pi](y) \right) \oplus h_i(y),$$

where, for every $j \leq N$, δ_{E_j} denotes the indicator function of the set E_j (defined by $\delta_{E_j}(y) = 1$ if $y \in E_j$ and $\delta_{E_j}(y) = 0$ otherwise), is at least t-resilient and admits $2^{n-1} - 2^{r-1}$ (resp. $2^{n-1} - 2^r$) for nonlinearity (the codes have been chosen having a pairwise intersection reduced to $\{0\}$ in order to ensure the injectivity (resp. the two-to-one property) of the functions $y \mapsto \sum_{j=1}^N \delta_{E_j}(y) \times [L_i^{[j]} \circ \pi](y)$, $i = 1, ..., m$).

In Construction 2, the value s can be larger than or equal to m. Clearly, the smaller is the value m and the easier is the definition of a large set of nonintersecting $[r, m, t+1]$ codes (Johansson and Pasalic give in [25] a lower bound on the cardinality of such a family of codes related to the parameters r, m and t). Notice that, when m is close to r and t is high, it is sometimes possible to define a unique $[r, m, t+1]$ code whereas it becomes impossible to define more than one nonintersecting code (in these cases, Construction 2 does not improve Construction 1, since the family of nonintersecting codes is reduced to one element). Johansson and Pasalic give in [25] numerical examples of t-resilient (n, m)-functions defined with Construction 2 for an injective function π, whose nonlinearity is strictly better than the nonlinearity of any (n, m)-function designed as in Construction 1 for the same parameters n and t. However, this construction is often worse than Construction 1 for large resiliency orders. Indeed, in order to define N nonintersecting $[r, m, t+1]$ codes in $\mathbb{F}_2^r$, the parameter r must be clearly greater than or equal to $N \times m$, whereas the parameter t satisfies $t < m$. Hence, the $(r + s, m)$-functions designed should be defined in a vector space of cardinality at least 2^{Nt} in order to have a resiliency order equal to t. Moreover, the difference $r - s$, that we want to minimize is lower bounded by $N \times m - \log_2(N \times (2^m - 1))$ (resp. $N \times m - \log_2(N \times (2^m - 1)) - 1$), that is approximatively by $(N - 1) \times m - \log_2(N)$, and this lower bound increases quickly with N (i.e. with the numbers of nonintersecting codes considered).

3.3 Other Constructions of Highly Nonlinear Resilient Vectorial Functions, Based on the Same Principle as the Maiorana-MacFarland Construction

Two constructions of highly nonlinear resilient vectorial functions respectively based on elliptic curves theory and on the trace of some power functions $x \mapsto x^d$ on finite fields, have been designed respectively by Cheon [13] and by Nyberg [34–36] (see also Khoo, Gong [26]). However, it is still an open problem to design highly nonlinear functions with high algebraic degree and high resiliency order with Cheon's method. On the other hand, the number of functions which can be designed by Nyberg and Cheon's methods are very small and the resiliency of the designed functions is difficult to study.

In fact, there exist actually only two main constructions of a large set of resilient highly nonlinear mappings: the Maiorana-MacFarland construction recalled in Section 3.2 and a second one proposed by Zhang and Zheng in [43, 44]. Zhang and Zheng's construction consists in the composition of a linear resilient (n, m, t)-function with a highly nonlinear permutation on $\mathbb{F}_2^m$ and it is based on the second part of Remark 1. Since it assumes the existence of a previously defined highly nonlinear function, it is a secondary construction.

Construction 3 *[44] Let L be a linear surjective (n, m, t)-function and let G be an (m, k)-function whose nonlinearity is denoted by N_G. Then the (n, k)-function $F = G \circ L$ is t-resilient, admits $2^{n-m} N_G$ for nonlinearity (where N_G denotes the nonlinearity of G) and its degree is the same as that of G.*

Remark 9. In [23], Gupta and Sarkar modified the Zhang, Zheng Construction 3 to design nonlinear functions with degree $d > m$. The first construction of highly nonlinear (n, m)-functions with algebraic degree strictly greater than m was obtained by Cheon in [13] by combining linear codes with inverse linear functions on finite field. But Gupta and Sarkar achieved this in an easier way by simply dropping some output bits in functions defined as in Construction 3.

Clearly, the nonlinearity of an (n, m, t)-function $F = G \circ L$ constructed using Construction 3 is maximal if the nonlinearity of G is maximal. Hence, if G is a permutation, then due to Chabaud-Vaudenay's bound recalled in section 2, the nonlinearity N_F of F is upper bounded by $2^{n-m}(2^{m-1} - 2^{\frac{m-1}{2}}) = 2^{n-1} - 2^{n-\frac{m+1}{2}}$, this bound being tight if and only if m is odd and G is almost bent. In [44], taking for function G the inverse function $x \mapsto x^{-1}$ on the finite Field $\mathbb{F}_{2^m}$ studied by Nyberg in [36], Zhang and Zheng obtained t-resilient functions having a nonlinearity greater than or equal to $2^{n-1} - 2^{n-m/2}$ and having $m - 1$ for algebraic degree.

The linear (n, m)-functions involved in the construction of Zhang and Zheng introduce a weakness which could be used to attack a system implementing them as cryptographic functions (for instance, it has been proved in [10] that their *unrestricted nonlinearity* is null and then, that this kind of functions cannot be used as a multi-output combination function in stream ciphers). However, this drawback can be avoided by concatenating such functions (obtained through Construction 3). Recall indeed that the concatenation of t-resilient functions is t-resilient. And a good nonlinearity can be obtained by concatenating functions with disjoint Walsh supports. We obtain this way a modified Maiorana-MacFarland's construction.

All the other primary constructions presented in $[25, 28, 29, 35]$ are based on Construction 1. As we show in what follows, the recent construction of (n, m, t)-functions defined by Gupta and Sarkar and presented as better than the previously known constructions are also obtained as a particular application of Construction 1 and is a secondary construction, since it assumes the existence of previously defined highly nonlinear functions. Les us recall Gupta and Sarkar's Construction.

Proposition 7. *[23] Let C be an $[r, m, t+1]$-linear code and let $L_1, \ldots, L_m$ be m linear functions defined as in Lemma 1 with respect to C and with a primitive element α of $\mathbb{F}_{2^m}$. Let s be an integer strictly lower than m and, for any integer p, denote $r + s + p$ by n. Define a function τ from $\mathbb{F}_2^s$ into $\mathbb{F}_{2^m}^*$ by $\tau(y) = \alpha^{\sum_{1 \leq i \leq s} y_i 2^{i-1}}$, then, for any (p, m)-function H, the (n, m)-function F whose coordinate functions f_i are defined by $f_i(x, y, y') = x \cdot [L_i \circ \tau](y) \oplus h_i(y')$ is t-resilient and admits $2^{n-1} - 2^{r-1}(2^p - 2N_H)$ for nonlinearity.*

In Proposition 7, let π denote the function defined from $\mathbb{F}_2^s \times \mathbb{F}_2^p$ into $\mathbb{F}_{2^m}$ by $\pi(y, y') = \tau(y)$. On the other hand, let H' be the $(s + p, m)$-function defined by $H'(y, y') = H(y')$. Then, the coordinate functions of the function F defined in Proposition 7 can be rewritten $f_i(x, (y, y')) = x \cdot [L_i \circ \pi](y, y') \oplus h_i'(y, y')$. Thus, the functions defined by Gupta and Sarkar belong to $\mathcal{M}_{n,m}$ and are particular cases of Relation (19). The resiliency of the constructed functions is clearly a direct consequence of Lemma 1 and one can already notice that Proposition 7 remains true if one assumes that the function τ is any injective or two-to-one function (in this case the value s in Proposition 7 can be equal to m) instead of the function $y \in \mathbb{F}_2^s \mapsto \alpha^{\sum_{1 \leq i \leq s} y_i 2^{i-1}}$. Applying Equation (20) to F, one obtains

$$N_F = 2^{r+s+p-1} - 2^{r-1} \max_{\substack{z \in \mathbb{F}_{2^m}, v \in \mathbb{F}_2^{m*} \\ (u', u'') \in \mathbb{F}_2^s \times \mathbb{F}_2^p}} \left| \sum_{y \in \pi^{-1}(z)} (-1)^{v \cdot H'(y,y') + u' \cdot y + u'' \cdot y'} \right|.$$

For every $z \in \mathbb{F}_{2^m}$, we have $\pi^{-1}(z) = \tau^{-1}(z) \times \mathbb{F}_2^p$, if $z \in \tau(\mathbb{F}_2^s)$ and $\tau^{-1}(z) = \emptyset$ otherwise. Since τ is injective (resp. two-to-one) from $\mathbb{F}_2^s$ into $\mathbb{F}_{2^m}^*$, we deduce that $\sum_{y \in \pi^{-1}(z)} (-1)^{v \cdot H'(y,y') + u' \cdot y + u'' \cdot y'}$ can be rewritten

$$\left(\sum_{y \in \tau^{-1}(z)} (-1)^{u' \cdot y} \right) \sum_{y' \in \mathbb{F}_2^p} (-1)^{v \cdot H(y') + u'' \cdot y'} \quad ,$$

which implies $N_F = 2^{r+s+p-1} - 2^{r-1}(2^p - 2N_H)$ (resp. $N_F = 2^{r+s+p-1} - 2^r(2^p - 2N_H)$).

3.4   Vectorial Maiorana-MacFarland (n, m)-Functions with $m > n/2$

Due to Remark 8 the construction of balanced highly nonlinear (n, m)-functions inside of $\mathcal{M}_{n,m}$ cannot be easily done using Corollary 1 and Proposition 5 when

m is strictly greater than $n/2$. The construction of Zhang and Zheng does not provide full satisfaction as we recalled in Section 3.3, and its modification by concatenating such functions extends the value of n and, in practice, makes m smaller than $n/2$. Hence, there is a need for a primary construction of highly nonlinear resilient (n, m)-functions such that $m > n/2$.

Definition 3. *We denote by $\mathcal{M}^\star_{n,m}$ the set of functions F defined from a product space $\mathbb{F}^n_2 = \mathbb{F}^r_2 \times \mathbb{F}^s_2$ into a space $\mathbb{F}^m_2$ and which can be written in the form*

$$F(x, y) = (F'(x, y), T(y)), \tag{21}$$

where F' is an $(r + s, p)$-function belonging to $\mathcal{M}_{r+s,p}$ (p is an integer strictly lower than m) and is defined as in (12), and T is any $(s, m - p)$-function.

According to Relation (15), the Walsh transform $\widehat{\chi_F}$ of F satisfies the relation

$$\widehat{\chi_F}\left((u, u'), (v, v')\right) = 2^r \sum_{y \in E_{u,v}} (-1)^{v \cdot H(y) + v' \cdot T(y) + u' \cdot y}. \tag{22}$$

We deduce the following proposition.

Proposition 8. *Let F be an element of $\mathcal{M}^\star_{r+s,m}$ defined as in Equation (21). Then, the nonlinearity of F is upper bounded by $2^r N_T$. Moreover, if the set $E_{u,v}$ has cardinality lower than or equal to 2 for every nonzero vector $v \in \mathbb{F}^p_2$ and every vector $u \in \mathbb{F}^r_2$, and if $s \geq 1$, then N_F equals $2^r N_T$.*

Proof. Let n denote the sum $r + s$. Due to Equation (8), we have

$$\max_{u' \in \mathbb{F}^s_2, v' \in \mathbb{F}^{m-p}_2{}^*} |\widehat{\chi_T}(u', v')| = 2^s - 2N_T$$

and

$$N_F = 2^{n-1} - \frac{1}{2} \max_{(u,u') \in \mathbb{F}^n_2, (v,v') \in \mathbb{F}^m_2{}^*} |\widehat{\chi_F}((u, u'), (v, v'))|.$$

We have

$$\max_{\substack{(u,u') \in \mathbb{F}^n_2, \\ (v,v') \in \mathbb{F}^m_2{}^*}} |\widehat{\chi_F}((u, u'), (v, v'))| \geq \max_{\substack{(u,u') \in \mathbb{F}^n_2, \\ v' \in \mathbb{F}^{m-p}_2{}^*}} |\widehat{\chi_F}((u, u'), (\mathbf{0}, v'))|$$

$$= 2^r \max_{u' \in \mathbb{F}^s_2, v' \in \mathbb{F}^{m-p}_2{}^*} |\widehat{\chi_T}(u', v')|$$

and, hence,

$$N_F = 2^{n-1} - \frac{1}{2} \max_{(u,u') \in \mathbb{F}^n_2, (v,v') \in \mathbb{F}^m_2{}^*} |\widehat{\chi_F}((u, u'), (v, v'))|$$

$$\leq 2^{n-1} - 2^{r-1} \max_{u' \in \mathbb{F}^s_2, v' \in \mathbb{F}^{m-p}_2{}^*} |\widehat{\chi_T}(u', v')| = 2^r N_T.$$

If the set $E_{u,v}$ has cardinality lower than or equal to 2 for every nonzero vector $v \in \mathbb{F}^p_2$ and every vector $u \in \mathbb{F}^r_2$, then Equation (22) implies that $|\widehat{\chi_F}|$ takes its values in the set $\{0, 2^r, 2^{r+1}\}$. Hence, the maximum value of $|\widehat{\chi_F}|$ under the condition $v \neq \mathbf{0}$ is lower than or equal to 2^r times the maximum value of $|\widehat{\chi_T}|$. Thus $N_F = 2^r N_T$.

A direct consequence of Relation (22) is

Proposition 9. *Let F be an element of $\mathcal{M}^*_{r+s,m}$ defined with respect to an $(r+s,p)$-function F' and an $(s, m-p, t')$-function T as in Equation (21). If F' satisfies the hypothesis of Corollary 1 (i.e. the family $(\phi_i(y))_{i\le p}$ is a basis of a p-dimensional subspace of $\mathbb{F}_2^r$ having $t+1$ for minimum Hamming weight) for some integer t, then the resiliency order of F is greater than or equal to $\min(t, t')$.*

Proposition 9 provides a secondary construction, because it uses a function T (supposed to be highly resilient and to have high nonlinearity). Obviously, T can be constructed according to the same principle as F. This leads to the following construction of (n, n)-functions (it would be also a simple matter to describe a construction of (n, m)-functions with $m \ne n$), obtained by applying recursively Proposition 9 with successive values of p equal to $n/2$, $n/4$, and so on.

Construction 4
Let k, n and n' be three integers such that $n = 2^k n'$. Let P be a permutation on $\mathbb{F}_2^{n'}$ and let F_i, $i = 0, 1, ..., k-1$, be balanced $(\frac{n}{2^i}, \frac{n}{2^{i+1}})$-functions designed as in Construction 1 (due to Lemma 1 such functions can always be defined). Then, the (n, n)-function $F = (F_0, F_1, ..., F_{k-1}, P)$ is balanced, admits $2^{n/2 + n/4 + ... + n/2^k} N_P$ for nonlinearity and its minimum algebraic degree $\deg_m$ is lower bounded by

$$\min\left(\deg_m(F_0), \deg_m(F_1), ..., \deg_m(F_{k-1}), \deg_m(P)\right)$$

(which is upper bounded by n').

4   Covering Sequences of Vectorial Functions

4.1   The Boolean Case

In [11], is introduced the notion of covering sequence of Boolean functions.

Definition 4. *[11] A covering sequence of a function $f : \mathbb{F}_2^n \mapsto \mathbb{F}_2$ is any real-valued function φ on $\mathbb{F}_2^n$ such that the real summation $\sum_{a \in \mathbb{F}_2^n} \varphi(a) D_a f$ is equal to a constant function ρ. The value of ρ is called the level of this sequence. If $\rho \ne 0$, then we say that the covering sequence is non-trivial.*

Notice that, for every $x \in \mathbb{F}_2^n$, the sum $\sum_{a \in \mathbb{F}_2^n} \varphi(a) D_a f(x)$ equals $\sum_{D_a f(x)=1} \varphi(a)$ and that, if it is constant, then the sum $\sum_{D_a f(x)=0} \varphi(a)$ is also constant. This will lead us to a natural generalization to (n, m)-functions. Before that, we recall in the next proposition the complete characterization given in [11] of the balancedness of Boolean functions by means of their covering sequences.

Proposition 10. *[11] Any Boolean function f on $\mathbb{F}_2^n$ is balanced if and only if it admits at least one non-trivial covering sequence. The same covering sequence – the constant sequence 1 – can be taken for all balanced functions. The level of this sequence with respect to any balanced function equals 2^{n-1}.*

The relevance of the notion of covering sequence to the study of Boolean functions is shown in [11], by characterizing it by means of the Walsh transform and by giving complete characterizations of the correlation-immunity and of the resiliency of Boolean functions with respect to their covering sequences (see generalizations in Theorem 2 and Corollary 2).

4.2   The Multiple Output Case

In the following definition, we extend the notion of covering sequence to the case of (n, m)-functions.

Definition 5. *We call covering sequence of an (n, m)-function F, a pair of functions (φ, ψ) from, respectively, $\mathbb{F}_2^n$ and $\mathbb{F}_2^m$ into $\mathbb{R}$, such that:*

$$\forall x \in \mathbb{F}_2^n, \forall b \in \mathbb{F}_2^m, \sum_{a \in \mathbb{F}_2^n;\ D_a F(x) = b} \varphi(a) = \psi(b). \tag{23}$$

Remark 10. A covering sequence (φ, ψ) of an (n, m)-function F satisfies the relation

$$\#Supp\ \psi \le \#Supp\ \varphi.$$

Indeed, let x be an element of $\mathbb{F}_2^n$. For every b in $\mathbb{F}_2^m$, we denote by $A_b(x)$ the set $\{a \in Supp\ \varphi; D_a F(x) = b\}$. According to Relation (23), if a vector $b \in \mathbb{F}_2^m$ belongs to $Supp\ \psi$ then, for every $x \in \mathbb{F}_2^n$, the set $A_b(x)$ is nonempty. By definition of covering sequences, for every x of $\mathbb{F}_2^n$, the family $(A_b(x))_{b \in Supp\ \psi}$ is a partition of a subset of $Supp\ \varphi$ and we conclude that the cardinality of the set $Supp\ \varphi$ is greater than or equal to the cardinality of the set $Supp\ \psi$.

The natural generalization of the notion of non-trivial covering sequence will appear below. To use the properties of the Walsh transform, we will need the following lemma.

Lemma 2. *Let F be an (n, m)-function. A pair of numerical functions φ, ψ from, respectively, $\mathbb{F}_2^n$ and $\mathbb{F}_2^m$ into $\mathbb{R}$ is a covering sequence of F if and only if, for every vector $v \in \mathbb{F}_2^{m*}$, we have*

$$\varphi \otimes \chi_F(\cdot, v) = \widehat{\psi}(v)\, \chi_F(\cdot, v), \tag{24}$$

where $\chi_F(\cdot, v)$ denotes the function $x \mapsto \chi_F(x, v) = (-1)^{v \cdot F(x)}$.

Proof. Relation (23) is equivalent to the fact that the functions

$$b \mapsto \sum_{a \in \mathbb{F}_2^n;\ D_a F(x)=b} \varphi(a)$$

and $b \mapsto \psi(b)$ are equal. Since two numerical functions are equal to each other if and only if their Fourier transforms are equal to each other, Relation (23) is equivalent to:

$$\forall v \in \mathbb{F}_2^{m*},\ \forall x \in \mathbb{F}_2^n,\ \sum_{a \in \mathbb{F}_2^n} \varphi(a)(-1)^{v \cdot D_a F(x)} = \sum_{b \in \mathbb{F}_2^m} \psi(b)(-1)^{v \cdot b},$$

that is to,

$$\forall v \in \mathbb{F}_2^{m*},\ \forall x \in \mathbb{F}_2^n,\ \sum_{a \in \mathbb{F}_2^n} \varphi(a)(-1)^{v \cdot F(x+a)} = (-1)^{v \cdot F(x)} \widehat{\psi}(v),$$

or equivalently,

$$\forall v \in \mathbb{F}_2^{m*},\ \varphi \otimes \chi_F(\cdot, v) = \widehat{\psi}(v)\, \chi_F(\cdot, v).$$

Remark 11. By summing up Relation (23) with b ranging over $\mathbb{F}_2^m$, we deduce $\widehat{\varphi}(\mathbf{0}) = \widehat{\psi}(\mathbf{0})$.

The following theorem generalizes Proposition 10 and leads to a natural definition of non-trivial covering sequence of vectorial functions.

Theorem 1. *An (n, m)-function F is balanced if and only if it admits at least one covering sequence (φ, ψ) satisfying $\widehat{\psi}(v) \neq \widehat{\varphi}(\mathbf{0})$ for every nonzero vector v of $\mathbb{F}_2^m$. Any balanced (n, m)-function F admits the pair of constant functions $(1, 2^{n-m})$ for covering sequence.*

Proof. Assume that (φ, ψ) is a covering sequence of F, then Equation (24) is satisfied:

$$\forall v \in \mathbb{F}_2^{m*},\ \forall x \in \mathbb{F}_2^n,\ [\varphi \otimes \chi_F(\cdot, v)](x) = \widehat{\psi}(v)\, \chi_F(x, v).$$

Summing up with x ranging over $\mathbb{F}_2^n$, that is, computing the values at $\mathbf{0}$ of the Fourier transform of both sides of this functional equality, and using Relation (7), we obtain:

$$\forall v \in \mathbb{F}_2^{m*},\ \widehat{\varphi}(\mathbf{0})\widehat{\chi_F}(\mathbf{0}, v) = \widehat{\psi}(v)\widehat{\chi_F}(\mathbf{0}, v),$$

that is, $(\widehat{\varphi}(\mathbf{0}) - \widehat{\psi}(v))\widehat{\chi_F}(\mathbf{0}, v) = 0$ for every nonzero vector $v \in \mathbb{F}_2^m$. If $\widehat{\varphi}(\mathbf{0}) - \widehat{\psi}(v)$ is nonzero for every $v \in \mathbb{F}_2^{m*}$, then the function $v \mapsto \widehat{\chi_F}(\mathbf{0}, v)$ is null on $\mathbb{F}_2^{m*}$, which implies that F is balanced (see Remark 1).
Conversely, if F is balanced, then, for every pair $(b, x) \in \mathbb{F}_2^m \times \mathbb{F}_2^n$, the cardinality of the set $\{a \in \mathbb{F}_2^n; D_a F(x) = b\}$ is constant equaling 2^{n-m} since the equation $D_a F(x) = b$ is equivalent to $F(x + a) = b + F(x)$. Let $\varphi : \mathbb{F}_2^n \mapsto \mathbb{R}$ and $\psi : \mathbb{F}_2^m \mapsto \mathbb{R}$ be respectively the constant function $x \mapsto 1$ and the constant function $y \mapsto 2^{n-m}$, then the pair (φ, ψ) is a covering sequence of F satisfying the relation $\widehat{\psi}(v) = 0 \neq \widehat{\psi}(\mathbf{0}) = \widehat{\varphi}(\mathbf{0}) = 2^{n-m}$ for every element v of $\mathbb{F}_2^{m*}$.

Definition 6. *A covering sequence* (φ, ψ) *of an* (n, m)-*function* F *is said to be non-trivial if* $\widehat{\psi}(v)$ *never equals* $\widehat{\varphi}(\mathbf{0})$ *(that is* $\widehat{\psi}(\mathbf{0})$*) when* v *ranges over* $\mathbb{F}_2^{m*}$.

Thus, according to Theorem 1, an (n, m)-function F is balanced if and only it admits a non-trivial covering sequence.

Remark 12. If ψ is a function from $\mathbb{F}_2^m$ into $\mathbb{R}_+$, then we have $\widehat{\psi}(v) \neq \widehat{\psi}(\mathbf{0})$ for every element v of $\mathbb{F}_2^{m*}$ if and only if the support of ψ has rank m (i.e. spans the whole vector space $\mathbb{F}_2^m$). Indeed, we have

$$\forall v \in \mathbb{F}_2^{m*}, \ \widehat{\psi}(v) \neq \widehat{\psi}(\mathbf{0}) \Longleftrightarrow \forall v \in (\mathbb{F}_2^m)^*, \ \sum_{b \in \mathbb{F}_2^m, b \in (v^\perp)^c} \psi(b) \neq 0$$

and, since $\psi(b) \geq 0, \forall b \in \mathbb{F}_2^m$, this relation is equivalent to saying that the support of ψ is not included in a linear hyperplane of $\mathbb{F}_2^m$.

To illustrate the fact that, in Theorem 1, the condition that the covering sequence is non-trivial is actually necessary, we give in the next proposition an example of an unbalanced function having a trivial covering sequence.

Proposition 11. *Let* F *be a balanced* (n, m)-*function admitting* (φ, ψ) *for covering sequence. Define an* $(n, m+1)$-*function* F^* *by* $F^*(x) = (F(x), 0)$. *Let* ψ^* *be real function on* $\mathbb{F}_2^{m+1}$ *such that* $\widehat{\psi^*}(v_1, ..., v_{m+1}) = \widehat{\psi}(v_1, ..., v_m)$ *if* $(v_1, ..., v_m) \neq \mathbf{0}$ *and* $\widehat{\psi^*}(\mathbf{0}, 1) = \widehat{\psi^*}(\mathbf{0}, 0) = \widehat{\psi}(\mathbf{0})$. *Then* F^* *admits* (φ, ψ^*) *for trivial covering sequence.*

Proof. For every vector $x \in \mathbb{F}_2^n$ and every nonzero vector $(v_1, ..., v_m, v_{m+1})$ of $\mathbb{F}_2^{m+1}$, we have, after denoting $(v_1, ..., v_m)$ by v,

$$[\varphi \otimes \chi_{F^*}(\cdot, (v, v_{m+1}))](x) = \sum_{a \in \mathbb{F}_2^n} \varphi(a) \, (-1)^{v \cdot F(x+a)} = [\varphi \otimes \chi_F(\cdot, v)](x),$$

since the $(m+1)$-th coordinate function of F^* is null. The pair (φ, ψ) being a covering sequence for F, the convolutional product $[\varphi \otimes \chi_F(\cdot, v)](x)$ equals $\widehat{\psi}(v) \chi_F(x, v)$ if $v \neq \mathbf{0}$ and equals $\widehat{\varphi}(\mathbf{0}) = \widehat{\psi}(\mathbf{0})$ if $v = \mathbf{0}$. Notice that, if v equals $\mathbf{0}$, then $v_{m+1} = 1$ since $(v, v_{m+1}) \neq \mathbf{0}$. If v differs from $\mathbf{0}$, then we have:

$$\varphi \otimes \chi_{F^*}(\cdot, (v, v_{m+1})) = \widehat{\psi}(v) \chi_F(\cdot, v) = \widehat{\psi^*}(v, v_{m+1}) \chi_{F^*}(\cdot, (v, v_{m+1})) \ .$$

Otherwise, if v equals $\mathbf{0}$, then we showed that $\varphi \otimes \chi_{F^*}(\cdot, (v, 1))$ equals $\widehat{\varphi}(\mathbf{0})$, that is $\widehat{\psi^*}(\mathbf{0}, 1)$, by construction of ψ^*. Hence, Equation (24) being satisfied by the function χ_{F^*} and the pair (φ, ψ^*), we deduce that (φ, ψ^*) is a covering sequence of F^* and, since $\widehat{\psi^*}(\mathbf{0}, 1)$ equals $\widehat{\varphi}(\mathbf{0})$, this covering sequence is trivial.

As we proved in Theorem 1, the same covering sequence $(1, 2^{n-m})$ can be taken for every balanced (n, m)-function. Finding a second covering sequence is a difficult problem in the general case. We show in the following proposition that the elements of $\mathcal{M}_{n,m}$ which satisfy the hypothesis of Corollary 1, admit several covering sequences.

Proposition 12. *Let F be an (n, m)-function derived from the (s, r)-functions $\phi_1, ..., \phi_m$ and the (s, m)-function H as in Equation (14). If the vectors $\phi_1(y), ..., \phi_m(y)$ are linearly independent for every $y \in \mathbb{F}_2^s$, then F admits the pair $(\delta_{\mathbb{F}_2^r \times \{\mathbf{0}\}}, 2^{r-m})$ for covering sequence.*

Proof. For every pair $(x, y) \in \mathbb{F}_2^r \times \mathbb{F}_2^s$, the function $a \mapsto D_{(a,\mathbf{0})}F(x, y)$ from $\mathbb{F}_2^r$ into $\mathbb{F}_2^m$ admits the Boolean functions $a \mapsto a \cdot \phi_1(y), \dots, a \mapsto a \cdot \phi_m(y)$ for coordinate functions. This implies that the function $a \mapsto D_{(a,\mathbf{0})}F(x, y)$ is a linear (r, m)-function and, since the vectors $\phi_1(y), ..., \phi_m(y)$ are assumed to be linearly independent for every $y \in \mathbb{F}_2^s$, then this linear function is balanced for every $y \in \mathbb{F}_2^s$. Hence, each element $b \in \mathbb{F}_2^m$ has the same number, 2^{r-m}, of pre-images by the function $a \mapsto D_{(a,\mathbf{0})}F(x, y)$ for any pair $(x, y) \in \mathbb{F}_2^r \times \mathbb{F}_2^s$ and we conclude that

$$\forall (x, y) \in \mathbb{F}_2^r \times \mathbb{F}_2^s, \forall b \in \mathbb{F}_2^m, \sum_{\substack{(a,a') \in \mathbb{F}_2^r \times \mathbb{F}_2^s \\ D_{(a,a')}F(x,y)=b}} \delta_{\mathbb{F}_2^r \times \{\mathbf{0}\}}(a, a') = 2^{r-m}.$$

In the following theorem, we generalize to a vectorial function F the characterization, given in [11], of its covering sequences by means of their Fourier transforms and of the Walsh support of F.

Theorem 2. *Let F be an (n, m)-function and let (φ, ψ) be any pair of real-valued functions respectively defined on $\mathbb{F}_2^n$ and on $\mathbb{F}_2^m$. Then, F admits (φ, ψ) for covering sequence if and only if, for every pair (u, v) belonging to $Supp\ \widehat{\chi_F}$, we have $\widehat{\varphi}(u) = \widehat{\psi}(v)$.*

Proof. The proof is very similar to the proof of Theorem 1: the pair (φ, ψ) is a covering sequence of F if and only if Relation (24) is satisfied; due to the bijectivity of the Fourier transform, for every nonzero vector $v \in \mathbb{F}_2^m$, the functions $\widehat{\psi}(v) \chi_F(\cdot, v)$ and $\varphi \otimes \chi_F(\cdot, v)$ are equal if and only if their Fourier transforms on $\mathbb{F}_2^n$ are equal. Thus, according to Relation (7), the pair (φ, ψ) is a covering sequence of F if and only if

$$\forall v \in \mathbb{F}_2^{m*}, \forall u \in \mathbb{F}_2^n, \widehat{\varphi}(u)\widehat{\chi_F}(u, v) = \widehat{\psi}(v)\,\widehat{\chi_F}(u, v),$$

that is, if and only if

$$((u, v) \in Supp\ \widehat{\chi_F}\ \text{and}\ v \neq \mathbf{0}) \Longrightarrow \left(\widehat{\varphi}(u) = \widehat{\psi}(v)\right).$$

Moreover, if $v = \mathbf{0}$, then $\widehat{\chi_F}(u, \mathbf{0})$ equals 0 for every nonzero vector $u \in \mathbb{F}_2^n$ and equals 2^n if $u = \mathbf{0}$. Since $\widehat{\varphi}(\mathbf{0}) = \widehat{\psi}(\mathbf{0})$, we obtain $((u, \mathbf{0}) \in Supp\ \widehat{\chi_F}) \Longrightarrow \widehat{\varphi}(u) = \widehat{\psi}(\mathbf{0})$ and we conclude that (φ, ψ) is a covering sequence of F if and only if

$$\forall (u, v) \in Supp\ \widehat{\chi_F}, \widehat{\varphi}(u) = \widehat{\psi}(v).$$

Corollary 2. *Let F be an (n, m)-function admitting (φ, ψ) for covering sequence. If the sets $\widehat{\varphi}(\{u \in \mathbb{F}_2^n / w_H(u) \leq t\})$ and $\widehat{\psi}(\mathbb{F}_2^{m*})$ are disjoint, then F is t-resilient.*

Remark 13. As in [11], a converse of Corollary 2 can be proven.

As a consequence of Theorem 2, we also deduce a property valid for a large class of functions including all balanced functions.

Corollary 3. *Let F be an (n, m)-function and let (φ, ψ) be a covering sequence for F. Assume that the functions φ and ψ are, respectively, different from the zero function on $\mathbb{F}_2^n$ and different from the zero function on $\mathbb{F}_2^{m*}$, then:*
1. For every vector $u \in \mathbb{F}_2^n$, there exists $v \in \mathbb{F}_2^{m*}$ such that $\widehat{\chi_F}(u, v) = 0$,
2. For every vector $v \in \mathbb{F}_2^m$, there exists a vector $u \in \mathbb{F}_2^n$ such that $\widehat{\chi_F}(u, v) = 0$.

Proof. 1. Suppose that there exists a vector $u \in \mathbb{F}_2^n$ such that $\widehat{\chi_F}(u, v) \neq 0$, $\forall v \in \mathbb{F}_2^{m*}$. Then, the set $\{u\} \times \mathbb{F}_2^{m*}$ being included in $Supp \, \widehat{\chi_F}$, we have, according to Theorem 2,

$$\forall v \in \mathbb{F}_2^{m*}, \widehat{\psi}(v) = \widehat{\varphi}(u).$$

Due to the inverse formula, this implies that $\psi(b)$ equals 0 for every nonzero vector b, which contradicts the hypothesis on ψ.
2. Suppose that there exists a vector $v \in \mathbb{F}_2^m$ such that $\widehat{\chi_F}(u, v) \neq 0$, $\forall u \in \mathbb{F}_2^n$. Then, the set $\mathbb{F}_2^n \times \{v\}$ being included in $Supp \, \widehat{\chi_F}$, we have, according to Theorem 2,

$$\forall u \in \mathbb{F}_2^n, \widehat{\varphi}(u) = \widehat{\psi}(v).$$

Due to the inverse formula, this implies that $\varphi(a)$ equals 0 for every vector $a \in \mathbb{F}_2^n$, which contradicts hypothesis on φ.

Remark 14. We excluded the case $v = \mathbf{0}$ in the first assumption of Corollary 3 since this value gives no information about the function F. Indeed, for any (n, m)-function F, we have $\widehat{\chi_F}(u, \mathbf{0}) = \mathbf{0}$ if $u \neq \mathbf{0}$ and $\widehat{\chi_F}(u, \mathbf{0}) = 2^n$ if $u = \mathbf{0}$, and the first assumption is trivially true for every nonzero vector u if we accept $v = \mathbf{0}$.

Due to Corollary 3, if an (n, m)-function F admits a covering sequence (φ, ψ) such that ψ is not the zero function on $\mathbb{F}_2^{m*}$, then for every vector $u \in \mathbb{F}_2^n$, there exists a nonzero linear combination $\sum_{i \leq m} v_i f_i$ of its coordinate functions whose sum with the Boolean linear function $x \mapsto u \cdot x$ is balanced. This is true in particular for any balanced function F, since, if (φ, ψ) is non-trivial, then ψ has this property.

4.3   Stability of Covering Sequences

In this section, we show that one can deduce, in some cases, a covering sequence of a function, either defined by composition or by concatenation, from covering sequences of the functions involved in the definition of the function. We show that this property is particularly relevant for the study of covering sequences of the elements of $\mathcal{M}_{n,m}^*$ and that it implies, for instance, that some permutations inside of $\mathcal{M}_{n,m}^*$ cannot be used as a round function in an iterated block cipher (we will explain why in Section 5).

Proposition 13. *Let* $F : \mathbb{F}_2^n \mapsto \mathbb{F}_2^m$ *and* $G : \mathbb{F}_2^m \mapsto \mathbb{F}_2^k$ *be two functions admitting respectively* (φ, ψ) *and* (ψ, θ) *for covering sequences. Then,* (φ, θ) *is a covering sequence of* $G \circ F$.

Proof. For every pair $(x, a) \in \mathbb{F}_2^n \times \mathbb{F}_2^n$, we have, denoting $D_a F(x)$ by b:

$$D_a[G \circ F](x) = G(F(x)) + G(F(x + a)) = G(F(x)) + G(F(x) + b) =$$

$$(D_b G)(F(x)).$$

Thus, for every pair $(x, c) \in \mathbb{F}_2^n \times \mathbb{F}_2^k$, we have:

$$\sum_{a \in \mathbb{F}_2^n, D_a[G \circ F](x) = c} \varphi(a) = \sum_{b \in \mathbb{F}_2^m, (D_b G)(F(x)) = c} \left(\sum_{a \in \mathbb{F}_2^n, D_a F(x) = b} \varphi(a) \right).$$

For every pair $(x, b) \in \mathbb{F}_2^n \times \mathbb{F}_2^m$, we have $\sum_{a \in \mathbb{F}_2^n, D_a F(x) = b} \varphi(a) = \psi(b)$ and thus, we get,

$$\sum_{a \in \mathbb{F}_2^n, D_a[G \circ F](x) = c} \varphi(a) = \sum_{b \in \mathbb{F}_2^m, (D_b G)(F(x)) = c} \psi(b).$$

Let y denote $F(x)$, then:

$$\sum_{a \in \mathbb{F}_2^n, D_a[G \circ F](x) = c} \varphi(a) = \sum_{b \in \mathbb{F}_2^m, D_b G(y) = c} \psi(b).$$

Since (ψ, θ) is a covering sequence of G, the sum $\sum_{b \in \mathbb{F}_2^m, D_b G(y) = c} \psi(b)$ takes constant value $\theta(c)$ for every pair $(y, c) \in \mathbb{F}_2^m \times \mathbb{F}_2^k$ and we deduce

$$\forall x \in \mathbb{F}_2^n, \ \forall c \in \mathbb{F}_2^k, \quad \sum_{a \in \mathbb{F}_2^n, D_a G \circ F(x) = c} \varphi(a) = \theta(c).$$

Remark 15. In particular, if a function F admits two covering sequences (φ_1, ψ_1) and (φ_2, ψ_2) such that $\psi_1 = \varphi_2$, then the function $F \circ F$ admits (φ_1, ψ_2) for covering sequence.

One can easily check that the $(r+s, m+k)$-function F obtained by concatenating the outputs of an (r, m)-function H having (φ, ψ) for covering sequence with the outputs of an independent (s, k)-function T having (φ', ψ') for covering sequence, defined by $F(x, y) = (H(x), T(y))$, admits (φ'', ψ'') for covering sequence, where $\varphi''(a, a') = \varphi(a)\varphi'(a')$ and $\psi''(b, b') = \psi(b)\psi'(b')$. When the inputs of H and T are not independent, it is generally not easy to deduce a covering sequence of F from those of H and T. The aim of the following proposition is to introduce a condition on the covering sequences of H and of T ensuring that the computation of a covering sequence of F is feasible. We will see in Corollary 4, that the hypothesis of this proposition, which seem to be very strong, are satisfied by the elements of $\mathcal{M}_{n,m}^*$.

Proposition 14. *Let F' be an $(r + s, p)$-function and let T be an $(s, m - p)$ function admitting a covering sequence (φ', ψ'). Assume that there exists φ : $\mathbb{F}_2^r \times \mathbb{F}_2^s \mapsto \mathbb{R}$ and $\psi : \mathbb{F}_2^p \mapsto \mathbb{R}$ such that, for every $a' \in Supp\ \varphi' \subseteq \mathbb{F}_2^s$, F' admits $(\varphi(\cdot, a'), \psi)$ for covering sequence. Let φ'' and ψ'' be the real-valued functions respectively defined on $\mathbb{F}_2^r \times \mathbb{F}_2^s$ and on $\mathbb{F}_2^p \times \mathbb{F}_2^{m-p}$ by $\varphi''(a, a') = \varphi(a, a')\varphi'(a')$ and by $\psi''(b, b') = \psi(b)\psi'(b')$. Then, the function F defined by $F(x, y) = (F'(x, y), T(y))$ admits (φ'', ψ'') for covering sequence.*

Proof. For every pair $(x, y) \in \mathbb{F}_2^r \times \mathbb{F}_2^s$ and for every element $(a, a') \in \mathbb{F}_2^r \times \mathbb{F}_2^s$, we have $D_{(a,a')}F(x, y) = (D_{(a,a')}F'(x, y), D_{a'}T(y))$. Then, for every pair $(x, y) \in \mathbb{F}_2^r \times \mathbb{F}_2^s$ and for every element $(b, b') \in \mathbb{F}_2^p \times \mathbb{F}_2^{m-p}$, we have

$$\sum_{\substack{(a,a')\in \mathbb{F}_2^r \times \mathbb{F}_2^s, \\ D_{(a,a')}F(x,y)=(b,b')}} \varphi(a, a')\varphi'(a') \quad = \quad \sum_{\substack{(a,a') \in \mathbb{F}_2^r \times \mathbb{F}_2^s, \\ D_{(a,a')}F'(x, y) = b, \\ D_{a'}T(y) = b'}} \varphi(a, a')\varphi'(a')$$

and the right-hand side can be rewritten

$$\sum_{\substack{a' \in \mathbb{F}_2^s, \\ D_{a'}T(y)=b'}} \varphi'(a') \left(\sum_{\substack{a \in \mathbb{F}_2^r, \\ D_{(a,a')}F'(x,y)=b}} \varphi(a, a') \right).$$

Since, for every $a' \in Supp\ \varphi'$, F' admits $(\varphi(\cdot, a'), \psi)$ for covering sequence, we deduce that $\sum_{\substack{a\in\mathbb{F}_2^r, \\ D_{(a,a')}F'(x,y)=b}} \varphi(a, a')$ equals $\psi(b)$. This implies

$$\sum_{\substack{(a,a')\in \mathbb{F}_2^r \times \mathbb{F}_2^s, \\ D_{(a,a')}F(x,y)=(b,b')}} \varphi(a, a')\varphi'(a') = \psi(b) \sum_{\substack{a' \in \mathbb{F}_2^s, \\ D_{a'}T(y)=b'}} \varphi'(a'). \qquad (25)$$

The function T admitting (φ', ψ') for covering sequence, the right-hand side of Equation (25) equals $\psi(b)\psi'(b')$ and the result follows.

Remark 16. Every $(s, m - p)$-function T admits clearly (δ_0, δ_0) for trivial covering sequence. Consequently, if F' admits $(\varphi(\cdot, 0), \psi)$ for non-trivial covering sequence, then F admits $(\varphi(\cdot, 0), \psi \times \delta_0)$ for non-trivial covering sequence, where $\psi \times \delta_0$ denotes the function $(b, b') \mapsto \psi(b) \times \delta_0(b')$.

We argued in Section 3.4 that there is a need for a construction providing a large set of balanced highly nonlinear (n, m)-functions with $m > n/2$ (and more particularly of balanced highly nonlinear permutations), and we proposed a natural extension, that we called $\mathcal{M}^*_{n,m}$, of the Maiorana-MacFarland construction to design such a set. In the following corollary, we show that, as a direct consequence of Proposition 12 and Proposition 14, the elements of $\mathcal{M}^*_{n,m}$ admit very particular covering sequences.

Corollary 4. *Let F be an $(r + s, m)$-function belonging to $\mathcal{M}^*_{r+s,m}$ and defined by $F(x, y) = (F'(x, y), T(y))$, where F' belongs to $\mathcal{M}_{r+s,p}$ and satisfies the hypothesis of Proposition 12, and T is an $(s, m - p)$-function. Then F admits the pair $(\delta_{\mathbb{F}_2^r \times \{0\}}, 2^{r-p} \delta_{\mathbb{F}_2^p \times \{0\}})$ for covering sequence. If p equals r and $m - p$ equals s (F is therefore an $(r + s, r + s)$-function), then F admits the pair $(\delta_{\mathbb{F}_2^r \times \{0\}}, \delta_{\mathbb{F}_2^r \times \{0\}})$ for covering sequence.*

Proof. Trivially, T admits $(\varphi', \psi') = (\delta_0, \delta_0)$ for covering sequence. According to Proposition 12, the function $F' \in \mathcal{M}_{r+s,p}$ admits the pair $(\delta_{\mathbb{F}_2^r \times \{0\}}, 2^{r-p}) = (\delta_{\mathbb{F}_2^r \times \mathbb{F}_2^s}(\cdot, \mathbf{0}), 2^{r-p})$ for covering sequence. This last covering sequence can be rewritten $(\varphi(\cdot, \mathbf{0}), \psi)$ and, hence, due to Proposition 14, F admits the pair of functions $(\delta_{\mathbb{F}_2^r \times \{0\}}, 2^{r-p} \delta_{\mathbb{F}_2^p \times \{0\}})$ for covering sequence.

5 Discriminators and Covering Sequences

5.1 On the Notion of Discriminator in Last-Round Attacks

To define an iterated block cipher in a formal way, we consider a family $(F_k)_{k \in \mathcal{K}}$ of (n, n)-functions, indexed by a value $k \in \mathcal{K}$ where $\mathcal{K}$ is called the *round key space*. The encryption function of the iterated block cipher with block size n, with R rounds and with round function F_k is defined by:

$$x^{(i)} = F_{k_i}\left(x^{(i-1)}\right) \text{ for } 1 \leq i \leq R, \tag{26}$$

where $x^{(0)}$ is the plaintext and $x^{(R)}$ is the ciphertext.

The vector $(k_1, \dots, k_R)$ is called the *key* and its coordinates are the *round keys*. The round keys may be derived from a unique master key which is shorter than the concatenation of all the round keys.

Most known or chosen plaintext attacks on iterated block ciphers (linear attacks [30], differential attacks [2, 7] or interpolations attacks [24] for instance) are divide-and-conquer techniques that find one round key after another in ascending order. Because these attacks try to recover the last-round key first, they are called *last-round attacks*. In a last-round attack, one considers the reduced cipher, i.e., the cipher obtained by removing the final round of the original cipher. The reduced cipher corresponds to the function:

$$G_{R-1} = G_{(k_1, \dots, k_{R-1})} = F_{k_{R-1}} \circ \dots \circ F_{k_1}. \tag{27}$$

The key point in a last-round attack is to be able to distinguish the reduced cipher from a random permutation for all possible values of the first $(R - 1)$ round keys $k_1, \dots, k_{R-1}$. Some information on k_R can be recovered by applying a discriminator to a subset of all functions H_k defined by $H_k\left(x^{(0)}\right) = F_k^{-1}\left(x^{(R)}\right)$ where k ranges over the set of all possible values of k_R. Notice that, in differential and linear attacks, the discriminator is not exactly used as a black box since it is combined with a more subtle approach to recover the last-round key. But it is always possible to use a discriminator more or less as a black box, that is, as an algorithm saying whether a given candidate key round can be accepted or not.

If the guessed k matches the actual last-round key k_R, then F_k^{-1} inverts the last encryption round and H_k corresponds to the reduced cipher, that is, to G_{R-1}. On the other hand, when k is a wrong guess, we get:

$$H_k = F_k^{-1} \circ F_{k_R} \circ F_{k_{R-1}} \circ \ldots \circ F_{k_1} = F_k^{-1} \circ F_{k_R} \circ G_{R-1} \ .$$

Since it essentially corresponds to the reduced cipher followed by two more encryption rounds, this function is supposed to act like a random permutation. This assumption is called the hypothesis of wrong-key randomization.

A discriminator of a subset of all the permutations is defined as follows:

Definition 7. *[7] Let $\mathcal{P}_n$ denote the set of all permutations of $\mathbb{F}_2^n$ and let $\mathcal{F}$ be a subset of $\mathcal{P}_n$. A discriminator for $\mathcal{F}$ with respect to a subset $\{x_0, \ldots, x_N\}$ of $\mathbb{F}_2^n$ is a function*

$$D : \quad (\mathbb{F}_2^n)^{N+1} \quad \longmapsto \quad \mathbb{F}_2$$
$$(y_0, \ldots, y_N) \mapsto D(y_0, \ldots, y_N)$$

for which there exists $\varepsilon > 0$ such that $|Pr_{f \in_R \mathcal{F}}[D(f(x_0), \ldots, f(x_N)) = 1] - Pr_{\pi \in_R \mathcal{P}_n}[D(\pi(x_0), \ldots, \pi(x_N)) = 1]|$ is greater than ε, where $\pi \in_R \mathcal{P}_n$ (resp. $f \in_R \mathcal{F}$) means that π (resp. f) is a randomly chosen permutation of $\mathbb{F}_2^n$ (resp. randomly chosen in $\mathcal{F}$).

When the discriminator is involved in a last-round attack, the vectors x_0, ..., x_N are the $N+1$ ciphertexts corresponding to $N+1$ chosen plaintexts (notice that, in the majority of last-round attacks, the plaintexts are supposed to be chosen in order to reveal a property of the reduced cipher that can be used to define a discriminator). The vectors y_0, ... , y_N designate the values $F_k^{-1}(x_0)$, ..., $F_k^{-1}(x_N)$, $k \in \mathcal{K}$, and the functions $f \in \mathcal{F}$ denote the functions $F_{k_R}^{-1}$, where k_R is an element of a subset of $\mathcal{K}$ (as small as possible) containing the last-round key used to cipher the plaintexts x_i, $i = 0, ..., N$.

Now the existence of a discriminator D for the family of reduced ciphers,

$$\mathcal{G} = \{G_K, K = (k_1, \ldots, k_{R-1}) \in \mathcal{K}^{R-1}\}$$

with respect to a set $\{x_0, \ldots, x_N\}$ leads to a last-round attack. The discriminator D should satisfy the hypothesis of *fixed-key equivalence*, i.e., it should return the same value for almost all reduced ciphers in $\mathcal{G}$.

In 1987, Evertse [20] introduces the notion of linear structure of vectorial functions and shows how linear structures can be used to define a discriminator in a last-round attack. Many attacks have been designed by generalizing the notion of linear structure. In particular, Canteaut and Videau [7] used a discriminator based on the notion of higher-order differential of vectorial functions, introduced by Knudsen (see [27]) as a generalization of the linear structures, and they succeeded in efficiently using this discriminator in last-round attacks against some iterated block ciphers: suppose that a reduced cipher G_{R-1} of an R-round iterated block cipher admits a higher-order differential of order N', that is, an N'-dimensional space E such that, for every $x \in \mathbb{F}_2^n$, we have $D_E G_{R-1}(x) = \sum_{e \in E} G_{R-1}(x + e) = \mathbf{0}$, then a discriminator can be defined by

$$D : \begin{array}{ccc} (\mathbb{F}_2^n)^N & \mapsto & \mathbb{F}_2 \\ (y_0, \ldots, y_N) & \mapsto & D(y_0, \ldots, y_N) \end{array} \tag{28}$$

where N equals $2^{N'}$ and where $D(y_0, \ldots, y_N)$ equals 1 if $\sum_{i=0}^{N}(y_0 + y_i) = \sum_{i=0}^{N} y_i = \mathbf{0}$ and equals 0 otherwise.

More generally, one can define a discriminator if there exists a set (not necessarily a space) V and a vector c such that the reduced cipher G_{R-1} satisfies $D_V G_{R-1}(x) = \sum_{v \in V} G_{R-1}(x + v) = c$ for every $x \in \mathbb{F}_2^n$. The pair (V, c) is called a *higher-order structure*.

The order of a higher-order differential can be proved to be related to the degree and to the nonlinearity of the vectorial functions (this is not the case of higher-order structures) and it is possible to prove that some reduced ciphers can be efficiently attacked using higher-order differentials. The notion of higher-order structure has never been used before (except for $\#V = 1$) to mount an attack on a cryptosystem because it is generally impossible to determine whether a given reduced cipher admits higher-order structures or not (if it is not a higher-order differential).

5.2 A Discriminator Using Covering Sequences

If a reduced cipher $G_{(k_1, \cdots, k_{R-1})}$ defined as in Equation (27) with respect to the (n, n)-functions $F_{k_1}, \ldots, F_{k_{R-1}}$ admits a same covering sequence (φ, ψ) for every choice of keys $(k_1, \cdots, k_{R-1})$, then one can define a discriminator D in the following way:

$$D : \begin{array}{ccc} (\mathbb{F}_2^n)^{N+1} & \mapsto & \mathbb{F}_2 \\ (y_0, \ldots, y_N) & \mapsto & D(y_0, \ldots, y_N) \end{array} \tag{29}$$

where N denotes the cardinality of the set $Supp\ \varphi = \{a_1, \cdots, a_N\}$ and D is defined by

$$D(y_0, \cdots, y_N) = 1 \iff \forall b \in Supp\ \psi, \quad \sum_{\substack{1 \le i \le N, \\ a_i \in Supp\ \varphi,\ y_0 + y_i = b}} \varphi(a_i) = \psi(b).$$

Indeed, let $x_0, x_1, \ldots, x_N$ be the $N+1$ ciphertexts corresponding to the plaintexts $x, x + a_1, \ldots, x + a_N$, where x can be any plaintext in $\mathbb{F}_2^n$. Since $G_{(k_1, \cdots, k_{R-1})}$

admits a same covering sequence (φ, ψ) for every choice of key $(k_1, \cdots, k_{R-1})$, then, for every choice of key $(k_1, \cdots, k_{R-1})$, we have

$$\forall b \in Supp\ \psi, \qquad \sum_{\substack{1 \le i \le N, \\ G_{k_1, \cdots, k_{R-1}}(x) + G_{k_1, \cdots, k_{R-1}}(x + a_i) = b}} \varphi(a_i) = \psi(b). \qquad (30)$$

In a last-round attack, the vectors y_0, ..., y_N in Relation (29) can be respectively defined as the values of $F_k^{-1}(x_0)$, ..., $F_k^{-1}(x_N)$. Hence, if the guessed round key k equals k_R, then $F_k^{-1}(x_0)$, $F_k^{-1}(x_1)$, ..., $F_k^{-1}(x_N)$ equal respectively $G_{k_1, \cdots, k_{R-1}}(x)$, $G_{k_1, \cdots, k_{R-1}}(x + a_1)$, ... , $G_{k_1, \cdots, k_{R-1}}(x + a_N)$ and Relation (30) can be rewritten as

$$\forall b \in Supp\ \psi, \qquad \sum_{\substack{1 \le i \le N, \\ a_i \in Supp\ \varphi,\ y_0 + y_i = b}} \varphi(a_i) = \psi(b).$$

We deduce that if y_0, ..., y_N are respectively defined as the values of $F_k^{-1}(x_0)$, ..., $F_k^{-1}(x_N)$ and if the guessed key k equals the round key k_R, then $D(y_0, \cdots, y_N) = 1$.

In many cryptosystems (like in AES or in DES) the round keys are introduced by addition and the round functions F_{k_i}, $1 \le i \le R$, are defined with respect to a unique function F by $F_{k_i}(x) = F(x + k_i)$.

In Theorem 1, we established the relationship between the existence of a covering sequence of a function F and the balancedness of this function. Let $(F_k)_{k \in \mathcal{K}}$ be a family of R (n, n)-functions and let $G_{k_1, \cdots, k_{R-1}}$ denote the function $F_{k_{R-1}} \circ \cdots \circ F_{k_1}$. Assume that each function F_{k_i} admits a same covering sequence (φ_i, ψ_i) for every choice of round key $k_i \in \mathcal{K}$ and that, for every index $i \le R - 1$, we have $\psi_i = \varphi_{i+1}$. Then, according to Proposition 13, the function $G_{k_1, \cdots, k_{R-1}}$ admits the same covering sequence (φ_0, ψ_{R-1}) for any choice of key $(k_1, \cdots, k_{R-1}, k_R)$.

Remark 17.
1. In order to be used in a cryptographic design, the (n, n)-functions F_{k_i} are necessarily balanced, therefore they admit at least one covering sequence. Moreover, if the round keys are introduced by addition and if there exists one of them, k', such that $F_{k'}$ admits a covering sequence (φ, ψ), then, for every other round key $k \in \mathcal{K}$, the pair (φ, ψ) is still a covering sequence of F_k. Indeed, if the round key is introduced by addition, then there exists an (n, n)-function F such that for every $k \in \mathcal{K} \subseteq \mathbb{F}_2^n$ and for every vector $x \in \mathbb{F}_2^n$, we have $F_k(x) = F(x + k)$. Since $F_{k'}$ admits (φ, ψ) for covering sequence, then it satisfies Relation (23) for every vector $x \in \mathbb{F}_2^n$, which implies that F satisfies also Relation (23) for every vector $x + k' \in \mathbb{F}_2^n$, that is, for every vector $x \in \mathbb{F}_2^n$. We deduce that F, as all the functions $F_k : x \mapsto F(x + k)$, admits also (φ, ψ) for covering sequence.
2. Let F be an (n, n)-function and, for every $k \in \mathbb{F}_2^n$, denote by F_k the (n, n)-function $x \mapsto F(x + k)$ (i.e. we assume that the round-key is introduced by addition). According to the remark above, if F admits the pairs (φ, ψ) and (ψ, φ) for

covering sequences, then, for any $(R-1)$-tuple $(k_1, ..., k_{R-1})$, the functions F_{k_1}, ..., $F_{k_{R-1}}$ admit the pairs (φ, ψ) and (ψ, φ) for covering sequences. Hence, for any $(R-1)$-tuple $(k_1, .., k_{R-1})$, these round functions satisfy the hypothesis of Proposition 13 and the function $G_{k_1, ..., k_{R-1}}$ defined as in Equation (27) admits (φ, ψ) and (ψ, φ) for covering sequences.

In the next corollary, we give examples of functions satisfying the hypothesis of the second part of the remark above.

Corollary 5. *Let F be a permutation on $\mathbb{F}_2^n = \mathbb{F}_2^r \times \mathbb{F}_2^s$ defined by $F(x, y) = (F'(x, y), P(y))$ where F' is an $(r+s, r)$-function belonging to $\mathcal{M}_{r+s,r}$ and where P is a permutation on $\mathbb{F}_2^s$. If $\mathcal{K}$ is included in $\mathbb{F}_2^n$ and if the round functions F_k of a R-round iterated block cipher are defined by $F_k(x, y) = F(x + k^{(L)}, y + k^{(R)})$ where $k = (k^{(L)}, k^{(R)}) \in \mathcal{K}$, then the reduced cipher $G_{k_1, \cdots, k_{R-1}}$ admits $(\delta_{\mathbb{F}_2^r \times \{0\}}, \delta_{\mathbb{F}_2^r \times \{0\}})$ for covering sequence.*

Proof. According to Corollary 4, the pair $(\delta_{\mathbb{F}_2^r \times \{0\}}, \delta_{\mathbb{F}_2^r \times \{0\}})$ is a covering sequence of F. Since the round keys k are introduced by addition (i.e. we have $F_{(k^{(L)}, k^{(R)})}(x, y) = F(x + k^{(L)}, y + k^{(R)})$), the pair $(\delta_{\mathbb{F}_2^r \times \{0\}}, \delta_{\mathbb{F}_2^r \times \{0\}})$ is a covering sequence of all the functions F_k for every choice of round key $k = (k^{(L)}, k^{(R)})$. After applying Proposition 13 to the round functions $F_{k_1}, \ldots, F_{k_{R-1}}$, where $(k_1, ..., k_{R-1})$ is any $(R-1)$-tuple of $\mathcal{K}$, we deduce that the reduced cipher $G_{k_1, \cdots, k_{R-1}}$ admits $(\delta_{\mathbb{F}_2^r \times \{0\}}, \delta_{\mathbb{F}_2^r \times \{0\}})$ for covering sequence.

Remark 18. The existence of a discriminator derived from covering sequences implies the definition of a new discriminator derived from a higher-order structure. Indeed, if an (n, m)-function G admits (φ, ψ) for covering sequence, then it admits a higher-order structure (V, c) such that $V = \{a \in \mathbb{F}_2^n; \varphi(a) \equiv 1 \mod 2\}$ and $c = \sum_{b \in \mathbb{F}_2^m} (\psi(b) \mod 2) \times b$, since one can always reduce modulo 2 the summations in (23) to obtain a higher-order structure. However, since covering sequences give more information about the function than the higher-order structures defined from them as above, a discriminator using covering sequences is more efficient than a discriminator derived from the corresponding higher structures.

5.3   On a Generalization of Covering Sequences of Vectorial Functions

Due to Proposition 18, the discrimination derived from covering sequences does not allow us to attack reduced ciphers which could not be attacked by using higher-order structures. A solution to design an attack on these functions which would work when no higher-order structure exists, would be to search for a generalization of covering sequences which still has a good behavior for the composition and the concatenation of vectorial functions. In what follows, we give a way to generalize the notion of covering sequence of vectorial functions.

Definition 8. *Let F be a function from $\mathbb{F}_2^n$ into $\mathbb{F}_2^m$. We call generalized covering sequence of F any pair of functions (Φ, Ψ) from, respectively, $(\mathbb{F}_2^n)^2$ and $\mathrm{Im}\, F \times \mathbb{F}_2^m$, into $\mathbb{R}$ such that:*

$$\forall x \in \mathbb{F}_2^n, \ \forall b \in \mathbb{F}_2^m, \quad \sum_{a \in \mathbb{F}_2^n, D_a F(x) = b} \varphi_x(a) = \psi_{F(x)}(b) \tag{31}$$

where $\varphi_x(a)$ and $\psi_{F(x)}(b)$ denote $\Phi(x, a)$ and $\Psi(F(x), b)$, respectively.

Remark 19. Any injective (n, m)-function F admits the generalized covering sequence (Φ, Ψ) where Φ is the constant function $(x, a) \in \mathbb{F}_2^{n^2} \mapsto 1$ and where $\Psi(F(x), b)$ denotes the cardinality of $\{a \in \mathbb{F}_2^n; D_a F(x) = b\}$.

Proposition 15. *Let $F : \mathbb{F}_2^n \mapsto \mathbb{F}_2^k$ and $G : \mathbb{F}_2^k \mapsto \mathbb{F}_2^m$ be two functions admitting respectively (Φ, Θ) and (Θ, Ψ) for generalized covering sequences. Then, (Φ, Ψ) is a generalized covering sequence of $G \circ F$.*

Proof. The proof is similar to that of Proposition 13.

Now, a discriminator can be derived from a generalized covering sequence of a round function F in a similar way as for covering sequences. Clearly, this new discriminator does not necessarily imply a second one involving higher-order structures. Such a discriminator will be efficient if the number of functions ψ_x which are different and the cardinality of the set $\bigcup_{x \in \mathbb{F}_2^n} Supp\, \varphi_x$ are small.

References

1. R. Anderson. Searching of the optimum correlation attack. In *Fast Software Encryption: Second International Workshop. Leuven, Belgium, 14-16 December 1994, Proceedings*, volume 1008 of *Lecture Notes in Computer Science*. Springer, 1995.
2. E. Biham and A. Shamir. Differential cryptanalysis of DES-like cryptosystems. *J. Cryptology*, 4(1):3–72, 1991.
3. A. Brouwer. http://www.win.tue.nl/~aeb/voorlincod.html.
4. P. Camion and A. Canteaut. Correlation-immune and resilient functions over a finite alphabet and their applications in cryptography. *Des. Codes Cryptogr.*, 16(2):121–149, 1999.
5. P. Camion, C. Carlet, P. Charpin, and N. Sendrier. On correlation-immune functions. In *Advances in Cryptology - CRYPTO' 91*, Lecture Notes in Computer Science, pages 86–100. Springer-Verlag, 1992.
6. A. Canteaut and M. Trabbia. Improved fast correlation attacks using parity-check equations of weight 4 and 5. In B. Preneel, editor, *Advances in Cryptology - EUROCRYPT 2000*, volume 1807 of *Lecture Notes in Computer Science*, pages 573–588. Springer, 2000.
7. A. Canteaut and M. Videau. Degree of composition of highly nonlinear functions and applications to higher order differential cryptanalysis. In *Advances in Cryptology - EUROCRYPT'2002*, volume 2332 of *Lecture Notes in Computer Science*, pages 518–533. Springer-Verlag, 2002.

8. C. Carlet. On cryptographic propogation criteria for boolean functions. In *Special Issue on Cryptology "Information and Computation"*, volume 151 1/2, pages 32–56, 1999.

9. C. Carlet. A larger class of cryptographic boolean functions via a study of the Maiorana-McFarland construction. In *Advances in Cryptology - CRYPTO 2002*, volume 2442 of *Lecture Notes in Computer Science*, pages 549–564. Springer-Verlag, 2002.

10. C. Carlet and E. Prouff. On a new notion of nonlinearity relevant to multi-ouput pseudo-random generators. *Selected Areas in Cryptography, 10th Annual International Workshop, SAC 2003*, 2003. to appear.

11. C. Carlet and Y. Tarannikov. Covering sequences of Boolean functions and their cryptographic significance. *Des. Codes Cryptogr.*, 25(3):263–279, 2002.

12. F. Chabaud and S. Vaudenay. Links between differential and linear cryptanalysis. In *Advances in cryptology—EUROCRYPT '94 (Perugia)*, volume 950 of *Lecture Notes in Comput. Sci.*, pages 356–365. Springer, Berlin, 1995.

13. J. H. Cheon. Nonlinear vector resilient funcions. In *Advances in Cryptology - CRYPTO 2001*, volume 26 of *Lecture Notes in Computer Science*, pages 396–407. Springer-Verlag, 1985.

14. Chor, Goldreich, Hastad, Freidmann, Rudich, and Smolensky. The bit extraction problem or t-resilient functions. In *FOCS: IEEE Symposium on Foundations of Computer Science (FOCS)*, 1985.

15. N. Courtois. Fast algebraic attacks on stream ciphers with linear feedback. In *Advances in cryptology—CRYPTO 2003 (Santa Barbara, CA)*, volume 2729 of *Lecture Notes in Computer Science*, pages 177–194. Springer, Berlin, 2003.

16. T. W. Cusick, C. Ding, and A. Renvall. *Stream Ciphers and Number Theory*, volume 55. North-Holland/Elsevier, 1998.

17. P. Delsarte. Bounds for unrestricted codes, by linear programming. *Philips Res. Rep.*, 27:272–289, 1972.

18. P. Delsarte. Four fundamental parameters of a code and their combinatorial significance. *Information and Control*, 23:407–438, 1973.

19. J. F. Dillon. *Elementary Hadamard Difference sets*. PhD thesis, University of Maryland, 1974.

20. J.-H. Evertse. Linear structures in blockciphers. In *Advances in Cryptology - EUROCRYPT'87*, volume 304 of *Lecture Notes in Computer Science*, pages 249–266. Springer-Verlag, 1988.

21. J.-C. Faugère and G. Ars. An algebraic cryptanalysis of nonlinear filter generators using gröbner bases. Rapport de Recherche 4739, INRIA, 2003.

22. R. Forré. A fast correlation attack on nonlinearly feedforward filtered shift register sequences. In *Advances in cryptology—EUROCRYPT '89 (Brighton, 1991)*, volume 434 of *Lecture Notes in Comput. Sci.*, pages 586–595. Springer, Berlin, 1990.

23. K. Gupta and P. Sarkar. Improved construction of nonlinear resilient s-boxes. In *Advances in Cryptology - ASIACRYPT 2002*, volume 2501 of *Lecture Notes in Computer Science*, pages 466–483. Springer-Verlag, 2002.

24. T. Jakobsen and L. R. Knudsen. The interpolation attack on block ciphers. *Lecture Notes in Computer Science*, 1267:28–40, 1997.

25. T. Johansson and E. Pasalic. A construction of resilient functions with high nonlinearity. In *Proceedings of the IEEE International Symposium on Information Theory*, 2000.

26. K. Khoo and G. Gong. Highly nonlinear sboxes with reduced bound on maximum correlation. In *Proceedings of IEEE International Symposium on Information Theory*, page 254, 2003.

27. L. Knudsen. Truncated and higher order differentials. In B. Preneel, editor, *Fast Software Encryption: Second International Workshop. Leuven, Belgium, 14-16 December 1994, Proceedings*, volume 1008 of *Lecture Notes in Computer Science*, pages 196–211. Springer, 1995.

28. K. Kurosawa, T. Satoh, and K. Yamamoto. Highly nonlinear t-resilient functions. *Journal of Universal Computer Science*, 3(6):721–729, June 1997.

29. S. Maitra and E. Pasalic. Linear codes in generalized construction of resilient functions with very high nonlinearity. In *Selected Areas in Cryptography, 8th Annual International Workshop, SAC 2001*, volume 2259 of *Lecture Notes in Computer Science*. Springer-Verlag, 2001. To appear in IEEE Transactions on Information Theory.

30. M. Matsui. The first experimental crytanalysis of the data encryption standard. In Y. Desmedt, editor, *Advances in Cryptology - CRYPTO '94, 14th Annual International Cryptology Conference, Santa Barbara, California, USA, August 21-25, 1994, Proceedings*, volume 839 of *Lecture Notes in Computer Science*, pages 1–11. Springer, 1994.

31. M. Matsui. Linear cryptanalysis method for DES cipher. In *Advances in Cryptology - EUROCRYPT'93*, volume 765 of *Lecture Notes in Computer Science*, pages 386–397. Springer-Verlag, 1994.

32. R. L. McFarland. A family of difference sets in non-cyclic groups. *J. Combinatorial Theory Ser. A*, 15:1–10, 1973.

33. M. J. Mihaljević and J. D. Golić. Convergence of a Bayesian iterative error-correction procedure on a noisy shift register sequence. In *Advances in cryptology—EUROCRYPT '92 (Balatonfüred, 1992)*, volume 658 of *Lecture Notes in Comput. Sci.*, pages 124–137. Springer, Berlin, 1993.

34. K. Nyberg. Perfect nonlinear S-boxes. In *Advances in cryptology—EUROCRYPT '91 (Brighton, 1991)*, volume 547 of *Lecture Notes in Comput. Sci.*, pages 378–386. Springer, Berlin, 1991.

35. K. Nyberg. On the construction of highly nonlinear permutations. In *Advances in cryptology—EUROCRYPT '92 (Balatonfüred, 1992)*, volume 658 of *Lecture Notes in Comput. Sci.*, pages 92–98. Springer, Berlin, 1993.

36. K. Nyberg. Differentially uniform mappings for cryptography. In *Advances in cryptology—EUROCRYPT '93 (Lofthus, 1993)*, volume 765 of *Lecture Notes in Comput. Sci.*, pages 55–64. Springer, Berlin, 1994.

37. O. S. Rothaus. On bent functions. In *Journal of Combinatorial Theory*, volume 20a, pages 300–305, 1976.

38. P. Sarkar and S. Maitra. Nonlinearity bounds and constructions of resilient Boolean functions. In *Advances in cryptology—CRYPTO 2000 (Santa Barbara, CA)*, volume 1880 of *Lecture Notes in Comput. Sci.*, pages 515–532. Springer, Berlin, 2000.

39. J. Seberry, M. Zhang, and Y. Zheng. On constructions and nonlinearity of correlation immune boolean functions. In *Advances in Cryptology - EUROCRYPT'93*, volume 1765 of *Lecture Notes in Computer Science*, pages 181–199. Springer-Verlag, 1993.

40. D. R. Stinson. Resilient functions and large sets of orthogonal arrays. *Congr. Numer.*, 92:105–110, 1993. Twenty-second Manitoba Conference on Numerical Mathematics and Computing (Winnipeg, MB, 1992).

41. D. R. Stinson and J. L. Massey. An infinite class of counterexamples to a conjecture concerning nonlinear resilient functions. *J. Cryptology*, 8(3):167–173, 1995.

42. G.-Z. Xiao and J. Massey. A spectral characterization of correlation-immune combining functions. In *IEEE Transactions on Information Theory*, volume IT 34, pages 569–571, May 1988.
43. X.-M. Zhang and Y. Zheng. On nonlinear resilient functions. In *Advances in Cryptology - EUROCRYPT '95*, volume 921 of *Lecture Notes in Computer Science*, pages 274–288. Springer, 1995.
44. X. M. Zhang and Y. Zheng. On crypographically resilient functions. In *IEEE Transactions on Information Theory*, volume 43, pages 1740–1747, 1997.

u_q-sharp subsets of a finite field

Pinaki Das

Department of Mathematics and Statistics
Penn State Altoona Altoona, PA 16601
das@math.psu.edu

Abstract. Let $f \in \mathbf{F}_q[x]$ where $\mathbf{F}_q$ is a finite field of characteristic p. Wan et al discovered a lower bound for the value set of f in terms of an invariant $u_q(f)$ associated to the polynomial f. We define a notion of u_q-sharp subsets of $\mathbf{F}_q$ and discuss related problems. We show how the notion of u_q-sharp sets may be used to give yet another proof of the classical Cauchy-Davenport theorem.

1 Introduction

In the Fq7 conference the author discussed a new proof of the Cauchy-Davenport theorem using ideas related to value sets of polynomials over finite fields, see [2]. The following problem (which is not discussed in [2]) arose during these investigations.

Let $\mathbf{F}_q$ be the finite field of $q = p^r$ elements, where p is a prime and r is a positive integer. The prime field of characteristic p is denoted by $\mathbf{F}_p$.

Definition 1. *Let $f(x_1, \cdots, x_n) \in \mathbf{F}_q[x_1, \cdots, x_n]$. The value set $V(f)$ of f is defined by $V(f) = \{f(\alpha_1, \cdots, \alpha_n) : (\alpha_1, \cdots, \alpha_n) \in \mathbf{F}_q^n\}$. In other words $V(f)$ is the range of f. We also define $u_q(f)$ to be the smallest positive integer i such that $\sum_{(\alpha_1, \cdots, \alpha_n) \in \mathbf{F}_q^n} f(\alpha_1, \cdots, \alpha_n)^i \neq 0$. If such an i does not exist then set $u_q(f) = \infty$.*

Note that in the above definition $u_q(f) = \infty$ if any of the variables x_i do not appear in the expression for f. To avoid confusion we may assume that all the variables x_i appear in the expression for f.

In [5] Wan et al. found a lower bound for the cardinality of the value set of a single variable polynomial $f(x)$ in terms of $u_q(f)$. In [2] their theorem was generalised to the multivariable case as follows. Note that in [5] and [2] the authors use the notation u_p instead of u_q for this invariant.

Theorem 1. *Let $f \in \mathbf{F}_q[x_1, x_2, \cdots, x_n]$ and let $V(f)$ be the value set of f. Let $u_q(f)$ be defined as in 1. If $u_q(f) < \infty$ then*

$$|V(f)| \geq u_q(f) + 1.$$

The results above are not enough to give a direct proof of the classical Cauchy-Davenport theorem whose statement follows after the following definition.

G. Mullen, A. Poli and H. Stichtenoth (Eds.): Fq7 2003, LNCS 2948, pp. 249–253, 2003.

Definition 2. *Let A and B be nonempty subsets of a field* $\mathbf{F}$*. The sumset* $A+B$ *is defined by*

$$A + B = \{a + b : a \in A \text{ and } b \in B\}. \tag{1}$$

The following is the classical Cauchy-Davenport theorem which has numerous applications in additive number theory (see [4] for example).

Theorem 2. *Let A, B be nonempty subsets of* $\mathbf{F}_p$*. Then*

$$|A + B| \geq \min\{p, |A| + |B| - 1\}.$$

In [2] we prove a more general version of the above theorem in the case of arbitrary fields, see also [1]. Here we introduce the notion of u_q-sharp sets. This gives yet another proof of the classical Cauchy-Davenport theorem.

2 u_q-sharp sets

Definition 3. *Let $f \in \mathbf{F}_q[x]$. We define f to be u_q-sharp if $V(f) = u_q(f) + 1$.*
  If A is a nonempty subset of $\mathbf{F}_q$ *we define A to be a u_q-sharp subset of* $\mathbf{F}_q$ *if there exists a polynomial $f \in \mathbf{F}_q[x]$ such that f is u_q-sharp and $V(f) = A$.*

It follows from the Hermite Dickson criterion (see [3, page 349]) that if f is a permutation polynomial then $u_q(f) = q - 1$. Hence permutation polynomials are u_q sharp.
  Since $u_q(x^n) = \frac{q-1}{(n,q-1)}$ and $V(x^n) = \frac{q-1}{(n,q-1)} + 1$, hence monomials are u_q-sharp.
  In [5] value sets of polynomials of the form $g_a(x) = x^r(x^{\frac{q-1}{d}} + a)$ where $d|q-1$ and $(r, q - 1) = 1$, are discussed at length. Thus it is well understood when such polynomials are u_q-sharp.
  It appears that u_q-sharp sets have not been investigated earlier. We can state the following.

Proposition 1.
  1) $\mathbf{F}_q$ *is u_q-sharp.*
  2) If $A \subset \mathbf{F}_q$ and $|A| = 2$ then A is u_q-sharp.
  3) If $A \subset \mathbf{F}_q$, $q \neq 3$ and $|A| = q - 1$ then A is not u_q-sharp.

  Proof. Statement 1) is obvious since permutation polynomials are u_q-sharp. The proofs of 2) and 3) follow easily from Proposition 2 below. $\square$

Definition 4. *Let $A = \{a_1, \cdots, a_K\} \subset \mathbf{F}_q$ with $|A| = K$. Let $f \in \mathbf{F}_q[x]$ such that $V(f) = A$. For $1 \leq i \leq K$, we let $m_i = |f^{-1}(a_i)| = |\{\alpha \in \mathbf{F}_q : f(\alpha) = a_i\}|$. We call $M(f) = \{m_i : i = 1, \cdots, K\}$ the set of multiplicities of f.*

Note that different sets of multiplicities give rise to different polynomials with the same value set A. Now let A be a u_q-sharp subset of $\mathbf{F}_q$ with $|A| = K$, and let $f \in \mathbf{F}_q[x]$ such that $V(f) = A$ and $|A| = u_q(f) + 1$. Let $\{m_1, \cdots, m_K\}$ be the set of multiplicities of f. The following proposition follows directly from the definition of $u_q(f)$ and the discussion above.

Proposition 2. *Let $A \subset \mathbf{F}_q$ with $|A| = K > 1$. Then A is a u_q-sharp subset of $\mathbf{F}_q$ if and only if there exists positive integers $m_1, \cdots, m_K$ with $\sum m_i = q$ such that m_i satisfies the following system of linear equations in $\mathbf{F}_q$.*

$$m_1 + m_2 + \cdots + m_K = 0 \qquad (2)$$
$$a_1 m_1 + a_2 m_2 + \cdots + a_K m_K = 0$$
$$\vdots$$
$$a_1^{K-1} m_1 + a_2^{K-1} m_2 + \cdots + a_K^{K-1} m_K = \beta,$$

where $\beta \neq 0$.

Discussion. Note that the matrix of coefficients in the above system is the Vandermonde matrix whose ij-th entry is (a_i^{j-1}) for $1 \leq i, j \leq K$. The determinant of the Vandermonde matrix is $\prod_{i>j}(a_i - a_j)$, which is nonzero since the a_i are distinct. Hence the system has a unique solution in $\mathbf{F}_q$. If this solution lies in the prime field $\mathbf{F}_p$, then we can find positive integers representing the residue classes and determine if the condition $\sum m_i = q$ holds in the ring of integers $\mathbf{Z}$.

We have seen in Proposition 1 that subsets of cardinality $q - 1$ are not u_q-sharp. We now give a nontrivial example of subsets of a prime field that are not u_q-sharp.

Example. Let $p \geq 5$. We use least positive integers to represent the residue classes in $\mathbf{F}_p$. Let c be any positive integer such that $2 \leq c \leq \frac{p-1}{2}$. Consider $A = \{1, c, p - c, p - 1\} \subset \mathbf{F}_p$. Then one can check that $(m_1, m_2, m_3, m_4) = (c, p - 1, 1, p - c)$ gives the unique solution of the following system of equations in $\mathbf{F}_p$. (Uniqueness follows from the discussion above).

$$m_1 + m_2 + m_3 + m_4 = 0$$
$$m_1 + cm_2 + (p - c)m_3 + (p - 1)m_4 = 0$$
$$m_1 + c^2 m_2 + (p - c)^2 m_3 + (p - 1)^2 m_4 = 0$$
$$m_1 + c^3 m_2 + (p - c)^3 m_3 + (p - 1)^3 m_4 = \beta,$$

where $\beta = 2c(1 - c^2) \neq 0 \pmod{p}$. Note that $\sum m_i = 2p$ in the ring of integers $\mathbf{Z}$. It is easy to check that $\sum m_i = 2p$ in the ring of integers, for all nonzero β. Hence from Proposition 2, A is not a u_p-sharp subset of $\mathbf{F}_p$.

One can however show the following for prime fields $\mathbf{F}_p$.

Theorem 3. *Let $A \subset \mathbf{F}_p$ such that $|A| > 1$ and let $q = p^i > p$. Then A is a u_q-sharp subset of $\mathbf{F}_q$.*

Proof: We prove it Let $A = \{a_1, \cdots, a_K\}$. Let $(m_1, \cdots, m_k)$ be the unique solution of the following system of equations in $\mathbf{F}_p$.

$$m_1 + m_2 + \cdots + m_K = 0$$
$$a_1 m_1 + a_2 m_2 + \cdots + a_K m_K = 0$$
$$\vdots$$
$$a_1^{K-1} m_1 + a_2^{K-1} m_2 + \cdots + a_K^{K-1} m_K = 1.$$

First observe that all the m_i have to be nonzero. Otherwise, without loss of generality assume that $m_1 = 0$. Then the first $K - 1$ equations form a system of homogeneous equations in the remaining $K - 1$ variables $m_2, \cdots, m_K$. The coefficient matrix of this system is a Vandermonde matrix with determinant $\prod_{2 \leq i < j \leq K} (a_j - a_i)$. The determinant is nonzero since the a_i are distinct. But then the homogeneous system must have only the zero solution. This contradicts the last equation.

We use least positive integer representatives for the m_i. If $\sum m_i = rp < q$ in the ring of integers $\mathbf{Z}$, where r is a positive integer such that $p \leq rp \leq K(p-1)$, we replace the positive integer representing the residue class m_1 by the positive integer $q - rp + m_1$. Then $\sum_i m_i = q$, and the proof is complete by Proposition 2. $\square$

Theorem 3 can be used to give a new proof of the classical Cauchy-Davenport theorem (Theorem 2) which is true for prime fields. For another proof, see [4] for example. As noted earlier the methods in [2] gives a more general version of the theorem.

Proof of Theorem 2: We can assume that $|A| > 1$ and $|B| > 1$. Also note that if A and B are subsets of $\mathbf{F}_p$ such that $|A| + |B| > p$ then $A \cap (\alpha - B) \neq \phi$ for each $\alpha \in \mathbf{F}_p$. Hence $|A+B| = p$ in this case. So we may assume that $|A| + |B| \leq p$. Let $q = p^2$. By Theorem 3, we can find $f, g \in \mathbf{F}_q[x]$ such that $V(f) = A$ and $V(g) = B$ with $|A| = u_q(f) + 1$ and $|B| = u_q(g) + 1$. Consider the polynomial $h(x,y) = f(x) + g(y) \in \mathbf{F}_q[x,y]$. Note that $V(h) = A + B$. Using the binomial theorem we get

$$\sum_{(x,y) \in \mathbf{F}_q^2} (h(x,y))^N = \sum_{i+j=N} \frac{N!}{i!j!} \sum_{x \in \mathbf{F}_q} (f(x))^i \sum_{y \in \mathbf{F}_q} (g(y))^j.$$

From the definition of the invariant u_q it follows that $\sum_{x \in \mathbf{F}_q} (f(x))^i = 0$ for $i < u_q(f)$ and $\sum_{y \in \mathbf{F}_q} (g(y))^j = 0$ for $j < u_q(g)$. Since $u_q(f) + u_q(g) = |A| + |B| - 2 \leq p - 2$, hence $\frac{(u_q(f) + u_q(g))!}{u_q(f)! u_q(g)!} \neq 0 \pmod{p}$. It follows that $u_q(h) = u_q(f) + u_q(g)$. From Theorem 1 we have $|A + B| = V(h) \geq u_q(h) + 1 = u_q(f) + u_q(g) + 1 = |A| + |B| - 1$. This proves the theorem. $\square$

We conclude with the following problem.

Problem: Study the structure of u_q-sharp subsets of prime and prime power fields. Find suitable conditions for a set to be a u_q-sharp subset of a finite field. Find new families of polynomials which are u_q-sharp. Find suitable conditions for a polynomial to be u_q-sharp.

Acknowledgments: I thank Gary Mullen and the referee for several helpful comments and suggestions.

References

1. N. Alon, *Combinatorial Nullstellensatz*, Combinatorics, Probability and Computing **8**(1999), 7-29.
2. P. Das, *Value sets of polynomials and the Cauchy Davenport theorem,* Finite Fields Appl. **10**(2004), 113-122.
3. R. Lidl and H. Niederreiter, *Finite Fields,* Encyclopedia Math. Appl. vol. 20, Cambridge Univ. Press, 1997.
4. M. B. Nathanson, *Additive Number Theory: Inverse Theorems and the Geometry of Sumsets,* Springer-Verlag, New York, 1996.
5. D. Wan, P. J.-S. Shiue, and C. S. Chen, *Value sets of polynomials over finite fields,* Proc. Amer. Math. Soc. **119**(1993), 711-717.

Cyclic Decomposition of Permutations of Finite Fields Obtained Using Monomials

Ivelisse M. Rubio[1,*] and Carlos J. Corrada-Bravo[2]

[1] Department of Mathematics
University of Puerto Rico at Humacao
ive@www.uprh.edu
[2] Department of Computer Science
University of Puerto Rico at Río Piedras
ccorrada@goliath.cnnet.clu.edu

Abstract. In this paper we study permutations of finite fields $\mathbf{F}_q$ that decompose as products of cycles of the same length, and are obtained using monomials $x^i \in \mathbf{F}_q[x]$. We give the necessary and sufficient conditions on the exponent i to obtain such permutations. We also present formulas for counting the number of this type of permutations. An application to the construction of encoders for turbo codes is also discussed.

1 Introduction

Consider $\mathbf{F}_q$, the finite field with q elements. It is well known that the function $\pi : \mathbf{F}_q \longrightarrow \mathbf{F}_q$ defined by $\pi(x) = x^i$ produces a permutation of the elements in $\mathbf{F}_q$ if and only if $gcd(i, q - 1) = 1$. Polynomials that produce permutations are called *permutation polynomials*. We are interested in permutations of $\mathbf{F}_q$ that decompose in cycles of the same length and are obtained using monomials x^i . When 0, 1 and -1 are the only elements fixed by the permutation, these monomials have been characterized in [6]. Here, we characterize the monomials that produce permutations of $\mathbf{F}_q$ that decompose in cycles of the same length and have any set of fixed elements. We also present formulas for counting the number of such monomials.

Applications of this type of permutations to the construction of encoders for turbo codes are being studied by the authors. Data obtained by Corrada-Bravo [2] suggests that the relation between the length of the cycles in the cyclic decomposition of the permutation and the length of the cycle of the convolutional code in the turbo code affects the performance of the code. In Section 4 we discuss the application of monomial permutations with cycles of the same length to turbo codes.

We first review some notation and present some results that will be used in the rest of the paper.

* This work was supported in part by the ADVANCE Institutional Transformation Program, NSF Grant SBE-0123654, and by the PR Space Grant IDEAS-ER Program, Grant NAGP5-40091.

G. Mullen, A. Poli and H. Stichtenoth (Eds.): Fq7 2003, LNCS 2948, pp. 254–261, 2003.

2   Preliminaries from Number Theory

Some of the following concepts and results are well known and can be found in almost any text in number theory. The other results are very easy to prove.

From now on p is a prime number, q is a power of a prime and n, k, l, j, h are positive integers. The *Euler function*, $\phi(n)$, denotes the number of positive integers not exceeding n that are relatively prime to n.

One of the most used concepts in this paper is the order of an element. The *order* of an integer i modulo n is the smallest positive integer j such that $i^j \equiv 1 \pmod{n}$ and it will be denoted by $j = ord_n(i)$.

Lemma 1. *If $i \equiv b \pmod{p^l}$, then $i^p \equiv b^p \pmod{p^{l+1}}$ for all $l \geq 1$.*

Lemma 2. *Let $j = ord_{p^l}(i)$. Then $j = ord_{p^{l+1}}(i)$ or $jp = ord_{p^{l+1}}(i)$.*

Proposition 1. *$j = ord_{p^k}(i)$ and $j|(p-1)$ if and only if $j = ord_{p^l}(i)$ for all $1 \leq l \leq k$.*

Lemma 3. *Let $p = ord_{p^k}(i)$ for some $k \geq 2$. Then either $2 = p = ord_{p^l}(i)$ for $2 \leq l \leq k$ or $i \equiv 1 \pmod{p^l}$ for $1 \leq l \leq k$.*

Lemma 4. *Let $j = ord_s(i)$, $j = ord_l(i)$ and $gcd(s, l) = 1$. Then $j = ord_{sl}(i)$.*

Lemma 5. *Let $j = ord_s(i)$, $i \equiv 1 \pmod{l}$ and $gcd(s, l) = 1$. Then $j = ord_{sl}(i)$.*

On Section 3.1 we will give formulas to count the number of permutation monomials $x^i \in \mathbf{F}_q[x]$ that decompose in cycles of the same length j. The next results will be helpful.

Proposition 2. *Let p be an odd prime and suppose that $j|\phi(p^n)$. Then, there are $\phi(j)$ incongruent elements of order j modulo p^n.*

Proposition 3. *The incongruent solutions of $x^2 \equiv 1 \pmod{2^k}$ are:*

$$\begin{cases} \pm 1, \ \pm\left(1 + 2^{k-1}\right) & \text{if } k \geq 3 \\ \pm 1 & \text{if } k = 2 \\ 1 & \text{if } k = 1 \ . \end{cases}$$

3   Cycles of the Same Length

The cycle structure of permutation monomials $x^i \in \mathbf{F}_q[x]$ was studied by Ahmad in [1]. The cycle structure for more general polynomials, specifically Dickson polynomials, was studied by Lidl and Mullen in [4]. Here, we present the necessary and sufficient conditions on the exponent i to obtain permutations of $\mathbf{F}_q$ that decompose in cycles of the same length. We will use the following result proven in [1].

Theorem 1. *The permutation of* $\mathbf{F}_q$ *given by* x^i *has a cycle of length* j *if and only if* $j = ord_t(i)$, *where* $t | (q - 1)$. *The number* N_j *of such cycles is*

$$jN_j = gcd\left(q - 1, i^j - 1\right) - \sum_{s|j, s<j} sN_s \ .$$

We say that a permutation has cycles of the same length j if the permutation decomposes in cycles of length j or 1. The next theorem characterizes the permutation monomials with this property.

Theorem 2. *Let* $q - 1 = p_0^{k_0} p_1^{k_1} \cdots p_r^{k_r}$. *The permutation of* $\mathbf{F}_q$ *given by* x^i *has cycles of the same length* j *if and only if one of the following holds for each* $l = 0, \cdots, r$:

1. $i \equiv 1 \left(mod\ p_l^{k_l}\right)$
2. $j = ord_{p_l^{k_l}}(i)$ *and* $j | (p_l - 1)$
3. $j = ord_{p_l^{k_l}}(i),\ k_l \geq 2$ *and* $j = p_l$.

Proof. ($\Longleftarrow$) If $i \equiv 1 \left(mod\ p_l^{k_l}\right)$ for all $l = 0, 1, \cdots, r$, then x^i is the identity permutation. Suppose that $1 < j = ord_{p_l^{k_l}}(i)$ for some of the l's and $i \equiv 1 \left(mod\ p_l^{k_l}\right)$ for the other. Proposition 1 and Lemma 3 guaranty that $j = ord_{p_l^k}(i)$ or $i \equiv 1 \left(mod\ p_l^k\right)$ for all $l = 0, 1, \cdots, r$ and $1 \leq k \leq k_l$. Now, if $t | (q - 1)$, then by Lemmas 4 and 5, we have that, $j = ord_t(i)$ or $i \equiv 1 (mod\ t)$. Hence, by Theorem 1, all the cycles have length j or 1.

($\Longrightarrow$) Suppose that all the cycles have the same length j. Then, by Theorem 1, $j = ord_t(i)$ or $i \equiv 1 (mod\ t)$ for all t that divides $q-1$. This holds in particular for $t = p_l^{k_l}$, $l = 0, 1, \cdots, r$. We only have to prove that, if $j = ord_{p_l^{k_l}}(i)$ then $j | (p_l - 1)$ or $j = p_l$, $k_l \geq 2$.

Suppose that $1 \neq j = ord_{p_l^{k_l}}(i)$. If $k_l = 1$ then $j | (p_l - 1)$ and we are done. If $k_l \geq 2$ and $j \nmid (p_l - 1)$, then Proposition 1 implies that $j \neq ord_{p_l^k}(i)$ for some $k < k_l$. Let s be the largest one such that $j \neq ord_{p_l^s}(i)$. Then $i \equiv 1 \ (mod\ p_l^s)$ because otherwise, by Theorem 1, there would be a cycle of length different from j. By Lemma 1, $i^{p_l} \equiv 1 \left(mod\ p_l^{s+1}\right)$. But $j = ord_{p_l^{s+1}}(i)$ implies that $j | p_l$ and hence $j = p_l$. □

The next results consider the special cases where $0, 1$ or $0, 1, -1$ are the only elements fixed by the permutation. These results were first presented on [6] but the proofs there did not use Theorem 1.

It is clear that 0 and 1 are always fixed by the permutation x^i. Fixed elements are the same as cycles of length 1, so, by Theorem 1, an element is fixed if and only if $i \equiv 1 (mod\ t)$, where $t | (q-1)$. Note that -1 is a fixed element if and only if $i \equiv 1 (mod\ 2)$. Hence, $0, 1, -1$ are the only elements fixed by the permutation if and only if $i \not\equiv 1 (mod\ t)$ for any $t \neq 2$ such that $t | (q - 1)$.

We first consider q being such that 4 does not divide $q - 1$.

Theorem 3. *Let $q - 1 = p_0^{k_0} p_1^{k_1} \cdots p_r^{k_r}$, $p_0 = 2$, $k_0 = 0, 1$. The permutation of $\mathbf{F}_q$ given by x^i decomposes in cycles with the same length j and $0, 1, -1$ or $0, 1$ are the only fixed elements if and only if $j = ord_{p_l^{k_l}}(i)$ and $j \mid (p_l - 1)$ for $p_l \neq 2$.*

Proof. ($\Longrightarrow$) Suppose that all the cycles have length j and $0, 1, -1$ or $0, 1$ are the only fixed elements. Then $j = ord_{p_l^k}(i)$ for $p_l \neq 2$, $k \leq k_l$ and, by Proposition 1, $j \mid (p_l - 1)$.

($\Longleftarrow$) Suppose that $j = ord_{p_l^{k_l}}(i)$ and $j \mid (p_l - 1)$ for $p_l \neq 2$. Then, Theorem 2 implies that all the cycles have the same length j. Also, Proposition 1 implies that $j = ord_{p_l^h}(i)$, $h \leq k_l$, $p_l \neq 2$. Hence $i \not\equiv 1 \pmod{t}$ for any $t \mid (q - 1), t \neq 2$ and the only possible fixed elements are $0, 1, -1$. $\square$

In the case where 4 divides $q - 1$ there are only two monomials that give permutations that decompose in cycles of the same length and have $0, 1, -1$ as the only fixed elements. Also, the length of the cycles on such permutations is always 2.

Theorem 4. *Let $q - 1 = p_0^{k_0} p_1^{k_1} \cdots p_r^{k_r}$, where $p_0 = 2$, $k_0 \geq 2$. The permutation of $\mathbf{F}_q$ given by x^i decomposes in cycles of the same length j and $0, 1, -1$ are the only fixed elements if and only if $j = ord_{p_l^{k_l}}(i)$ for $p_l \neq 2$, $j = ord_{2^h}(i)$ for $2 \leq h \leq k_0$, and $j = 2$.*

Proof. By the arguments given before Theorem 3, $0, 1, -1$ are the only fixed elements if and only if $i \not\equiv 1 \pmod{p_l^h}$ for $p_l \neq 2$, $h \leq k_l$, and for $p_0 = 2$, $2 \leq h \leq k_0$. Since $j \mid (p_0 - 1)$ would imply that $j = 1$, the result now follows applying Theorem 2. $\square$

Corollary 1. *Let $q - 1 = p_0^{k_0} p_1^{k_1} \cdots p_r^{k_r}$, where $p_0 = 2$, $k_0 > 2$. The permutation of $\mathbf{F}_q$ given by x^i decomposes in cycles of the same length j and $0, 1, -1$ are the only fixed elements if and only if $j = 2$ and $i = q - 2$ or $i = \frac{q-3}{2}$.*

Proof. By Proposition 3, for $k_0 > 2$, $2 = ord_{2^{k_0}}(i)$ if and only if $i \equiv -1$ or $i \equiv \pm (1 + 2^{k_0 - 1}) \pmod{2^{k_0}}$. But $2 = ord_4(i)$ if and only if $i \equiv -1$ or $i \equiv -1 - 2^{k_0 - 1} \pmod{2^{k_0}}$.

Then, by the previous theorem and Proposition 3, we have cycles of the same length j and $0, 1, -1$ are the only fixed elements if and only if $i \equiv -1$ or $i \equiv -1 - 2^{k_0 - 1} \pmod{2^{k_0}}$, and $i \equiv -1 \pmod{p_l^{k_l}}$. Hence, there are only two i's such that the permutation x^i decomposes in cycles of the same length $j = 2$ and have $0, 1, -1$ as the only fixed elements. Noting that $\frac{q-3}{2} + 1 + 2^{k_0 - 1} = \frac{2^{k_0} \left(p_1^{k_1} \cdots p_r^{k_r} + 1 \right)}{2} = 2^{k_0} s$, for some $s \in \mathbb{Z}$, we get that $i \equiv -1 \pmod{q - 1}$ and $i \equiv \frac{q-3}{2} \pmod{q - 1}$ are the only solutions. $\square$

Corollary 2. *Let $q - 1 = 4 p_1^{k_1} \cdots p_r^{k_r}$. The permutation of $\mathbf{F}_q$ given by x^i decomposes in cycles of the same length j and $0, 1, -1$ are the only fixed elements if and only if $j = 2$ and $i = q - 2$.*

Proof. By arguments similar to those given in the previous proof, the only i such that x^i gives a permutation with cycles of the same length $j = 2$ and have $0, 1, -1$ as the only fixed elements is $i \equiv -1 \pmod{q-1}$.   □

3.1   Counting the Number of Permutation Monomials that Decompose in Cycles of the Same Length

In this section we give formulas for counting the number of permutation mono-mials that decompose in cycles of the same length j. To do this, we define a bijection between the set of all the i such that x^i decompose in cycles of the same length j and another set. Let $q - 1 = p_0^{k_0} p_1^{k_1} \cdots p_r^{k_r}$, and define

$$U_j = \left\{ i \mid x^i \text{ is a permutation of } \mathbf{F}_q \text{ that decomposes in cycles of length } j \right\} .$$

and

$$W_j = \left\{ (w_0, w_1, \cdots, w_r) \mid w_n \in \mathbb{Z}_{p_n^{k_n}}, \; j = ord_{p_n^{k_n}}(w_n) \text{ for } j \mid (p_n - 1), \right.$$

$$\left. \text{or } j = ord_{p_n^{k_n}}(w_n) \text{ for } k_n \geq 2 \text{ and } j = p_n, \text{ or } w_n \equiv 1 \pmod{p_n^{k_n}} \right\} .$$

Lemma 6. *Let* $q - 1 = p_0^{k_0} p_1^{k_1} \cdots p_r^{k_r}$ *and* U_j *and* W_j *be defined as above. Let* $f_j : U_j \longrightarrow W_j$ *be defined by* $f_j(i) = (w_0, w_1, \cdots, w_r)$, *where* $i \equiv w_n \pmod{p_n^{k_n}}$ *for* $0 \leq n \leq r$. *Then* f_j *is a bijection.*

Proof. Note that by Theorem 2, if x^i decomposes in cycles of length j if and only if for each $0 \leq n \leq r$, we have that $j = ord_{p_n^{k_n}}(i)$ for $j \mid (p_n - 1)$, or $j = ord_{p_n^{k_n}}(i)$ for $k_n \geq 2$ and $j = p_n$, or $i \equiv 1 \pmod{p_n^{k_n}}$. Since $i \equiv w_n \pmod{p_n^{k_n}}$ for $0 \leq n \leq r$ we have that $f_j(U_j) \subseteq W_j$.

To see that f_j is onto, let $(w_0, w_1, \cdots, w_r) \in W_j$. We need to find i such that x^i decomposes in cycles of length j and $i \equiv w_n \pmod{p_n^{k_n}}$ for $0 \leq n \leq r$. By the definition of W_j and Theorem 2, we just have to find a solution to the system

$$\begin{cases} i \equiv w_0 \pmod{p_0^{k_0}} \\ i \equiv w_1 \pmod{p_1^{k_1}} \\ \quad\vdots \\ i \equiv w_r \pmod{p_r^{k_r}} \end{cases} .$$

The Chinese Remainder Theorem guaranties that there is a unique solution modulo $q - 1 = p_0^{k_0} p_1^{k_1} \cdots p_r^{k_r}$.   □

Theorem 5. *Let $q - 1 = p_0^{k_0} p_1^{k_1} \cdots p_r^{k_r}$. Then, the number of permutations x^i of $\mathbf{F}_q$ with cycles of the same length $j \neq 1$ is*

$$\prod_{n=0}^{r} f\left(j, p_n^{k_n}\right) - 1, \tag{1}$$

where, for p_n odd,

$$f\left(j, p_n^{k_n}\right) = \begin{cases} 1 + \phi(j) & \text{if } j \mid (p_n - 1) \\ 1 + \phi(j) & \text{if } j = p_n \text{ and } k_n \geq 2 \\ 1 & \text{otherwise,} \end{cases} \tag{2}$$

and, for $p_n = 2$,

$$f\left(j, 2^k\right) = \begin{cases} 4 & \text{if } j = 2, \ k \geq 3 \\ 2 & \text{if } j = 2, \ k = 2 \\ 1 & \text{if } j = 2, \ k = 1, \quad \text{or } j > 2 \ . \end{cases} \tag{3}$$

Proof. From the previous lemma we have that counting the x^i with cycles of length j is the same as counting the elements in W_j. For each $0 \leq n \leq r$, we have to count the number $f\left(j, p_n^{k_n}\right)$ of elements in $\mathbb{Z}_{p_n^{k_n}}$ of order 1, or of order j if $j \mid (p_n - 1)$ or $j = p_n$. Formula 1 give us the number of all possible x^i with cycles of length j; we subtract 1 for the case where $i \equiv 1 \pmod{q - 1}$, that is when all the elements are fixed.

By Proposition 2, there are $\phi(j)$ elements of order j and one element congruent to 1 in $\mathbb{Z}_{p_n^{k_n}}$ for each p_n odd. This give us (2). For the case $p_n = 2$ and $j > 2$, by Theorem 2, one must have that $i \equiv 1 \pmod{2^k}$ and hence $f\left(j, 2^k\right) = 1$ for $j > 2$. The other cases on (3) follow from Proposition 3. □

Now consider the case where the permutation x^i has cycles of length j and $0, 1, -1$ are the only elements fixed by the permutation. This case is of particular interest for the application to turbo codes as we will explain in the next section.

Corollary 2 says that, when $q - 1 = 4p_1^{k_1} \cdots p_r^{k_r}$, the only permutation $x^i \in \mathbf{F}_q[x]$ that decomposes in cycles of the same length and has $0, 1, -1$ as the only fixed elements is x^{q-2}. For the case when $q - 1 = 2^k p_1^{k_1} \cdots p_r^{k_r}$, $k > 2$, Corollary 1 give us two permutation monomials with this property: x^{q-2} and $x^{\frac{q-3}{2}}$. The following proposition give us the number of monomials with this property for the other cases.

Proposition 4. *Let $q - 1 = p_0^k p_1^{k_1} \cdots p_r^{k_r}$, $k = 0, 1$. The number of monomials $x^i \in \mathbf{F}_q[x]$ with cycles of length j and have $0, 1$ or $0, 1, -1$ as the only fixed elements is $\phi(j)^r$ if $j \mid (p_n - 1)$ for all $1 \leq n \leq r$, and 0 otherwise.*

Proof. By Theorem 3, x^i has cycles of length j and $0, 1$ or $0, 1, -1$ are the only fixed elements only if $j \mid (p_n - 1)$ for all $1 \leq n \leq r$. Suppose that $j \mid (p_n - 1)$ for all $1 \leq n \leq r$. As in in Lemma 6, we can construct a bijection between the set of monomials x^i that decompose in cycles of length j and have $0, 1$ or $0, 1, -1$ as the only fixed elements and the set

$$W_j = \left\{ (w_1, w_2, \cdots, w_r) \mid w_n \in \mathbb{Z}_{p_n^{k_n}}, \ j = ord_{p_n^{k_n}}(w_n) \right\} \ .$$

Again, counting the number of such monomials is the same as counting the number of elements in W_j. Hence, by Proposition 2, there are $\phi(j)^r$ monomials $x^i \in \mathbf{F}_q[x]$ that have cycles of length j and have $0, 1$ or $0, 1, -1$ as the only fixed elements. □

4   Application to Turbo Codes

Error control codes are used in digital communication systems to protect information from errors that might occur during transmission. Turbo codes are specially suitable for satellite communication systems since they provide error control performance with a good reduction in the transmitter power levels.

One of the main components of a turbo encoder is the interleaver, which permutes the information symbols. The current practice to construct interleavers is to choose them randomly. The fact that these interleavers are found by computer search implies that they have to be stored in memory. Although good performance can be obtained with this type of construction, it is bad for implementation as well as for performance analysis. To avoid this problem, researchers have considered deterministic constructions that can be generated on the fly and that perform as well as random interleavers.

Most of the known methods for constructing interleavers algebraically do not produce interleavers that perform well. Some of the properties associated to the interleaver that are important to obtain "good" turbo encoders are the spreading and the dispersion factors. An article by Takeshita and Costello [7] as well as data obtained by Corrada-Bravo [2] suggested that another important property of an interleaver is the length of the cycles in the cyclic decomposition of the permutation in relation to the length of the cycle of the convolutional code in the turbo code.

We are constructing interleavers using permutation monomials that give permutations with a fixed length of cycles and studying the spreading and dispersion properties as well as the performance of the codes. In particular, we are studying permutations that only fix $0, 1, -1$ because, usually, permutations with few fixed elements have good dispersion. Also, it is very simple to construct monomials $x^i \in \mathbf{F}_q[x]$ that give permutations that decompose in cycles of the same length j and only fix $0, 1, -1$: if $4 \mid (q-1)$ and $j = 2$, x^{q-2} and $x^{\frac{q-3}{2}}$ are the only choices; for $q - 1 = p_0^k p_1^{k_1} \cdots p_r^{k_r}$, $k = 0, 1$ and $j \mid (p_n - 1)$ for all $1 \leq n \leq r$, we only have to find i such that

$$\begin{cases} i \equiv 1 \quad (mod\ 2) \\ i \equiv w_1 \ \left(mod\ p_1^{k_1}\right) \\ \quad \vdots \\ i \equiv w_r \ (mod\ p_r^{k_r}) \ . \end{cases}$$

The proof of the Chinese Remainder Theorem gives an easy way to construct these i's.

Our simulations show that although our interleavers do not have better dispersion or spreading than random interleavers, interleavers with certain length of the cycles perform as well or better than them. More details on this can be found on [3]

Graphs with large girth have been used for the construction of regular and irregular low density parity check (LDPC) codes and recently, in [8] the author derived interleavers for turbo codes from graphs which have large girth. The *girth* (the length of the shortest cycle) of the turbo code graph, captures the relation between the cycle length of interleavers and the cycle length of the convolutional codes. We are carrying further studies on this relation in an attempt to answer the question as to which other parameters are necessary to established how an interleaver is going to perform. With this approach, we hope to be able to predict the performance of a turbo code with a particular interleaver, based on the cycle length of the convolutional code and the cycle structure of the interleaver. This would remove of the analysis (up to a degree) the painstaking and time consuming task of simulation.

Acknowledgments

The authors want to thank Professor Gary Mullen for suggesting the use of Theorem 1 to prove some of the results in this paper. We also appreciate the contribution of the following students from the University of Puerto Rico at Humacao: Marian Hernández-Viera, Yara Luis, Luis Medina-Rivera, Aida Navarro, Liannette Passapera, and Everilis Santana-Vega.

References

1. S. Ahmad, Cycle structure of automorphisms of finite cyclic groups, *J. Comb. Thy.* **6** (1969), 370-374.
2. C. J. Corrada-Bravo, *Sequence Designs for Applications in Ultra-wideband Systems and Turbo Codes*, Ph.D. Thesis, Univesity of Southern California, August, 2002.
3. C. J. Corrada-Bravo and I. Rubio, "Deterministic Interleavers for Turbo Codes with Random-like Performance and Simple Implementation", *Proceedings of the International Symposium on Turbo Codes*, September, 2003, pp 555-558.
4. R. Lidl, G.L. Mullen, Cycle Structure of Dickson Permutation Polynomials, *Mathematical Journal of Okayama University* **33** (1991), 1-11.
5. K. Rosen, *Elementary Number Theory and its Applications*, Fourth Edition, Addison Wesley, 1999.
6. I. Rubio, *Cyclic Decomposition of Monomial Permutations*, M.S. Thesis, University of Puerto Rico, Río Piedras, 1988.
7. O. Takeshita and D. Costello "New deterministic interleaver designs for turbo Codes", *IEEE-IT*, Vol. 46, pp. 1988-2006, Sept. 2000.
8. P. O. Vontobel, "On the Construction of Turbo Code Interleavers Based on Graphs with Large Girth", *Proc. IEEE Intern. Conf. Communications*, Vol.3, pp.1408-1412, New York, NY, USA, Apr. 28-May 2, 2002.

Author Index